# THE EVOLUTIONARY ECOLOGY OF INVASIVE SPECIES

# THE EVOLUTIONARY ECOLOGY OF INVASIVE SPECIES

**JOHANNES LE ROUX**
Associate Professor, Department of Biological Sciences,
Macquarie University, Sydney, NSW, Australia

**ACADEMIC PRESS**
An imprint of Elsevier

Academic Press is an imprint of Elsevier
125 London Wall, London EC2Y 5AS, United Kingdom
525 B Street, Suite 1650, San Diego, CA 92101, United States
50 Hampshire Street, 5th Floor, Cambridge, MA 02139, United States
The Boulevard, Langford Lane, Kidlington, Oxford OX5 1GB, United Kingdom

**Notices**
Knowledge and best practice in this field are constantly changing. As new research and experience
broaden our understanding, changes in research methods, professional practices, or medical
treatment may become necessary.

Practitioners and researchers must always rely on their own experience and knowledge in evaluating
and using any information, methods, compounds, or experiments described herein. In using such
information or methods they should be mindful of their own safety and the safety of others, including
parties for whom they have a professional responsibility.

To the fullest extent of the law, neither the Publisher nor the authors, contributors, or editors, assume
any liability for any injury and/or damage to persons or property as a matter of products liability,
negligence or otherwise, or from any use or operation of any methods, products, instructions, or ideas
contained in the material herein.

**Library of Congress Cataloging-in-Publication Data**
A catalog record for this book is available from the Library of Congress

**British Library Cataloguing-in-Publication Data**
A catalogue record for this book is available from the British Library

ISBN 978-0-12-818378-6

For information on all Academic Press publications
visit our website at https://www.elsevier.com/books-and-journals

*Publisher:* Charlotte Cockle
*Acquisitions Editor:* Anna Valutkevich
*Editorial Project Manager:* John Leonard
*Production Project Manager:* Selvaraj Raviraj
*Cover Designer:* Greg Harris

Typeset by STRAIVE, India

Working together
to grow libraries in
developing countries

www.elsevier.com • www.bookaid.org

# For Adeline and Benjamin

*May the wonders of the natural world never cease to amaze you.*

# Contents

*Preface*                                                                    *xi*

## 1. Setting the scene                                                        1

   1.1 A brave new world                                                       1
   1.2 Invasive species are major features of Anthropocene habitats            2
   1.3 A brief overview of rapid evolution in invasive species                 3
   1.4 About this book                                                         5
   References                                                                  7

## 2. Life in the fast lane                                                   11

   2.1 Introduction                                                           11
   2.2 Microevolution during biological invasions                            15
   2.3 Epigenetic variation                                                   25
   2.4 Holobionts and hologenomes                                             26
   2.5 Conclusions                                                            28
   References                                                                 28

## 3. A home away from home: The role of eco-evolutionary
   experience in establishment and invasion success                         35

   3.1 Introduction                                                          36
   3.2 Eco-evolutionary experience and invasiveness                          40
   3.3 How does eco-evolutionary experience influence the pace of
       contemporary evolution and the ecological impacts caused by
       invasive species?                                                     45
   3.4 Conclusions and future directions                                     48
   References                                                                 49

## 4. Phenotypic plasticity and the emerging field of 'invasion
   epigenetics'                                                              55

   4.1 Introduction                                                          56
   4.2 Mechanisms of epigenetic variation                                     59
   4.3 The ecological consequences of epigenetic variation                   63
   4.4 The evolutionary consequences of epigenetic variation                 64
   4.5 The role of epigenetic variation in establishment success             64
   4.6 Epigenetic variation during range expansions                          68

**4.7** Epigenetics, hybridisation, and polyploidization: Insights from
invasive *Spartina anglica*                                         69
**4.8** Conclusions                                                 71
References                                                          72

## 5. Drivers of rapid evolution during biological invasions    79
**5.1** Introduction                                                80
**5.2** Changes in abiotic conditions                              82
**5.3** Changes in biotic conditions                               88
**5.4** Conclusions                                                 92
References                                                          93

## 6. The current state of research on the evolutionary ecology of invasive species    99
**6.1** Introduction                                               100
**6.2** Causes of rapid evolution                                  102
**6.3** Traits that frequently undergo rapid evolution during invasion    118
**6.4** Introduction history                                       119
**6.5** Experimental design                                        122
**6.6** Conclusions                                                123
References                                                          125

## 7. Evolutionary impacts of invasive species on native species    135
**7.1** Introduction                                               136
**7.2** Direct evolutionary impacts                                140
**7.3** Genetic impacts via hybridisation                          147
**7.4** Indirect impacts                                           148
**7.5** Invasive species as drivers of extinction and speciation   149
**7.6** Conclusion                                                 152
References                                                          153

## 8. Invasion genetics: Molecular genetic insights into the spatial and temporal dynamics of biological invasions    159
**8.1** Introduction                                               160
**8.2** The genetics of invasive species                           165
**8.3** Genetic signatures of adaptation                           174
**8.4** Concluding remarks and future directions                   178
References                                                          179

**9. Incorporating evolutionary biology into invasive species management**     **189**

9.1 Introduction     190

9.2 Evolutionary approaches to enhance the success of classical biological control     192

9.3 The use of native enemies as biological control agents     195

9.4 Engineering genomes to manage invasive species     199

9.5 Gene drives: Frankenstein species or promising tools for the genetic control of invasive species?     200

9.6 Conclusions     203

References     203

*Index*     *209*

# Preface

A few significant changes came up in quick succession in my personal life towards the end of 2017. We found out that my wife, Kate, was expecting our second child, I accepted a new position at Macquarie University in Sydney, and I agreed to writing this book. I had some experience in raising my own offspring and moving between jobs on different continents, but virtually none in book writing. It has been a steep learning curve to say the least.

This book was partly inspired by the work of others that has influenced my own scientific thinking, particularly during my early days in science. For instance, I fondly remember 'stumbling' upon papers by Scott Carroll during my PhD, on his beloved soapberry bugs and balloon vines (also see the cover of this book!) and reading these with great enthusiasm. Research such as Scott's instilled an ongoing fascination with contemporary evolution in me. Many years later, Scott and I would become close friends and study balloon vines and soapberry bugs together. George Cox's (2004) book, *Alien Species and Evolution—The Evolutionary Ecology of Exotic Plants, Animals, Microbes, and Interacting Native Species*, is another example of work that greatly inspired me as a student. I still regard this book as one of the most important published in invasion biology. Since then, many excellent books covering various topics related to the ecology, impacts, and spatial and temporal dynamics of biological invasions have appeared. Primary research on rapid evolution in invasive species also blossomed since the early 2000s, yet, no synthesis of this impressive body of work has been attempted. This book is an effort to do so. I am certainly not an expert on all the topics discussed in this book and cannot pretend that all there is to know will be found in its pages. Therefore, while my hope is that this book will inspire students and researchers alike, I suspect that not everyone will find it wholly satisfactory.

I have received generous and critical feedback from colleagues and friends on various aspects of this book. Scott Carroll, Allan Ellis, Heinrich Kroukamp, Llewellyn Foxcroft, Dick Frankham, Nicola Hawes, Heidi Hirsch, Anthony Manea, Daniel Montesinos, Ethan Newman, Ana Novoa, Dave Richardson, Daniel Rubinoff, Wolf Saul, John Terblanche, and Mark Wright, all reviewed specific chapters of the book. I would also like to thank Chris Lochner for the stunning cover image that he painted for the book. Last, but not least, I would like to say a massive thank you to my wife, Kate, without whose love and support this project would simply not have been possible.

# CHAPTER 1

# Setting the scene

## Chapter outline

1.1 A brave new world   1
1.2 Invasive species are major features of Anthropocene habitats   2
1.3 A brief overview of rapid evolution in invasive species   3
1.4 About this book   5
References   7

> *The second half of the twentieth century is unique in the entire history of human existence on Earth. Many human activities reached take-off points sometime in the twentieth century and have accelerated sharply towards the end of the century. The last 50 years have without doubt seen the most rapid transformation of the human relationship with the natural world in the history of humankind.*
>
> **Steffen et al. (2004)**

## 1.1 A brave new world

The geological timescale provides us with an overall idea of chronological events that shaped our planets history, patched together from snapshots of observable changes in the Earth System. These snapshots, preserved in strata of rock, glacier ice, and marine sediments, tell us about our planet's past climate conditions and when different organisms appeared, evolved, and became extinct. Human activities now impact every corner of the globe, to the extent where, similar to meteor strikes and previous ice ages, we are impacting the Earth System in ways that will be detectable for millions of years to come, defining the Anthropocene (Waters et al., 2016). It is worth putting the coming into being of this latest epoch into perspective. If we were to condense the Earth's 4.5-billion-year-old history into a single calendar year, then the first single-celled life evolved somewhere in September, dinosaurs appeared around Christmas, Angiosperms evolved 3 days later, and the Anthropocene starts during the last nanosecond of New Year's Eve. Man-made changes to the Earth System are happening at unprecedented rates!

*The Evolutionary Ecology of Invasive Species*
http://doi.org/10.1016/B978-0-12-818378-6.00009-7

Biodiversity in the Anthropocene will leave distinct footprints in the geological record. It is estimated that around 43% of the planet's ice- and desert-free land is now under farming (Poore and Nemecek, 2018) while, more broadly, upwards of three-quarters of all terrestrial habitats are in one way or another transformed by human activities (Sage, 2020; Ellis et al., 2021). These changes, among others, are causing a major shuffling of global biodiversity. Anthropocene landscapes are dominated by domestic and exotic species, while native species are going extinct hundreds to thousands of times faster than in pre-human times (Ceballos and Ehrlich, 2002; Blackburn et al., 2019; Humphreys et al., 2019; Le Roux et al., 2019). Some has argued that the Earth is facing a looming sixth mass extinction (Ceballos et al., 2015).

Rapid human-mediated changes to the Earth System have been referred to as the 'Great Acceleration': a surge in socio-economic activities since the 1950s that coincides with an acceleration in Earth System impacts such as levels of greenhouse gasses, ocean acidification, and deforestation (Steffen et al., 2015). There is no indication that these impacts are slowing down. For example, human-made material outweighed all living biomass for the first time in 2020 (Elhacham et al., 2020) while more recently, in April 2021, the concentration of atmospheric $CO_2$ exceeded 420 ppm for the first time in recorded history, the midpoint towards double pre-industrial $CO_2$ levels. We are living in a brave new world.

## 1.2 Invasive species are major features of Anthropocene habitats

For thousands of years humans have intentionally or accidentally moved species beyond their natural geographic ranges, often over large distances. In many instances these translocations have resulted in "invasive" (i.e. spreading) populations. Biological invasions are key components of Anthropocene habitats (Sage, 2020) and interact with many other Anthropocene features such as accelerated native species extinctions (Blackburn et al., 2019). Thousands of exotic species are now found across the globe (Seebens et al., 2017). For instance, van Kleunen et al. (2015a) estimated that 13,168 plant species, or 3.9% of the world's vascular flora, have one or more naturalised populations. Biological invasions also fit the general trends of the Great Acceleration, with marked increases in the introduction of exotic species from the 1950s onwards (Seebens et al., 2017), a symptom of globalisation and the associated increased movement of people and goods (Hulme, 2009). The resulting, and ever-increasing, number of invasive species are causing tremendous economic impacts, with a recent study suggesting that these costs amount to US$20 billion per year globally (Diagne et al., 2021), a likely gross underestimate.

## 1.3 A brief overview of rapid evolution in invasive species

Another central feature of the Anthropocene is that humans are a dominant evolutionary force. The inadvertent side effects of our ecological disruptions, such as land conversion, pollution, urbanisation, wild harvest, climate change, and invasive species introductions, are causing rapid evolutionary responses in most wild populations (Palumbi, 2001; Wood et al., 2021). From a biodiversity perspective the challenges arising from human-mediated evolution in the Anthropocene can be characterised as those stemming from unwanted species (e.g. invasive species) adapting too quickly, and valued species (e.g. endemic species) adapting too slowly (Carroll et al., 2014). Biologists have traditionally viewed evolution as a slow process and have only recently embraced the idea that it can operate over much shorter timescales. Research on invasive species, in particular, has helped to facilitate this paradigm shift (Reznick et al., 2019).

Darwin studied the geological record to formulate his theory of evolution by natural selection and this told him that evolution was a slow process. Exotic species also did not escape Darwin's attention (Ludsin and Wolfe, 2001) and he was one of the first naturalists to think about the role of evolution in the context of biological invasions. Most famous maybe was his idea that relatedness between exotic and resident species will determine establishment success, in what is now commonly referred to as Darwin's naturalisation hypothesis (Cadotte et al., 2018). He observed that 'naturalised plants are of a highly diversified nature. They differ, moreover, to a large extent from the indigenes ...', which he interpreted as exotic species being less likely to establish in areas where closely related natives occur, since competition for available resources would be high (Darwin, 1859; Cadotte et al., 2018). Of course, at the time, exotic species were rare curiosities rather than commonly observed occurrences.

Many have argued that it was Elton's (1958) book, *The Ecology of Invasions by Animals and Plants*, that paved the way for invasion biology as an independent scientific discipline (Richardson and Pyšek, 2008). In his book, Elton (1958) provided the first global synthesis of human introductions of exotic species and the impact of these translocations on biotic interactions and communities. Despite his detailed treatment of the evolution of biogeographic realms (chapter 2 in *The Ecology of Invasions by Animals and Plants*) he only peripherally discussed contemporary evolution, probably because rapid evolution was still largely unknown at the time. For example, he only briefly mentioned the role of genetics in the decline of invasive Canadian pondweed (*Elodea canadensis*) in Europe, resistance evolution in insects and fungi, and hybridisation and polyploidy in invasive *Spartina* cordgrasses.

The publication of *The Genetics of Colonizing Species* (Baker and Stebbins, 1965) was a pioneering moment in studies on the evolutionary biology of invasive species. This volume resulted from a symposium that was held on in California in 1964 that brought together geneticists, ecologists, taxonomists, and applied researchers interested in invasive species biology from various parts of the world. The diverse topics covered in this publication have since become central theses in invasion biology, such as Baker's ideal weed (van Kleunen et al., 2015b), the impact of demographic processes on genetic variation (Dlugosch and Parker, 2008; Vicente et al., 2021), or how different mating systems impact exotic plant colonisation success (Barrett et al., 2008). However, the notion of rapid evolution was not discussed in great detail in *The Genetics of Colonizing Species*. Instead, the volume focussed on how past evolution (i.e. preadaptation) and demographic processes like founder events and genetic bottlenecks influence the ecological dynamics of invasive populations. Following the publication of *The Genetics of Colonizing Species* a handful of studies described the role of rapid evolution in the success of exotic species (Baker, 1974; Williamson, 1996; Sax et al., 2005), yet, it seemed like an interest in evolutionary biology never really caught on in invasion biology (Callaway and Maron, 2006). This all changed somewhere between the late 1990s and early 2000s, when evidence for contemporary evolution started to accumulate, often from studies involving invasive species (e.g. Reznick and Ghalambor, 2001; Strauss et al., 2006; Reznick et al., 2019). At the same time, an appreciation for invasive species as drivers of evolutionary change in native species emerged (e.g. Carroll et al., 2005). While many books covering diverse topics in invasion biology have appeared since the 2000s (e.g. Sax et al., 2005; Richardson, 2011; Hui and Richardson, 2017), most only briefly treated rapid evolution. Cox's (2004) book, *Alien Species and Evolution—The Evolutionary Ecology of Exotic Plants, Animals, Microbes, and Interacting Native Species*, was the first comprehensive synthesis of rapid evolution in invasive species.

## 1.4 About this book

Research on the evolutionary ecology of invasive species has exploded since the publication of Cox's (2004) book. The current book aims to synthesise some of this recent research and to discuss new and emerging insights gained from these studies. We start in Chapter 2 by outlining why invasive species represent high levels of evolutionary opportunity, whether due to strong selection in their new ranges or release from selection pressures experienced in their native ranges. The individual drivers of microevolution—assortative mating, gene flow, genetic drift, mutation, selection—and how these relate to the circumstances that typically accompany the introduction, establishment, and spread of exotic species will be discussed.

As we mentioned, Darwin (1859) was one of the first naturalists to suggest that past evolutionary history may influence the establishment success of exotic species. When exotic species are closely related to resident species then competition for available resources may be high. On the other hand, close relatedness to resident natives may also equip exotic species with pre-adaptations to their new environments (Cadotte et al., 2018). Evolutionary history is therefore a potentially important yardstick for invasion success (Divíšek et al., 2018). In Chapter 3 we discuss the role of eco-evolutionary experience (EEE)—the exposure of both exotic and native species to biotic interactions over historical evolutionary timescales—in the establishment success and adaptability of exotic species, and the evolutionary responses by native species to invaders (Saul et al., 2013; Saul and Jeschke, 2015). Numerous ecological and evolutionary hypotheses have been put forward to explain establishment and invasion success by exotic species (Jeschke and Heger, 2018; Enders et al., 2020), and we will discuss how EEE with various interaction types and abiotic conditions fits into these hypotheses.

Phenotypic responses by individuals under novel environmental conditions often stems from plasticity. High levels of plasticity generally equip species for survival and reproduction under heterogenous environmental conditions and, therefore, selection on plasticity may be strong in invasive populations (Colautti et al., 2017). It is only recently that we have begun to understand the inner workings, and evolutionary significance of, phenotypic plasticity. In particular, epigenetic variation is now widely recognised for its role in plastic responses (Ashe et al., 2021). In Chapter 4 we focus on 'invasion epigenetics' as an emerging research field in the study of rapid evolution in invasive species. We will make a convincing case for the importance of epigenetic mechanisms as a bet hedging strategy in invasive populations, especially those that lack appreciable levels standing genetic diversity.

The abiotic and biotic conditions that promote rapid evolution have received much research attention (Reznick et al., 2019). From a biotic perspective, most exotic species will be liberated from interactions in their native ranges, simultaneously experiencing a whole suite of novel interactions in their new ranges (Le Roux et al., 2020). The evolutionary component of such interaction rewiring has been repeatedly shown (e.g. Strauss et al., 2006) and is one of the topics of discussion in Chapter 5. We will also discuss rapid evolutionary responses in invasive species to abrupt changes in abiotic conditions.

Given the ever-growing number of studies documenting rapid evolution in invasive species, are any general patterns emerging? Chapter 6 aims to answer this question by reviewing research over the last decade on rapid evolution in invasive species to establish whether certain traits consistently undergo rapid evolution in specific taxa and, if so, whether similar selection pressures underlie such evolution. As studies on rapid evolution in invasive species took off in the late 1990s, a parallel appreciation for exotic species as drivers of evolutionary change in native species has emerged. A particularly dramatic example comes from Nevada in the United States, where the native Edith's checkerspot butterfly (*Euphydryas editha*) evolved host preference for invasive ribwort plantain (*Plantago lanceolata*) (Singer and Parmesan, 2018). Subsequent changes in land use led to the recovery of grasslands and smothering of ribwort plantain plants, in turn, creating microclimate conditions that were unsuitable for the development of Edith's checkerspots. The evolved preference for ribwort plantain meant that the butterfly was unable to switch back to its original native host plant and the Nevadan population went extinct. Examples like these and other evolutionary impacts on native species by invasive species are discussed in Chapter 7.

During the 1970s–80s, after the publication of *The Genetics of Colonizing Species* (Baker and Stebbins, 1965), researchers started to investigate the ecological and evolutionary genetics of invasive species (Barrett, 2015). However, it is only later that population genetic studies on invasive species became common, driven by an exponential growth in molecular techniques and analytical tools from the early 1990s onwards. These studies have contributed immensely to our understanding of rapid evolution and it is no coincidence that the field of 'invasion genetics' is prospering. In Chapter 8 we focus on the lessons learned from invasion genetics studies and how these have contributed to our understanding of rapid evolution in general.

Historically, the ecology of invasive species has enjoyed much more research attention than their evolutionary biology. This may reflect concerns for the ecological causes and consequences of invasive species and, therefore, the need to manage them in order to minimise impacts on native species, habitats, and ecosystem services. It is now widely accepted that we can longer ignore the solutions offered by evolutionary biology to manage invasive species. Chapter 9 discusses the use of evolutionary principles to manage invasive species.

# References

Ashe, A., Colot, V., Oldroyd, B.P., 2021. How does epigenetics influence the course of evolution? Philos. Trans. R. Soc. B Biol. Sci. 376, 20200111.

Baker, H.G., 1974. The evolution of weeds. Annu. Rev. Ecol. Syst. 5, 1–24.

Baker, H.G., Stebbins, G.L. (Eds.), 1965. The Genetics of Colonizing Species. Academic Press, New York.

Barrett, S.C.H., 2015. Foundations of invasion genetics: the Baker and Stebbins legacy. Mol. Ecol. 24, 1927–1941.

Barrett, S.C.H., Colautti, R.I., Eckert, C.G., 2008. Plant reproductive systems and evolution during biological invasion. Mol. Ecol. 17, 373–383.

Blackburn, T.M., Bellard, C., Ricciardi, A., 2019. Alien versus native species as drivers of recent extinctions. Front. Ecol. Environ. 17, 203–207.

Cadotte, M.W., Campbell, S.E., Li, S.-.P., Sodhi, D.S., Mandrak, N.E., 2018. Preadaptation and naturalization of nonnative species: Darwin's two fundamental insights into species invasion. Annu. Rev. Plant Biol. 69, 661–684.

Callaway, R.M., Maron, J.L., 2006. What have exotic plant invasions taught us over the past 20 years? Trends Ecol. Evol. 21, 369–374.

Carroll, S.P., Loye, J.E., Dingle, H., Mathieson, M., Famula, T.R., Zalucki, M.P., 2005. And the beak shall inherit—evolution in response to invasion. Ecol. Lett. 8, 944–951.

Carroll, S.P., Jørgensen, P.S., Kinnison, M.T., Bergstrom, C.T., Denison, R.F., Gluckman, P., Smith, T.B., Strauss, S.Y., Tabashnik, B.E., 2014. Applying evolutionary biology to address global challenges. Science 346, 1245993.

Ceballos, G., Ehrlich, P.R., 2002. Mammal population losses and the extinction crisis. Science 296, 904–907.

Ceballos, G., Ehrlich, P.R., Barnosky, A.D., García, A., Pringle, R.M., Palmer, T.M., 2015. Accelerated modern human-induced species losses: entering the sixth mass extinction. Sci. Adv. 1, e1400253.

Colautti, R.I., Alexander, J.M., Dlugosch, K.M., Keller, S.R., Sultan, S.E., 2017. Invasions and extinctions through the looking glass of evolutionary ecology. Philos. Trans. R. Soc. B Biol. Sci. 372, 20160031.

Cox, G.W., 2004. Alien Species and Evolution: The Evolutionary Ecology of Exotic Plants, Animals, Microbes, and Interacting Native Species. Island Press, Washington, DC.

Darwin, C.R., 1859. The Origin of Species. John Murray, London.

Diagne, C., Leroy, B., Vaissière, A.C., Gozlan, R.E., Roiz, D., Jarić, I., Salles, J.M., Bradshaw, C.J.A., Courchamp, F., 2021. High and rising economic costs of biological invasions worldwide. Nature 592, 571–576.

Divíšek, J., Chytrý, M., Beckage, B., Gotelli, N.J., Lososová, Z., Pyšek, P., Richardson, D.-M., Molofsky, J., 2018. Similarity of introduced plant species to native ones facilitates naturalization, but differences enhance invasion success. Nat. Commun. 9, 4631.

Dlugosch, K.M., Parker, I.M., 2008. Founding events in species invasion: genetic variation, adaptive evolution, and the role of multiple introductions. Mol. Ecol. 17, 431–449.

Elhacham, E., Ben-Uri, L., Grozovski, J., Bar-On, Y.M., Milo, R., 2020. Global human-made mass exceeds all living biomass. Nature 588, 442–444.

Ellis, E.C., Gauthier, N., Klein Goldewijk, K., Bird, R., Boivin, N., Diaz, S., Fuller, D., Gill, J., Kaplan, J., Kingston, N., Locke, H., McMichael, C., Ranco, D., Rick, T., Shaw, M.R., Stephens, L., Svenning, J.C., Watson, J.E.M., 2021. People have shaped most of terrestrial nature for at least 12,000 years. Proc. Natl. Acad. Sci. U. S. A. https://doi.org/10.1073/pnas.2023483118.

Elton, C.S., 1958. The Ecology of Invasions by Animals and Plants. Methuen, London.

Enders, M., Havemann, F., Ruland, F., Bernard-Verdier, M., Catford, J.A., Gómez-Aparicio, L., Haider, S., Heger, T., Kueffer, C., Kühn, I., Meyerson, L.A., Musseau, C., Novoa, A., Ricciardi, A., Sagouis, A., Schittko, C., Strayer, D.L., Vilà, M., Essl, F., Hulme, P.E., van Kleunen, M., Kumschick, S., Lockwood, J.L., Mabey, A.L., McGeoch, M.A., Palma, E., Pyšek, P., Saul, W.-C., Yannelli, F.A., Jeschke, J.M., 2020. A conceptual map of invasion biology: integrating hypotheses into a consensus network. Glob. Ecol. Biogeogr. 29, 978–991.

Hui, C., Richardson, D.M., 2017. Invasion Dynamics. Oxford University Press, Oxford.

Hulme, P.E., 2009. Trade, transport and trouble: managing invasive species pathways in an era of globalization. J. Appl. Ecol. 46, 10–18.

Humphreys, A.M., Govaerts, R., Ficinski, S.Z., Nic Lughadha, E., Vorontsova, M.S., 2019. Global dataset shows geography and life form predict modern plant extinction and rediscovery. Nat. Ecol. Evol. 3, 1043–1047.

Jeschke, J.M., Heger, T. (Eds.), 2018. Invasion Biology: Hypotheses and Evidence. CABI, Wallingford, UK.

Le Roux, J.J., Hui, C., Castillo, M.L., Iriondo, J.M., Keet, J.-.H., Khapugin, A.A., Médail, F., Rejmánek, M., Theron, G., Yannelli, F.A., Hirsch, H., 2019. Recent anthropogenic plant extinctions differ in biodiversity hotspots and coldspots. Curr. Biol. 29, 2912–2918.

Le Roux, J.J., Clusella-Trullas, S., Mokotjomela, T.M., Mairal, M., Richardson, D.M., Skein, L., Wilson, J.R., Weyl, O.L.F., Geerts, S., 2020. Biotic interactions as mediators of biological invasions: insights from South Africa. In: van Wilgen, B.W., Measey, J., Richardson, D.M., Wilson, J.R., Zengeya, T.A. (Eds.), Biological Invasions in South Africa. Springer, Berlin, pp. 385–428.

Ludsin, S.A., Wolfe, A.D., 2001. Biological invasion theory: Darwin's contributions from the origin of species. BioScience 51, 780–789.

Palumbi, S.R., 2001. Humans as the world's greatest evolutionary force. Science 293, 1786–1790.

Poore, J., Nemecek, T., 2018. Reducing food's environmental impacts through producers and consumers. Science 360, 987–992.

Reznick, D.N., Ghalambor, C.K., 2001. The population ecology of contemporary adaptations: what empirical studies reveal about the conditions that promote adaptive evolution. Genetica 112, 183–198.

Reznick, D.N., Losos, J., Travis, J., 2019. From low to high gear: there has been a paradigm shift in our understanding of evolution. Ecol. Lett. 22, 233–244.

Richardson, D.M. (Ed.), 2011. Fifty Years of Invasion Ecology. The Legacy of Charles Elton. Wiley-Blackwell, Oxford.

Richardson, D.M., Pyšek, P., 2008. Fifty years of invasion ecology—the legacy of Charles Elton. Divers. Distrib. 14, 161–168.

Sage, R.F., 2020. Global change biology: a primer. Glob. Chang. Biol. 26, 3–30.

Saul, W.-C., Jeschke, J.M., 2015. Eco-evolutionary experience in novel species interactions. Ecol. Lett. 18, 236–245.

Saul, W.-C., Jeschke, J.M., Heger, T., 2013. The role of eco-evolutionary experience in invasion success. NeoBiota 17, 57–74.

Sax, D.F., Stachowicz, J.J., Gaines, S.D. (Eds.), 2005. Species Invasions: Insights into Ecology, Evolution, and Biogeography. Sinauer Associates, Sunderland, MA, USA.

Seebens, H., Blackburn, T.M., Dyer, E.E., Genovesi, P., Hulme, P.E., Jeschke, J.M., Pagad, S., Pyšek, P., Winter, M., Arianoutsou, M., Bacher, S., Blasius, B., Brundu, G., Capinha, C., Celesti-Grapow, L., Dawson, W., Dullinger, S., Fuentes, N., Jäger, H., Kartesz, J., Essl, F., 2017. No saturation in the accumulation of alien species worldwide. Nat. Commun. 8, 14435. https://doi.org/10.1038/ncomms14435.

Singer, M.C., Parmesan, C., 2018. Lethal trap created by adaptive evolutionary response to an exotic resource. Nature 557, 238–241.

Steffen, W., Sanderson, A., Tyson, P.D., 2004. Global Change and the Earth System: A Planet under Pressure. The IGBP Book Series. Springer-Verlag, Berlin, Heidelberg, New York, p. 18.

Steffen, W., Broadgate, W., Deutsch, L., Gaffney, O., Ludwig, C., 2015. The trajectory of the Anthropocene: the great acceleration. Anthr. Rev. 2, 81–98.

Strauss, S.Y., Lau, J.A., Carroll, S.P., 2006. Evolutionary responses of natives to introduced species: what do introductions tell us about natural communities? Ecol. Lett. 9, 357–374.

van Kleunen, M., Dawson, W., Essl, F., Pergl, J., Winter, M., Weber, E., Kreft, H., Weigelt, P., Kartesz, J., Nishino, M., Antonova, L.A., Barcelona, J.F., Cabezas, F.J., Cárdenas, D., Cárdenas-Toro, J., Castaño, N., Chacón, E., Chatelain, C., Ebel, A.L., Figueiredo, E., Pyšek, P., 2015a. Global exchange and accumulation of non-native plants. Nature 525, 100–103. https://doi.org/10.1038/nature14910.

van Kleunen, M., Dawson, W., Maurel, N., 2015b. Characteristics of successful alien plants. Mol. Ecol. 24, 1954–1968.

Vicente, S., Máguas, C., Richardson, D.M., Trindade, H., Wilson, J.R.U., Le Roux, J.J., 2021. Highly diverse and highly successful: invasive Australian acacias have not experienced genetic bottlenecks globally. Ann. Bot. 128, 149–157.

Waters, C.N., Zalasiewicz, J., Summerhayes, C., Barnosky, A.D., Poirier, C., Gałuszka, A., Cearreta, A., Edgeworth, M., Ellis, E.C., Ellis, M., Jeandel, C., Leinfelder, R., McNeill, J.R., Richter, D.d., Steffen, W., Syvitski, J., Vidas, D., Wagreich, M., Williams, M., Zhisheng, A., Wolfe, A.P., 2016. The Anthropocene is functionally and stratigraphically distinct from the Holocene. Science 351. https://doi.org/10.1126/science.aad2622, aad2622.

Williamson, M., 1996. Biological Invasions. Chapman and Hall, London.

Wood, Z.T., Palkovacs, E.P., Olsen, B.J., Kinnison, M.T., 2021. Humans as the world's greatest eco-evolutionary force. EcoEvoRxiv. https://doi.org/10.1093/biosci/biab010.

# CHAPTER 2

# Life in the fast lane

## Contents

| | |
|---|---|
| **2.1** Introduction | 11 |
| **2.2** Microevolution during biological invasions | 15 |
|     **2.2.1** Founder events, genetic drift and inbreeding | 16 |
|     **2.2.2** Mutation, hybridisation, and whole-genome duplication | 19 |
|     **2.2.3** Assortative mating | 23 |
|     **2.2.4** Natural selection | 23 |
|     **2.2.5** Artificial selection | 25 |
| **2.3** Epigenetic variation | 25 |
| **2.4** Holobionts and hologenomes | 26 |
| **2.5** Conclusions | 28 |
| References | 28 |

## Abstract

This chapter summarises the ecological conditions and demographic processes that underlie most biological invasions and how these create unique opportunities for rapid evolution to occur. The chapter discusses the roles of founder events, inbreeding, genetic drift, migration, mutation, natural selection, and non-assortative mating in causing rapid evolutionary change in invasive species, and provides examples of each process. The chapter also briefly discusses how non-genetic (epigenetic) and hologenomic diversity impact on the evolutionary trajectories of invasive species.

**Keywords:** Biological invasions, Genetic drift, Inbreeding, Inbreeding depression, Invasive species, Local adaptation, Migration, Natural selection, Rapid evolution, Spatial sorting

## 2.1 Introduction

*...we can see that when a plant or animal is placed in a new country amongst new competitors, though the climate may be exactly the same as in its former home, ... the conditions of its life will generally be changed in an essential manner...*

**Darwin (1859)**

The intentional introduction of cane toads (*Rhinella marina*) into Australia in 1935 to control sugarcane pests will go down in history as one of the most

*The Evolutionary Ecology of Invasive Species*
https://doi.org/10.1016/B978-0-12-818378-6.00001-2

spectacular failures of 'biological control'. Cane toads were introduced to control the French's cane beetle (*Lepidiota frenchi*) and the grey-back cane beetle (*Dermolepida albohirtum*), pests whose larvae feed on the roots of sugarcane plants (Shine et al., 2020). This introduction was wholly misguided because of the different ecologies of cane toads and beetles. The idea was for cane toads to control soil-dwelling beetle larvae by preying on their airborne parents. However, cane toads only forage on the ground and are nocturnal while their intended prey is diurnal. Sugarcane plantations in Australia were also too dry for cane toads and soon it was time for them to move on. The scene was set for what would become one of the most extraordinary invasions by an exotic animal. Today, cane toads are found over more than one million square kilometres of tropical and subtropical Australia (Urban et al., 2007). Their ecological impacts have been severe, e.g., population numbers of many native invertebrates and predators dramatically declined because of cane toad predation and poisoning, respectively (Shine et al., 2020). In some instances, these impacts have resulted in rapid evolutionary responses in native species (Shine, 2010), such as snakes evolving resistance to toad toxins or behaviours to avoid them (Phillips et al., 2003, 2004; Phillips and Shine, 2004, 2006a,b,c).

The impressive range expansion of cane toads in Australia not only tells the story of a highly fecund and mobile species being at the right place at the right time, but also one of a species that underwent rapid evolution to attain such phenomenal ecological success. For example, within decades of their arrival in Australia, dispersal at the invasion front of toads accelerated dramatically (Phillips et al., 2006; Phillips et al., 2007; Alford et al., 2009; Fig. 2.1). This accelerated dispersal has been linked to evolutionary changes in morphology (longer legs, Phillips et al., 2006), behaviour (more frequent and further movements along straight paths; Alford et al., 2009), and physiology (increased endurance; Llewelyn et al., 2010).

Our extraordinary ability and propensity to redistribute myriad life forms around the world is increasingly homogenising the planet's biota. Cane toads in Australia are just one example of the consequences of these human-mediated translocations. Whether deliberate or unintentional, the continuous introduction of exotic species has led to an astonishing, and seemingly never-ending, build-up of invasive species globally (Seebens et al., 2017, 2021). The impacts of these species ramify deeply into the ecosystems they invade, and biological invasions are now one of the main drivers of global change (Pyšek et al., 2020; Sage, 2020). Biological invasions not only impact

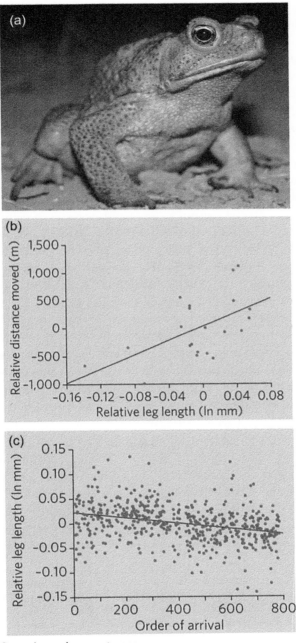

**Fig. 2.1** See figure legend on next page.

*Continued*

biodiversity and ecosystems but also the livelihoods of people and economies of the world (Pimentel et al., 2005; Diagne et al., 2021; Linders et al., 2021).

Despite their ecological and economic impacts, invasive species do represent unique 'natural experiments' to study fundamental questions in ecology and evolution. The quest to understand why some exotic species become successful invaders and others not has been at the forefront of invasion biology research (e.g. Rejmanek and Richardson, 1996; Pyšek and Richardson, 2007). Decades of ecological research have not yielded satisfying or general answers to this question, probably because of the highly context dependency of biological invasions (Catford et al., 2019; Pyšek et al., 2020). However, it is clear that, irrespective of life form, biogeographic origin, or habitat type, all invasive species share one common attribute: rapid population growth and expansion (i.e. rapid increases in density and range sizes) under novel ecological conditions. It is therefore surprising that the evolutionary biology of invasive species has historically received little research attention (Barrett, 2015). This has now changed, and over recent years an appreciation for biological invasions as excellent systems to study contemporary evolution has emerged (Reznick et al., 2019).

As for all organisms, the fitness of invasive populations is an attribute of the genotypes they hold, being condition dependent, i.e. on genotype x environment interactions (GxE). Importantly, evolutionary responses are not always the result of selection but may also stem from non-adaptive processes (Colautti and Lau, 2015; Hodgins et al., 2018). Rapid range expansions in new environments create distinct opportunities for microevolution to occur, whether via genetic drift, migration, non-random mating, or natural selection. Even when standing genetic diversity is low (Sherpa and Després, 2021), many exotic species survive and, in some instances, undergo rapid local adaption (Allendorf and Lundquist, 2003; Estoup et al., 2016). Low standing genetic diversity may also lead to strong stochastic fluctuations in diversity, i.e. genetic drift, with important consequences for evolutionary change (Dlugosch and Parker, 2008). On the other hand, genetic variation in invasive populations may be augmented by new mutations. The role of

---

**Fig. 2.1, cont'd**    (a) Invasive cane toads (*Rhinella marina*) rapidly evolved longer legs and mobility as they spread across Australia. (b) Compared with their shorter-legged conspecifics, toads that evolved longer hind limbs also move further and (c) they evolved at the leading edge of the invasion (based on order of arrival at the study site). *(Panels b and c: Modified from Phillips, B.L., Brown, G.P., Webb, J.K., Shine, R., 2006. Invasion and the evolution of speed in toads. Nature 439, 803; Photograph (panel a): Matthew Greenlees.)*

new mutations in the evolution of invasive species is contentious, given their infrequent occurrence, slow paths to fixation, and because they often have no (i.e. being neutral) or detrimental effects (Bock et al., 2015). High propagule pressure (i.e. the number, frequency, and size of introduction events), particularly when stemming from multiple introductions, is often linked with the establishment success of exotic species (Simberloff, 2009). Similar to migration between native range populations, multiple introductions can create novel genetic variation in invasive populations via admixture, which may increase trait variation (Irimia et al., 2021) or rescue severely bottlenecked populations. Rapid range expansions also provide opportunities for assortative mating to occur in invasive populations, e.g., via spatial sorting (Shine et al., 2011; Phillips and Perkins, 2019). Lastly, the amount of standing genetic variation in invasive populations will ultimately determine how these populations respond to natural selection (Hodgins et al., 2018; Sherpa and Després, 2021), and many studies have demonstrated local adaption in invasive populations (e.g. Colautti and Barrett, 2013; Mackin et al., 2021). This chapter provides an overview of the main drivers of microevolution and how these operate during invasion. We also briefly discuss the roles of phenotypic plasticity and hologenomic variation in rapid evolution during invasion.

## 2.2 Microevolution during biological invasions

To understand the relative importance and dynamics of the factors that underlie the evolutionary responses of invasive species, we need to consider the overall process that accompanies the transition of an exotic species from being introduced to becoming invasive. All exotic species must negotiate a series of barriers (geography, possibly cultivation, survival, reproduction, dispersal, environment) in order to become invasive, what is commonly referred as the introduction–naturalisation–invasion-continuum (INIC, Blackburn et al., 2011; Richardson and Pyšek, 2012). In short, the INIC conceptualises that a species has to be introduced (deliberately or accidently) by humans into a new environment, survive in that new environment, reproduce, disperse, and survive any additional novel conditions it may encounter (Blackburn et al., 2011; Richardson and Pyšek, 2012). The transitioning of exotic species across these barriers lies at the heart of biogeography, population demography, abiotic and biotic interactions, and therefore evolutionary ecology. It is also important to remember that historical evolution (i.e. preadaptations) play an important role in determining the success of exotic species under novel environmental conditions. That is, colonisation is to some extent dependent on the 'eco-evolutionary experience' (EEE) of both the introduced exotic species and members of the recipient community (Saul et al., 2013; Saul and Jeschke, 2015). Originally defined as the historical exposure of organisms to biotic interactions, EEE emphasises the role of traits that have been previously selected for in both exotic and resident species in driving the establishment success and adaptability of exotic species. As we will discuss in Chapter 3, EEE is a good yardstick for gauging the ease by which introduced species integrate into novel ecological contexts.

## 2.2.1 Founder events, genetic drift and inbreeding

Allele frequencies often fluctuate at random between successive generations. Such genetic drift is an important driver of microevolutionary change, especially in small populations. Many invasive populations, particularly those stemming from unintentional introductions, are founded by a few individuals (Hodgins et al., 2018). However, ascribing shifts in trait values to genetic drift rather than natural selection can be challenging (Keller and Taylor, 2008; Liao et al., 2020). One example comes from the flowering rush, *Butomus umbellatus*. Native European populations of this aquatic plant consist of two cytotypes: triploids that only reproduce asexually via rhizome fragmentation and diploids that can reproduce via rhizome fragments and sexually via seeds. Seventy-one percent of invasive populations in North America are diploid, while only 16% of European populations have this cytotype (Kliber and Eckert, 2005). This makes it tempting to conclude that natural selection favoured sexual reproduction in North America as it would increase genetic diversity and dispersal (Brown and Eckert, 2004). This notion is supported by data showing that North American populations have higher investment in both sexual flowers and asexual rhizome fragments than European populations (Brown and Eckert, 2004). However, genetic analyses indicated that North American invasive populations have experienced a severe genetic bottleneck, with 95% of all plants sharing a single genotype (Kliber and Eckert, 2005). This suggests that reproduction is almost exclusively asexual in invasive populations despite a high investment in seed production. Therefore what appears to be adaptive evolution in the invasive range is likely the consequence of a severe founder event of a small number of asexual genotypes, biased towards diploids that happen to have high capacities for seed production (Brown and Eckert, 2004; Kliber and Eckert, 2005).

Another example of a founder event causing rapid trait differentiation between native and invasive populations comes from South American water hyacinth (*Eichhornia crassipes*). This species has a tristylous breeding system (Ornduff, 1987), so that sexual reproduction only occurs when individuals of different style morphs (long-, mid-, or short) cross-fertilise, an adaptation for the maintenance of high genetic variation. Water hyacinth can also reproduce asexually. Founder events have dramatically influenced the distribution of style morphs throughout the species' invasive ranges globally, essentially nullifying the functioning of the tristylous breeding. All invasive populations consist predominantly, or exclusively, of mid-styled morphs and reproduction is asexually only (Barrett, 1989; Barrett et al., 2008).

Genetic bottlenecks often lead to inbreeding depression (i.e. the expression of deleterious recessive alleles). So why are so many bottlenecked populations of exotic species able to successfully establish and invade new habitats? One possible answer is the interaction between inbreeding depression and environmental conditions experienced by invasive species (Schrieber and Lachmuth, 2017). A growing body of evidence suggests that the effects of inbreeding are often less severe under benign environmental conditions than under harsh and stressful conditions (e.g. Fox and Reed, 2011; Yun and Agrawal, 2014; Schrieber and Lachmuth, 2017). This will of course only be true if deleterious alleles are involved in metabolic pathways associated with stress responses (Schrieber and Lachmuth, 2017). Moreover, inbreeding depression in itself is a stress, as it regularly causes dysfunctional physiological responses, such as an increased synthesis of heat shock proteins (Kristensen et al., 2010). This may create feedback conditions whereby inbred individuals are further compromised in their ability to cope with additional stress (Schrieber and Lachmuth, 2017). Insights from plant-herbivore interactions provide compelling evidence for the link between inbreeding depression and levels of environmental stress. Campbell et al. (2013) investigated the impacts of herbivory on inbreeding depression experienced by inbred and outbred lines of Carolina horsenettle (*Solanum carolinense*). These authors found that "herbivory accounted for a majority of the observed inbreeding depression, with inbreeding loads associated with asexual reproduction [i.e. inbred plants] being consistently high across most families under herbivory, but variable, and indistinguishable from zero overall, when protected [from herbivores]". Indeed, outbred Carolina horsenettle lines had twice the growth performance and fitness of inbred lines under herbivory, but these differences disappeared in the absence of herbivory (Campbell et al., 2013). This example illustrates that selection may only act on deleterious alleles and cause inbreeding depression under stressful environmental conditions. Under benign environmental conditions deleterious alleles may essentially become neutral and drift towards high frequencies, even in inbred populations (Schrieber and Lachmuth, 2017).

Biological invasions often occur in benign environments. For example, exotic species are often liberated from their specialist native enemies (Keane and Crawley, 2002; Heger and Jeschke, 2018), introduced into highly disturbed habitats, or may experience low levels of competition from natives (Catford et al., 2012; Montesinos, 2021). Under these circumstances, it is conceivable that inbreeding depression linked to genes involved in enemy defence, nutrient stress, and competition will be negligible, even in populations with low genetic diversity. For example, Daehler (1999) compared inbreeding depression between outcrossed and selfed progeny of invasive smooth cordgrass (*Spartina alterniflora*) under conditions of high competition and high or low nutrients. Inbreeding depression was consistently higher when plants were grown under competition than under both nutrient conditions. This indicates a larger portion of smooth cordgrass' inbreeding depression is associated with deleterious alleles linked to interspecific competition but not nutrient stress (Daehler, 1999).

Inbreeding may also "benefit" exotic species when they experience weak or moderate genetic bottlenecks, which may allow selection to purge deleterious alleles. This is eloquently illustrated by the globally invasive ladybird, *Harmonia axyridis* (Fig. 2.2; Facon et al., 2011). By comparing fitness of inbred or outbred populations of this species, Facon et al. (2011) showed that invasive populations suffered lower levels of inbreeding depression than some native populations. Specifically, they found inbred and outbred invasive populations to have shorter generation times and higher lifetime performances (i.e. higher fitness) than inbred, but not outbred, native range populations (Fig. 2.2). Inbred native range populations suffered from severe inbreeding depression. These findings suggest that invasive populations have been purged of deleterious alleles. Such purging might be particularly prominent during the initial stages of the invasion (i.e. during establishment) when populations are small. Purging of deleterious alleles would also benefit the establishment of new outlying foci, founded by one or a few individuals, during range expansion. This may have been the case for invasive ladybirds in Europe. Within 8 years of their introduction into Europe these insects evolved higher flying speeds at the leading edge (Lombaert et al., 2014). The purging of deleterious alleles (Facon et al., 2011) may have contributed this rapid evolution since small outlying populations are expected to

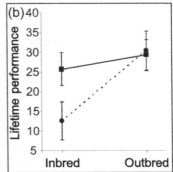

**Fig. 2.2** (a) The ladybird (*Harmonia axyridis*) has undergone rapid evolution in its invasive range in Europe. (b) Founder effects led to the purging of deleterious alleles, so that inbred and outbred invasive populations (*squares*) have similar levels of fitness (measured as the product of hatching rate, larval survival, and subsequent fecundity; see Facon et al., 2011 for details). On the other hand, genetically diverse native range populations (*circles*) suffer marked inbreeding depression when inbred, while outbred populations have levels of fitness that are similar to both inbred and outbred invasive populations. *(Adapted from Facon, B., Hufbauer, R.A., Tayeh, A., Loiseau, A., Lombaert, E., Vitalis, R., et al., 2011. Inbreeding depression is purged in the invasive insect* Harmonia axyridis. *Curr. Biol. 21, 424–427; Photograph (panel a): Ingrid Minnaar.)*

maintain high relative fitness. That is, ladybirds would benefit from low inbreeding depression following long-distance dispersal by a few individuals and consecutive founder events. Similar examples exist in other taxonomic groups. For example, a study on invasive slender false brome (*Brachypodium sylvaticum*) in the USA found core populations to harbour more genetic diversity, but also to suffer from higher inbreeding depression, than leading-edge populations (Marchini et al., 2015).

## 2.2.2 Mutation, hybridisation, and whole-genome duplication

New mutations potentially provide novel phenotypic variation. However, it is reasonable to expect new mutations to contribute little, if anything, to rapid evolutionary responses in invasive species. Mutations occur infrequently and often have deleterious or neutral effects and, therefore, are eliminated by selection or drift. Even when mutations have tangible fitness benefits they usually take time to reach high frequencies. Despite this, a handful of studies have demonstrated new mutations to underlie evolutionary responses in invasive populations. For example, Rollins et al. (2016) found a new mitochondrial mutation in invasive European starlings

(*Sturnus vulgaris*) in Australia to have quickly reached high frequencies in leading-edge populations, while being absent from core populations. This suggests that strong selection, rather than drift, is acting on this mutation, a notion that is supported by simulation models (Rollins et al., 2016).

Hybridisation can rapidly alter the genetic background of organisms, and often in profound ways. Unlike mutations, hybridisation will usually not contribute dominant deleterious alleles to hybrid genomes. Many invasive species, especially plants, are hybrids between native and exotic species (e.g. Manzoor et al., 2020, for review see Ellstrand and Schierenbeck, 2000; Gaskin, 2017). Hybridisation can occur at various levels of relatedness. For example, distinct native range populations of the same species may be co-introduced into the same area. This can lead to genetic admixture, a common feature of invasive populations (e.g. Vicente et al., 2021). The benefits of genetic admixture are well known and include, among others, increased genetic diversity and novel trait variation (e.g. Qiao et al., 2019). Keller et al. (2014) used trait and population genetic data to determine whether genetic admixture contributed to the invasion success of bladder campion (*Silene vulgaris*) in North America. Interestingly, this species also consists of admixed populations in its native range in Europe (Keller et al., 2009). Molecular data suggested that European populations became naturally admixed sometime between the late Pleistocene and the early Holocene, as a result of post-glacial population expansion and secondary contact (Keller et al., 2014). Human-mediated admixture in the invasive North American range of the species only occurred in the last century (Keller et al., 2014). Keller et al. (2014) found levels of admixture to be positively correlated with genetic diversity (heterozygosity) and fitness (number of fruits produced) in North America, but not in Europe. This suggests that recent admixture between distinct bladder campion lineages likely masked deleterious alleles and increased fitness in North American populations.

Closely related species may also hybridise to form invasive hybrids (see Schierenbeck and Ellstrand, 2009). The novel genetic variation gained from hybridisation has various implications for rapid evolution and has been repeatedly linked to invasiveness (Ellstrand and Schierenbeck, 2000; Schierenbeck and Ellstrand, 2009; te Beest et al., 2012). Despite often being maladaptive, some hybrid genotypes may generate highly adaptive trait values. For instance, hybrids often express trait values that transgress those of both parental species, often referred to as heterosis, transgression, or hybrid vigour (te Beest et al., 2012). For example, hybrids between the native sea fig (*Carpobrotus chilensis*) and invasive Hottentots fig (*Carpobrotus edulis*) in

California show higher herbivore resistance and, in some environments, higher vegetative growth than both parental species (Vilà and D'Antonio, 1998). Distinguishing between adaptive evolution and the effects of heterosis can be challenging and calls for experiments that compare the fitness of hybrids and their parental species under different environmental conditions or in reciprocal transplant experiments. Hybridisation will also benefit exotic populations by increasing standing genetic variation.

Hybrids are often inviable or sterile, especially when the genomes of parental species are incompatible (te Beest et al., 2012). In some instances, whole-genome processes like polyploidization can restore fertility while, in other, changes in reproductive strategies, such as the adoption of agamospermy, may occur (Runemark et al., 2019). The latter may lead to the fixation of genotypes and therefore heterosis (e.g. Welles and Ellstrand, 2020; also see Schierenbeck and Ellstrand, 2009; te Beest et al., 2012 for reviews). Humans also deliberately create hybrids and polyploids, e.g., for genetic improvement of forestry species, conferring the genetic advantages discussed before to these species. For example, Griffin et al. (2012) found tetraploid *Acacia mangium* to have lower seed set than diploid genotypes, but a higher capacity for self-fertilisation, a trait often associated with invasiveness (Rambuda and Johnson, 2004; Rodger et al., 2013). Similarly, human-created hybrids between different *Casuarina* species (Ho et al., 2002) and various *Prosopis* species (van Klinken et al., 2006) are now considered some of the most damaging invasive trees globally.

The recent invasion of freshwater habitats around the world by the marbled crayfish (*Procambarus virginalis*) represents one of the only examples that I am aware of where polyploid speciation preceded invasion (also see Welles and Ellstrand, 2020). The coming into being of marbled crayfish is a truly remarkable story. The ancestral species, slough crayfish (*P. fallax*), is native to Florida in the USA. This species underwent autopolyploidization to give rise to the all-female triploid marbled crayfish (Scholtz et al., 2003; Lyko, 2017; Vogt et al., 2015). Marbled crayfish reproduces parthenogenetically, grows bigger, and are more fecund than the slough crayfish, which made it an attractive species in the pet trade (Fig. 2.3; Vogt, 2017). The species first appeared in the German aquarium trade in 1995, and it remains unknown whether it originated in Germany following the introduction of slough crayfish from the USA or whether it was introduced directly from the USA. The current restriction of marbled crayfish to Africa, Asia, and Europe supports the intriguing hypothesis of a post-introduction genesis. Irrespective of where and when marbled crayfish originated, it is astonishing to think that

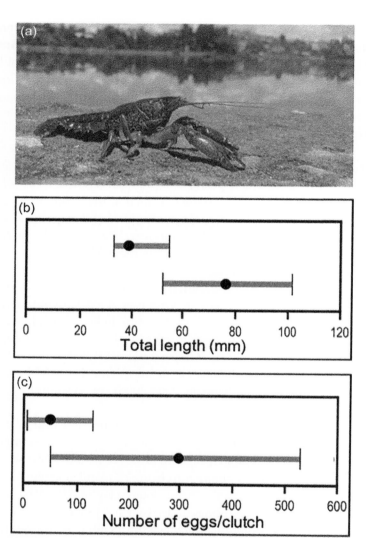

**Fig. 2.3** (a) Marble crayfish (*Procambarus virginalis*) is a clonal polyploid descendant of the slough crayfish (*P. fallax*). Compared to the ancestral species, marble crayfishes grow bigger (b) and are more fecund (c). *Red (dark grey in the printed version) bars* indicate trait values for marble crayfish and *blue (light grey in the printed version) bars* those for slough crayfish (*dots and horizontal bars* indicate means and ranges, respectively). The marble crayfish is now invasive in 12 countries. *(Panels b and c: Modified from Vogt, G., 2017. Facilitation of environmental adaptation and evolution by epigenetic phenotype variation: insights from clonal, invasive, polyploid, and domesticated animals. Environ. Epigenet. 3, 1–17; Photograph (panel a): Ranja Andriantsoa.)*

all invasive populations, now found in 12 countries on three continents, are the descendants of a single female once held captive in an aquarium in Germany (Gutekunst et al., 2018).

### 2.2.3 Assortative mating

The spatial and temporal dynamics of rapidly expanding invasive populations differ substantially from that of populations under ecological equilibria. For example, spatial sorting theory predicts the accrual of highly dispersive phenotypes at the leading edge of expanding invasive populations (Shine et al., 2011; Phillips and Perkins, 2019). This, in turn, will increase the probability of highly dispersive individuals mating with each other, rather than with slow-dispersing individuals, at the leading edge (Shine et al., 2011; Phillips and Perkins, 2019). It follows that higher dispersal ability will evolve in leading-edge populations. Spatial sorting therefore represents an intriguing alternative to natural selection, whereby selection on phenotypes acts through space rather than time (Phillips and Perkins, 2019). The rapid evolution of increased dispersal at the leading edge of cane toads in Australia resulted from spatial sorting (Phillips et al., 2006). Leading-edge toads also have higher feeding rates, grow faster, and store more energy than toads at the core (Brown et al., 2013). However, spatial sorting can lead to trade-offs so that higher-dispersing phenotypes may not necessarily have high local fitness. In Australia, for example, more dispersive cane toads at the leading edge are more prone to severe spinal injuries (Brown et al., 2007; Shilton et al., 2008) and have higher mortality rates (Phillips et al., 2008), compared with less dispersive toads at the core. In this instance the perceived benefits of spatially sorted traits at the leading edge may therefore be balanced by the negative effects on fitness in the long run (Brown et al., 2013).

### 2.2.4 Natural selection

Some have argued that evolutionary responses in invasive species are at least as common than they are in native species (Oduor et al., 2016). However, whether these responses are adaptive often remains unknown (Colautti and Lau, 2015, but see Colautti and Barrett, 2013). As invasive species spread, strong selection may stem from changes in abiotic attributes of their physiological niches, such as temperature, precipitation, and soil chemistry. These changes often manifest as clinal trait variation. For example, many plant invaders show latitudinal clines in size and phenology (e.g. Novy et al., 2013, also see Chapter 6). Because of the usually strong covariance between latitude and abiotic conditions, such clines may reflect plastic responses, patterns of isolation by distance, or local adaptation. As we will discuss in Chapter 6, local adaptation in clinal variation has been demonstrated for some invasive species.

The selection pressures experienced by species in their historical native ranges may also be significantly relaxed upon introduction into new areas or may disappear entirely. For example, exotic species are often liberated from their specialist natural enemies (Heger and Jeschke, 2018), creating opportunities for the reallocation of (often costly) resources used for defence towards performance, in what has been referred to as the 'evolution of increased competitive ability' (EICA) (Blossey and Nötzold, 1995). An emerging view is that EICA may involve various trade-offs (i.e. not only specialist enemy defence) or, sometimes, no trade-offs (e.g. Ridenour et al., 2008). How common EICA is remains an open question in ecology, with mixed support from empirical data (Bossdorf et al., 2005). One example comes from invasive Chinese tallow (*Sapium sebiferum*) in the USA. Siemann and Rogers (2001) compared the performance and chemical defences of recently introduced, long-established, and native populations of this tree. They found tannin concentrations to be highest in native range populations, intermediate in old invasive populations, and completely absent in recently established populations. This suggests that the reallocation of defence resources towards growth and reproduction may only be temporary, as native resident herbivores and pathogens 'catch up' with invasive populations over time. Similarly, Zangerl and Berenbaum (2005) analysed concentrations of defence chemicals in herbarium specimens of the parsnip (*Pastinaca sativa*), spanning 152 years of collections in the species' invasive North American range. They found a spike in defence phytotoxins to coincide with the accidental introduction of a major specialist herbivore of the species, the parsnip webworm (*Depressaria pastinacella*), again suggesting that invasive species may re-evolve defence strategies against specialist enemies.

It was recently suggested that transport vectors impose strong selection on exotic species while in transit (Pettit et al., 2017; Briski et al., 2018). For example, population numbers of plankton often plummet during transport in ballast water because of physico-chemical conditions related to temperature, salinity, dissolved oxygen levels, exposure to rust and metal-based paints, etc. (Seiden et al., 2011; Briski et al., 2014; Chan et al., 2015). Selection is therefore expected to increase the frequency of genotypes that are preadapted to these conditions which, in turn, may increase colonisation success in highly polluted anthropic habitats like harbours (Briski et al.,

2018). Similarly, fouling organisms that are attached to ship hulls may experience selection for high attachment strength and/or low drag coefficients, which may benefit invasiveness and facilitate secondary introductions (e.g. Clarke Murray et al., 2012). The main environmental drivers that underlie rapid adaptive evolution, and the various evolutionary responses by invasive species, are discussed in detail in Chapters 5 and 6, respectively.

## 2.2.5  Artificial selection

Exotic species that have been intentionally introduced for human usage, such as in the ornamental trade or for forestry, are sometimes subjected to artificial selection. Artificial selection may, in turn, inadvertently select for traits that benefit invasiveness. For instance, resistance selection against resident pathogens in exotic forestry trees is a common practice. Forestry species are also often subjected to artificial selection in traits that enhance yield (growth rates and biomass production, e.g. Wang et al., 1999). No empirical evidence exists, to my knowledge, that directly links the invasiveness of exotic crop or forestry species to artificial selection prior to their escape and spread.

## 2.3  Epigenetic variation

The study of phenotypic variation due to changes in gene expression that are modulated by non-genetic mechanisms is referred to as epigenetics. The nature of epigenetic modifications is often contingent upon environmental conditions and thus integral in modulating phenotypic plasticity (see Chapter 4 for details). It is now widely accepted that epigenetic variation may hold important consequences for evolution (Feinberg and Irizarry, 2010; Rey et al., 2020). From cancer tumours (Feinberg et al., 2006) to recently diverged species complexes (e.g. Paun et al., 2010), the effects of epigenetic modifications on individual fitness, population divergence, and even speciation are evident. Yet, the link between epigenetic variation and heritable phenotypic variation remains poorly understood. Some studies have shown that certain types of epigenetic variation, such as DNA methylation, impact trait variation (e.g. Herrera and Bazaga, 2010, 2013). While it is tempting to interpret these correlations as epigenetic variation being able to contribute to adaptive evolution, they may simply reflect genetic differences, i.e. epigenetic variation that is under genetic control. Moreover, environmentally induced epigenetic variation is often not heritable, and therefore only contributes to phenotypic plasticity rather than the heritable variation thereof (Richards, 2006). Despite these challenges, it is now widely accepted that epigenetic variation plays an important role in the establishment success of exotic species (see Chapter 4).

Some epigenetic variation has been found to be 'partially' heritable, i.e. over a few generations, before being reversed (Rey et al., 2020). This may serve as a sort of 'molecular memory' to optimise the environmental compatibility of genotypes through plasticity (Wilschut et al., 2016). Richards et al. (2012) measured population epigenetic structure (as differences in DNA methylation) in a few genotypes of Japanese knotweed (*Reynoutria japonica* and the hybrid taxon *R. × bohemica*), a highly successful invasive plant. They grew knotweed plants from different invaded habitats in the USA under common garden conditions and found that a significant proportion of epigenetic variation could be explained by population origin. This supports the idea that epigenetic variation is strongly influenced by local environmental conditions and, more importantly, that it can be inherited via clonal propagation, even in the absence of the environmental factors that created it.

As we discussed earlier, interspecific hybridisation has been repeatedly linked with invasiveness (Ellstrand and Schierenbeck, 2000). The union of divergent parental genomes in hybrids often demands major epigenetic remodelling (Rieseberg, 2001). For example, in invasive allopolyploid cordgrass (*Spartina anglica*) in Europe, changes in DNA methylation are predominantly associated with interspecific hybridisation and not with intraspecific genome doubling (Salmon et al., 2005). Research on the roles of epigenetic variation in the establishment success and spread of exotic species is emerging rapidly and Chapter 4 discusses the field of 'invasion epigenetics' in detail.

## 2.4 Holobionts and hologenomes

Higher organisms are not autonomous individuals but represent various biological components, each comprised of many closely associated organisms, or so-called holobionts (Bordenstein and Theis, 2015; Zenni et al., 2017; Carthey et al., 2020). Host-associated microbial communities (or microbiomes) can have significant impacts on the physiology, anatomy, behaviour, and reproduction of their hosts, and therefore represent a significant source of phenotypic variation (Carthey et al., 2020; Zenni et al., 2017). It is important to keep in mind that microbiome diversity and composition can be driven by deterministic processes, either through host selection or vertical transmission, or by stochastic processes such as ecological drift or random dispersal (Bordenstein and Theis, 2015). Our current understanding of how the genomes of hosts and their microbiota, or hologenomes, influence evolution remains rudimentary, with limited insights from model organisms. For example, changes in the gut microbiome of fruit flies, associated with food sources, can cause heritable changes in mating behaviour, suggesting that diet could act as a precursor to reproductive isolation (Sharon et al., 2010).

The fruit fly example illustrates that hologenomic variation can facilitate rapid trait diversification in response to novel environmental conditions. Exotic species will experience substantial release from their microbiomes upon introduction, in the same way they are freed from their macro symbionts (Le Roux, 2020). This can lead to 'hologenomic bottlenecks' with consequences for the survival, reproduction, and adaptive capacities of exotic species. Losses in both ecological (i.e. loss of microbiome richness) and genetic diversity (i.e. loss of microbial genetic diversity) can cause hologenomic bottlenecks. However, unlike genetic bottlenecks, hologenomic bottlenecks are expected to quickly recover as exotic species accumulate new microbiota over time. Invasive red turpentine beetles (*Dendroctonus valens*) in China provide an excellent example of how microbiome reassembly could impact evolutionary responses in invasive species. This North American beetle was co-introduced into China with its symbiotic fungus, *Leptographium procerum*, and has had devastating impacts on native Chinese red pines (*Pinus tabuliformis*) (Sun et al., 2013). The beetle acts as a vector for the fungus which, upon infection of the pine, stimulates the release of volatile pine terpenoids. These terpenoids, in turn, attract red turpentine beetles to pine trees (Lu et al., 2010, 2011). However, soon after their introduction into China, the beetle also formed associations with resident fungi. These Chinese fungi cause the expression of a defensive phenolic compound, naringenin, by pines, which strongly suppresses the co-introduced invasive beetle-fungus complex (Cheng et al., 2016). Subsequently, a tripartite symbiosis developed between the co-introduced beetle and fungus and the microbiome of the beetle gallery (space between bark and tree wood where beetle larvae develop). These new associations have dramatically increased the beetle's fitness via bacterial-mediated degradation of naringenin (Cheng et al., 2018). An appreciation of the role of hologenomic variation during biological invasions, their impacts on native communities, and the evolutionary responses of invasive species is only just emerging (Le Roux, 2020) and, as the invasion of red turpentine beetles in China illustrates, provides fascinating opportunities for future research.

## 2.5 Conclusions

The ecology of invasive species has traditionally received more research attention than their evolutionary biology. Ecologists have been primarily interested in understanding why certain species become invasive and others not, why some habitats are more vulnerable to invasion than others, what the ecological impacts of invasive species are, and how to best manage invasive populations. The evolutionary underpinnings of many of these aspects have since emerged. The demographic and ecological conditions during the introduction, establishment, and spread of exotic species create unique opportunities for microevolution, making biological invasions excellent systems to study rapid evolution.

## References

Alford, R.A., Brown, G.P., Schwarzkopf, L., Phillips, B.L., Shine, R., 2009. Comparisons through time and space suggest rapid evolution of dispersal behaviour in an invasive species. Wildl. Res. 36, 23–28.

Allendorf, F.W., Lundquist, L.L., 2003. Introduction: population biology, evolution, and control of invasive species. Conserv. Biol. 17, 24–30.

Barrett, S.C.H., 1989. Waterweed invasions. Sci. Am. 260, 90–97.

Barrett, S.C.H., 2015. Foundations of invasion genetics: the Baker and Stebbins legacy. Mol. Ecol. 24, 1927–1941.

Barrett, S.C.H., Colautti, R.I., Eckert, C.G., 2008. Plant reproductive systems and evolution during biological invasion. Mol. Ecol. 17, 373–383.

Blackburn, T.M., Pyšek, P., Bacher, S., Carlton, J.T., Duncan, R.P., Jarošík, V., Wilson, J.R.U., Richardson, D.M., 2011. A proposed unified framework for biological invasions. Trends Ecol. Evol. 26, 333–339.

Blossey, B., Nötzold, R., 1995. Evolution of increased competitive ability in plants: a hypothesis. J. Ecol. 83, 887–889.

Bock, D.G., Caseys, C., Cousens, R.D., Hahn, M.A., Heredia, S.M., Hübner, S., Turner, K.G., Whitney, K.D., Rieseberg, L.H., 2015. What we still don't know about invasion genetics. Mol. Ecol. 24, 2277–2298.

Bordenstein, S.R., Theis, K.R., 2015. Host biology in light of the microbiome: ten principles of holobionts and hologenomes. PLoS Biol. 13, 1–23.

Bossdorf, O., Auge, H., Lafuma, L., Rogers, W.E., Siemann, E., Prati, D., 2005. Phenotypic and genetic differentiation between native and introduced plant populations. Oecologia 144, 1–11.

Briski, E., Chan, F., MacIsaac, H.J., Bailey, S.A., 2014. A conceptual model of community dynamics during the transport stage of the invasion process: a case study of ships' ballast. Divers. Distrib. 20, 236–244.

Briski, E., Chan, F.T., Darling, J.A., Lauringson, V., Macisaac, H.J., Zhan, A., Bailey, S.A., 2018. Beyond propagule pressure: importance of selection during the transport stage of bio- logical invasions. Front. Ecol. 16, 345–353.

Brown, J.S., Eckert, C.G., 2004. Evolutionary increase in sexual and clonal reproductive capacity during biological invasion in an aquatic plant *Butomus umbellatus* (Butomaceae). Am. J. Bot. 92, 495–502.

Brown, G.P., Shilton, C., Phillips, B.L., Shine, R., 2007. Invasion, stress, and spinal arthritis in cane toads. Proc. Natl. Acad. Sci. U. S. A. 104, 17698–17700.

Brown, G.P., Kelehear, C., Shine, R., 2013. The early toad gets the worm: cane toads at an invasion front benefit from higher prey availability. J. Anim. Ecol. 82, 854–862.

Campbell, S.A., Thaler, J.S., Kessler, A., 2013. Plant chemistry underlies herbivore-mediated inbreeding depression in nature. Ecol. Lett. 16, 252–260.

Carthey, A.J.R., Blumstein, D.T., Gallagher, R.V., Tetu, S.G., Gillings, M.R., 2020. Conserving the holobiont. Funct. Ecol. 34, 764–776.

Catford, J.A., Daehler, C.C., Murphy, H.T., Sheppard, A.W., Hardesty, B.D., Westcott, D. A., Rejmánek, M., Bellingham, P.J., Pergl, J., Horvitz, C.C., Hulme, P.E., 2012. The intermediate disturbance hypothesis and plant invasions: implications for species richness and management. Perspect. Plant Ecol. Evol. Syst. 14, 231–241.

Catford, J.A., Smith, A.L., Wragg, P.D., Clark, A.T., Kosmala, M., Cavender-Bares, J., Reich, P.B., Tilman, D., 2019. Traits linked with species invasiveness and community invasibility vary with time, stage and indicator of invasion in a long-term grassland experiment. Ecol. Lett. 22, 593–604.

Chan, F.T., Bradie, J., Briski, E., Bailey, S.A., Simard, N., MacIsaac, H.J., 2015. Assessing introduction risk using species' rank-abundance distributions. Proc. R. Soc. B Biol. Sci. 282, 20141517.

Cheng, C., Xu, L., Xu, D., Lou, Q., Lu, M., Sun, J., 2016. Does cryptic microbiota mitigate pine resistance to an invasive beetle–fungus complex? Implications for invasion potential. Sci. Rep. 6, 33110.

Cheng, C., Wickham, J.D., Chen, L., Xu, D., Lu, M., Sun, J., 2018. Bacterial microbiota protect an invasive bark beetle from a pine defensive compound. Microbiome 6, 132.

Clarke Murray, C., Therriault, T.W., Martone, P.T., 2012. Adapted for invasion? Comparing attachment, drag and dislodgment of native and nonindigenous hull fouling species. Biol. Invasions 14, 1651–1663.

Colautti, R.I., Barrett, S.C.H., 2013. Rapid adaptation to climate facilitates range expansion of an invasive plant. Science 342, 364–366.

Colautti, R.I., Lau, J.A., 2015. Contemporary evolution during invasion: evidence for differentiation, natural selection, and local adaptation. Mol. Ecol. 24, 1999–2017.

Daehler, C.C., 1999. Inbreeding depression in smooth cordgrass (Spartina alterniflora, Poaceae) invading San Francisco Bay. Am. J. Bot. 86, 131–139.

Darwin, C.R., 1859. The Origin of Species. John Murray, London.

Diagne, C., Leroy, B., Vaissière, A.C., Gozlan, R.E., Roiz, D., Jarić, I., Salles, J.M., Bradshaw, C.J.A., Courchamp, F., 2021. High and rising economic costs of biological invasions worldwide. Nature 592, 571–576.

Dlugosch, K.M., Parker, I.M., 2008. Founding events in species invasions: genetic variation, adaptive evolution, and the role of multiple introductions. Mol. Ecol. 17, 431–449.

Ellstrand, N.C., Schierenbeck, K.A., 2000. Hybridization as a stimulus for the evolution of invasiveness in plants? Proc. Natl. Acad. Sci. U. S. A. 97, 7043–7050.

Estoup, A., Ravigné, V., Hufbauer, R., Vitalis, R., Gautier, M., Facon, B., 2016. Is there a genetic paradox of biological invasion? Annu. Rev. Ecol. Evol. Syst. 47, 51–72.

Facon, B., Hufbauer, R.A., Tayeh, A., Loiseau, A., Lombaert, E., Vitalis, R., Guillemaud, T., Lundgren, J.G., Estoup, A., 2011. Inbreeding depression is purged in the invasive insect Harmonia axyridis. Curr. Biol. 21, 424–427.

Feinberg, A.P., Irizarry, R.A., 2010. Stochastic epigenetic variation as a driving force of development evolutionary adaptation and disease. Proc. Natl. Acad. Sci. U. S. A. 107, 1757–1764.

Feinberg, A.P., Ohlsson, R., Henikoff, S., 2006. The epigenetic progenitor origin of human cancer. Nat. Rev. Genet. 7, 21–33.

Fox, C.W., Reed, D.H., 2011. Inbreeding depression increases with environmental stress: an experimental study and meta-analysis. Evolution 65, 246–258.

Gaskin, J.F., 2017. The role of hybridization in facilitating tree invasion. AoB Plants 9, plw079.

Griffin, A.R., Vuong, T.D., Vaillancourt, R.E., Harbard, J.L., Harwood, C.E., Nghiem, C.Q., Thinh, H.H., 2012. The breeding systems of diploid and neoautotetraploid clones of Acacia mangium Willd. in a synthetic sympatric population in Vietnam. Sex. Plant Reprod. 25, 1–9.

Gutekunst, J., Andriantsoa, R., Falckenhayn, C., Hanna, K., Stein, W., Rasamy, J., Lyko, F., 2018. Clonal genome evolution and rapid invasive spread of the marbled crayfish. Nat. Ecol. Evol. 2, 567–573.

Heger, T., Jeschke, J.M., 2018. Enemy release hypothesis. In: Jeschke, J.M., Heger, T. (Eds.), Invasion Biology: Hypotheses and Evidence. CABI, pp. 92–102.

Herrera, C.M., Bazaga, P., 2010. Epigenetic differentiation and relationship to adaptive genetic divergence in discrete populations of the violet Viola cazorlensis. New Phytol. 187, 867–876.

Herrera, C.M., Bazaga, P., 2013. Epigenetic correlates of plant phenotypic plasticity: DNA methylation differs between prickly and nonprickly leaves in heterophyllous Ilex aquifolium (Aquifoliaceae) trees. Bot. J. Linn. Soc. 171, 441–452.

Ho, K.Y., Ou, C.H., Yang, J.C., Hsiao, J.Y., 2002. An assessment of DNA polymorphisms and genetic relationships of Casuarina equisetifolia using RAPD markers. Bot. Bull. Acad. Sin. 43, 93–98.

Hodgins, K.A., Bock, D.G., Rieseberg, L.H., 2018. Trait evolution in invasive species. Annu. Plant Rev. Online 1, 1–37. https://doi.org/10.1002/9781119312994.apr0643.

Irimia, R.E., Hierro, J.L., Branco, S., Sotes, G., Cavieres, L.A., Eren, O., Lortie, C.J., French, K., Callaway, R.M., Montesinos, D., 2021. Experimental admixture among geographically disjunct populations of an invasive plant yields a global mosaic of reproductive incompatibility and heterosis. J. Ecol. 109, 2152–2162.

Keane, R.M., Crawley, M.J., 2002. Exotic plant invasions and the enemy release hypothesis. Trends Ecol. Evol. 17, 164–170.

Keller, S.R., Taylor, D.R., 2008. History, chance and adaptation during biological invasion: separating stochastic phenotypic evolution from response to selection. Ecol. Lett. 11, 852–866.

Keller, S.R., Sowell, D.R., Neiman, M., Wolfe, L.M., Taylor, D.R., 2009. Adaptation and colonization history affect the evolution of clines in two introduced species. New Phytol. 183, 678–690.

Keller, S.R., Fields, P.D., Berardi, A.E., Taylor, D.R., 2014. Recent admixture generates heterozygosity-fitness correlations during the range expansion of an invading species. J. Evol. Biol. 27, 616–627.

Kliber, A., Eckert, C.G., 2005. Interaction between founder effect and selection during biological invasion in an aquatic plant. Evolution 59, 1900–1913.

Kristensen, T.N., Pedersen, K.S., Vermeulen, C.J., Loeschcke, V., 2010. Research on inbreeding in the 'omic' era. Trends Ecol. Evol. 25, 44–52.

Le Roux, J.J., 2020. Molecular ecology of plant–microbial interactions during invasions: progress and challenges. In: Traveset, A., Richardson, D.M. (Eds.), Plant Invasions: The Role of Biotic Interactions. Invasive Species Series, CABI, UK, pp. 340–362.

Liao, Z.Y., Scheepens, J.F., Li, Q.M., Wang, W.-B., Feng, Y.-L., Zheng, Y.-L., 2020. Founder effects, post-introduction evolution and phenotypic plasticity contribute to invasion success of a genetically impoverished invader. Oecologia 192, 105–118.

Linders, T.E.W., Schaffner, U., Alamirew, T., Allan, E., Choge, S.K., Eschen, R., Shiferaw, H., Manning, P., 2021. Stakeholder priorities determine the impact of an alien tree invasion on ecosystem multifunctionality. People Nat. 3, 658–672.

Llewelyn, J., Phillips, B.L., Alford, R.A., Schwarzkopf, L., Shine, R., 2010. Locomotor performance in an invasive species: cane toads from the invasion front have greater endurance, but not speed, compared to conspecifics from a long-colonised area. Oecologia 162, 343–348.

Lombaert, E., Estoup, A., Facon, B., Joubard, B., Grégoire, J.-.C., Jannin, A., Blin, A., Guillemaud, T., 2014. Rapid increase in dispersal during range expansion in the invasive ladybird *Harmonia axyridis*. J. Evol. Biol. 27, 508–517.

Lu, M., Wingfield, M.J., Gillette, N.E., Mori, S.R., Sun, J.-H., 2010. Complex interactions among host pines and fungi vectored by an invasive bark beetle. New Phytol. 187, 859–866.

Lu, M., Wingfield, M.J., Gillette, N., Sun, J.H., 2011. Do novel genotypes drive the success of an invasive bark beetle–fungus complex? Implications for potential reinvasion. Ecology 92, 2013–2019.

Lyko, F., 2017. The marbled crayfish (Decapoda: Cambaridae) represents an independent new species. Zootaxa 4363, 544–552.

Mackin, C.R., Peña, J.F., Blanco, M.A., Balfour, N.J., Castellanos, M.C., 2021. Rapid evolution of a floral trait following acquisition of novel pollinators. J. Ecol. 109, 2234–2246.

Manzoor, S.A., Grffiths, G., Obiakara, M.C., Esparza-Estrada, C.E., Lukac, M., 2020. Evidence of ecological niche shift in *Rhododendron ponticum* (L.) in Britain: hybridization as a possible cause of rapid niche expansion. Ecol. Evol. 10, 2040–2050.

Marchini, G.L., Sherlock, N.C., Ramakrishnan, A.P., Rosenthal, D.M., Cruzan, M.B., 2015. Rapid purging of genetic load in a metapopulation and consequences for range expansion in an invasive plant. Biol. Invasions 18, 183–196.

Montesinos, D., 2021. Fast invasives fastly become faster: invasive plants align largely with the fast side of the plant economics spectrum. J. Ecol., 1–5. https://doi.org/10.1111/1365-2745.13616. In press.

Novy, A., Flory, S.L., Hartman, J.M., 2013. Evidence for rapid evolution of phenology in an invasive grass. J. Evol. Biol. 26, 443–450.

Oduor, A.M.O., Leimu, R., van Kleunen, M., 2016. Invasive plant species are locally adapted just as frequently and at least as strongly as native plant species. J. Ecol. 104, 957–968.

Ornduff, R., 1987. Reproductive systems and chromossome races of *Oxalis pes-caprae* L. and their bearing on the genesis of a noxious weed. Ann. Mo. Bot. Gard. 74, 79–84.

Paun, O., Bateman, R.M., Fay, M.F., Hedren, M., Civeyrel, L., Chase, M.W., 2010. Stable epigenetic effects impact adaptation in allopolyploid orchids (Dactylorhiza: Orchidaceae). Mol. Biol. Evol. 27, 2465–2473.

Pettit, L., Greenlees, M., Shine, R., 2017. The impact of transportation and translocation on dispersal behaviour in the invasive cane toad. Oecologia 184, 411–422.

Phillips, B.L., Perkins, T.A., 2019. Spatial sorting as the spatial analogue of natural selection. Theor. Ecol. 12, 155–163.

Phillips, B.L., Shine, R., 2004. Adapting to an invasive species: toxic cane toads induce morphological change in Australian snakes. Proc. Natl. Acad. Sci. U. S. A. 101, 17150–17155.

Phillips, B.L., Shine, R., 2006a. Allometry and selection in a novel predator-prey system: Australian snakes and the invading cane toad. Oikos 112, 122–130.

Phillips, B.L., Shine, R., 2006b. Spatial and temporal variation in the morphology (and thus, predicted impact) of an invasive species in Australia. Ecography 29, 205–212.

Phillips, B.L., Shine, R., 2006c. An invasive species induces rapid adaptive change in a native predator: cane toads and black snakes in Australia. Proc. R. Soc. B Biol. Sci. 273, 1545–1550.

Phillips, B.L., Brown, G.P., Shine, R., 2003. Assessing the potential impact of cane toads on Australian snakes. Conserv. Biol. 17, 1738–1747.

Phillips, B.L., Brown, G.P., Shine, R., 2004. Assessing the potential for an evolutionary response to rapid environmental change: invasive toads and an Australian snake. Evol. Ecol. Res. 6, 799–811.

Phillips, B.L., Brown, G.P., Webb, J.K., Shine, R., 2006. Invasion and the evolution of speed in toads. Nature 439, 803.

Phillips, B.L., Brown, G.P., Greenlees, M., Webb, J.K., Shine, R., 2007. Rapid expansion of the cane toad (*Bufo marinus*) invasion front in tropical Australia. Austral Ecol. 32, 169–176.

Phillips, B.L., Brown, G.P., Travis, J.M.J., Shine, R., 2008. Reid's paradox revisited: the evolution of dispersal in range-shifting populations. Am. Nat. 172, S34–S48.

Pimentel, D., Zuniga, R., Morrison, D., 2005. Update on the environmental and economic costs associated with alien-invasive species in the United States. Ecol. Econ. 52, 273–288.

Pyšek, P., Richardson, D.M., 2007. Traits associated with invasiveness in alien plants: where do we stand? In: Nentwig, W. (Ed.), Biological Invasions. Springer, pp. 97–125.

Pyšek, P., Hulme, P.E., Simberloff, D., Bacher, S., Blackburn, T.M., Carlton, J.T., Dawson, W., Essl, F., Foxcroft, L.C., Genovesi, P., Jeschke, J.M., Kühn, I., Liebhold, A.M., Mandrak, N.E., Meyerson, L.A., Pauchard, A., Pergl, J., Roy, H.E., Seebens, H., van Kleunen, M., Vilà, M., Wingfield, M.J., Richardson, D.M., 2020. Scientists' warning on invasive alien species. Biol. Rev. 95, 1511–1534.

Qiao, H., Liu, W., Zhang, Y., Zhang, Y.-.Y., Li, Q.Q., 2019. Genetic admixture accelerates invasion via provisioning rapid adaptive evolution. Mol. Ecol. 28, 4012–4027.

Rambuda, T.D., Johnson, S.D., 2004. Breeding systems of invasive alien plants in South Africa: does Baker's rule apply? Divers. Distrib. 10, 409–416.

Rejmanek, M., Richardson, D.M., 1996. What attributes make some plant species more invasive? Ecology 77, 1655–1661.

Rey, O., Eizaguirre, C., Angers, B., Baltazar-Soares, M., Sagonas, K., Prunier, J.G., Blanchet, S., 2020. Linking epigenetics and biological conservation: towards a conservation epigenetics perspective. Funct. Ecol. 34, 414–427.

Reznick, D.N., Losos, J., Travis, J., 2019. From low to high gear: there has been a paradigm shift in our understanding of evolution. Ecol. Lett. 22, 233–244.

Richards, E.J., 2006. Inherited epigenetic variation—revisiting soft inheritance. Nat. Rev. Genet. 7, 395–401.

Richards, C.L., Schrey, A.W., Pigliucci, M., 2012. Invasion of diverse habitats by few Japanese knotweed genotypes is correlated with epigenetic differentiation. Ecol. Lett. 15, 1016–1025.

Richardson, D.M., Pyšek, P., 2012. Naturalization of introduced plants: ecological drivers of biogeographic patterns. New Phytol. 196, 383–396.

Ridenour, W.M., Vivanco, J.M., Feng, Y., Horiuchi, J., Callaway, R.M., 2008. No evidence for trade-offs: *Centaurea* plants from America are better competitors and defenders. Ecol. Monogr. 78, 369–386.

Rieseberg, L.H., 2001. Chromosomal rearrangements and speciation. Trends Ecol. Evol. 16, 351–358.

Rodger, J.G., van Kleunen, M., Johnson, S.D., 2013. Pollinators, mates and Allee effects: the importance of self-pollination for fecundity in an invasive lily. Funct. Ecol. 27, 1023–1033.

Rollins, L.A., Woolnough, A.P., Fanson, B.G., Cummins, M.L., 2016. Selection on mitochondrial variants occurs between and within individuals in an expanding invasion. Mol. Biol. Evol. 33, 995–1007.

Runemark, A., Vallejo-Marin, M., Meier, J.I., 2019. Eukaryote hybrid genomes. PLoS Genet. 15, e1008404.

Sage, R.F., 2020. Global change biology: a primer. Glob. Chang. Biol. 26, 3–30.

Salmon, A., Ainouche, M.L., Wendel, J.F., 2005. Genetic and epigenetic consequences of recent hybridization and polyploidy in *Spartina* (Poaceae). Mol. Ecol. 14, 1163–1175.

Saul, W.C., Jeschke, J.M., 2015. Eco-evolutionary experience in novel species interactions. Ecol. Lett. 18, 236–245.

Saul, W.C., Jeschke, J., Heger, T., 2013. The role of eco-evolutionary experience in invasion success. NeoBiota 17, 57–74.

Schierenbeck, K.A., Ellstrand, N.C., 2009. Hybridization and the evolution of invasiveness in plants and other organisms. Biol. Invasions 11, 1093–1105.

Scholtz, G., Braband, A., Tolley, L., Reimann, A., Mittmann, B., Lukhaup, C., Steuerwald, F., Vogt, G., 2003. Parthenogenesis in an outsider crayfish. Nature 421, 806.

Schrieber, K., Lachmuth, S., 2017. The genetic paradox of invasions revisited: the potential role of inbreeding × environment interactions in invasion success. Biol. Rev. 92, 939–952.

Seebens, H., Blackburn, T.M., Dyer, E.E., Genovesi, P., Hulme, P.E., Jeschke, J.M., Pagad, S., Pyšek, P., Winter, M., Arianoutsou, M., Bacher, S., Blasius, B., Brundu, G., Capinha, C., Celesti-Grapow, L., Dawson, W., Dullinger, S., Fuentes, N., Jäger, H., Kartesz, J., Essl, F., 2017. No saturation in the accumulation of alien species worldwide. Nat. Commun. 8, 14435. https://doi.org/10.1038/ncomms14435.

Seebens, H., Bacher, S., Blackburn, T.M., Capinha, C., Dawson, W., Dullinger, S., Genovesi, P., Hulme, P.E., van Kleunen, M., Kühn, I., Jeschke, J.M., Lenzner, B., Liebhold, A.M., Pattison, Z., Pergl, J., Pyšek, P., Winter, M., Essl, F., 2021. Projecting the continental accumulation of alien species through to 2050. Glob. Chang. Biol. 27, 970–982.

Seiden, J.M., Way, C.J., Rivkin, R.B., 2011. Bacterial dynamics in ballast water during transoceanic voyages of bulk carriers: environmental controls. Mar. Ecol. Prog. Ser. 436, 145–159.

Sharon, G., Segal, D., Ringo, J.M., Hefetz, A., Zilber-Rosenberg, I., Rosenberg, E., 2010. Commensal bacteria play a role in mating preference of *Drosophila melanogaster*. Proc. Natl. Acad. Sci. U. S. A. 107, 20051–20056.

Sherpa, S., Després, L., 2021. The evolutionary dynamics of biological invasions: a multi-approach perspective. Evol. Appl. 14, 1463–1484.

Shilton, C.M., Brown, G.P., Benedict, S., Shine, R., 2008. Spinal arthropathy associated with *Ochrobactrum anthropi* in free-ranging cane toads (*Chaunus [Bufo] marinus)* in Australia. Vet. Pathol. 45, 85–94.

Shine, R., 2010. The ecological impact of invasive cane toads (*Bufo marinus*) in Australia. Q. Rev. Biol. 85, 253–291.

Shine, R., Brown, G.P., Phillips, B.L., 2011. An evolutionary process that assembles phenotypes through space rather than through time. Proc. Natl. Acad. Sci. U. S. A. 108, 5708–5711.

Shine, R., Ward-Fear, G., Brown, G.P., 2020. A famous failure: why were cane toads an ineffective biocontrol in Australia? Conserv. Sci. Pract. 2, e296.

Siemann, E., Rogers, W.E., 2001. Genetic differences in growth of an invasive tree species. Ecol. Lett. 4, 514–518.

Simberloff, D., 2009. The role of propagule pressure in biological invasions. Annu. Rev. Ecol. Evol. Syst. 40, 81–102.

Sun, J.H., Lu, M., Gillette, N.E., Wingfield, M.J., 2013. Red turpentine beetle: innocuous native becomes invasive tree killer in China. Annu. Rev. Entomol. 58, 293–311.

te Beest, M., Le Roux, J.J., Richardson, D.M., Brysting, A.K., Suda, J., Kubesová, M., Pysek, P., 2012. The more the better? The role of polyploidy in facilitating plant invasions. Ann. Bot. 109, 19–45.

Urban, M.C., Phillips, B.L., Skelly, D.K., Shine, R., 2007. The cane toad's (*Chaunus [Bufo] marinus*) increasing ability to invade Australia is revealed by a dynamically updated range model. Proc. R. Soc. B Biol. Sci. 274, 1413–1419.

van Klinken, R.D., Graham, J., Flack, L.K., 2006. Population ecology of hybrid mesquite (*Prosopis* species) in Western Australia: how does it differ from native range invasions and what are the implications for impacts and management? Biol. Invasions 8, 727–741.

Vicente, S., Máguas, C., Richardson, D.M., Trindade, H., Wilson, J.R.U., Le Roux, J.J., 2021. Highly diverse and highly successful: invasive Australian acacias have not experienced genetic bottlenecks globally. Ann. Bot. 128, 149–157.

Vilà, M., D'Antonio, C.M., 1998. Hybrid vigor for clonal growth in *Carpobrotus* (Aizoaceae) in coastal California. Ecol. Appl. 8, 1196–1205.

Vogt, G., 2017. Facilitation of environmental adaptation and evolution by epigenetic phenotype variation: insights from clonal, invasive, polyploid, and domesticated animals. Environ. Epigenet. 3, 1–17.

Vogt, G., Falckenhayn, C., Schrimpf, A., Schmid, K., Hanna, K., Panteleit, J., Helm, M., Schulz, R., Lyko, F., 2015. The marbled crayfish as a paradigm for saltational speciation by autopolyploidy and parthenogenesis in animals. Biol. Open 4, 583–594.

Wang, T., Aitken, S.N., Rozenberg, P., Carlson, M.R., 1999. Selection for height growth and Pilodyn pin penetration in lodgepole pine: effects on growth traits, wood properties, and their relationships. Can. J. For. Res. 29, 434–445.

Welles, S.R., Ellstrand, N.C., 2020. Evolution of increased vigour associated with allopolyploidization in the newly formed invasive species *Salsola ryanii*. AoB Plants 12, plz039.

Wilschut, R.A., Oplaat, C., Snoek, L.B., Kirschner, J., Verhoeven, K.J.F., 2016. Natural epigenetic variation contributes to heritable flowering divergence in a widespread asexual dandelion lineage. Mol. Ecol. 25, 1759–1768.

Yun, L., Agrawal, A.F., 2014. Variation in the strength of inbreeding depression across environments: effects of stress and density dependence. Evolution 68, 3599–3606.

Zangerl, A.R., Berenbaum, M.R., 2005. Increase in toxicity of an invasive weed after reassociation with its coevolved herbivore. Proc. Natl. Acad. Sci. U. S. A. 102, 15529–15532.

Zenni, R.D., Dickie, I.A., Wingfield, M.J., Hirsch, H., Crous, C.J., Meyerson, L.A., Burgess, T.I., Zimmermann, T.G., Klock, M.M., Siemann, E., Erfmeier, A., Aragon, R., Montti, L., Le Roux, J.J., 2017. Evolutionary dynamics of tree invasions: complementing the unified framework for biological invasions. AoB Plants 9, lw085.

# CHAPTER 3

# A home away from home: The role of eco-evolutionary experience in establishment and invasion success

## Chapter Outline

3.1 Introduction 36
3.2 Eco-evolutionary experience and invasiveness 40
   3.2.1 Exotic species x native herbivore/predator/pathogen interactions 40
   3.2.2 Exotic species x native prey/resource interactions 42
   3.2.3 Exotic species x native competitor interactions 43
   3.2.4 Exotic species x native mutualist interactions 44
3.3 How does eco-evolutionary experience influence the pace of contemporary evolution and the ecological impacts caused by invasive species? 45
3.4 Conclusions and future directions 48
References 49

## Abstract

This chapter evaluates the historical contexts within which individual biotic interactions have evolved and how these influence the establishment success of exotic species. The effects of past evolution on different biotic interactions, or so-called eco-evolutionary experience, emphasise the role of preadaptations to various interaction types in predicting the invasiveness of exotic species and the invasibility of communities. This chapter summarises examples of eco-evolutionary experience with different interaction types (including predation, competition, parasitism, and mutualism) and how these relate to hypotheses frequently tested in invasion ecology that invoke biotic interactions. The chapter also discusses how eco-evolutionary experience of exotic and native species affects the rate of contemporary evolution during invasion and the ecological impacts caused by invasive species. Suggestions for future research avenues to understand the role of eco-evolutionary experience in the establishment success of exotic species are provided throughout the chapter.

**Keywords:** Biotic interactions, Competition, Eco-evolutionary experience, Herbivory, Historical evolution, Mutualism, Parasitism, Phylogenetic relatedness, Preadaptation, Predation

*The Evolutionary Ecology of Invasive Species*
https://doi.org/10.1016/B978-0-12-818378-6.00005-X

*During invasions, species reach areas where they are not native and interact with species that they have not evolved with. Such settings lead to 'novelty' in biotic interactions in invaded areas, which may likely be decisive for the success or failure of invasions.*

**Saul et al. (2013)**

## 3.1 Introduction

Darwin was one of the first naturalists to think about the role of past evolution in the establishment success of exotic species (Darwin, 1859). On the one hand, he thought that exotic species will be less likely to establish in areas where congeneric or closely related natives are present because of high competition for available resources (a.k.a. Darwin's naturalisation hypothesis; Darwin, 1859; Daehler, 2001). However, he also thought that if adaptations to local conditions are important for survival and colonisation then exotic species will be better preadapted to conditions in new environments that harbour close relatives (a.k.a. Darwin's preadaptation hypothesis; Ricciardi and Mottiar, 2006; Cadotte et al., 2018).

If Darwin was correct, then one would expect the establishment success of exotic species to depend, to some degree, on their phylogenetic relatedness to resident species. Empirical data support both Darwin's viewpoints, in some instances indicating that native and exotic species are phylogenetically more closely related to one another than expected by chance alone (e.g. Park and Potter, 2013) while, in other instances, this is not the case (Strauss et al., 2006). However, it is important to consider spatial scale when interpreting the role of relatedness to natives in establishment success of exotic species (Mitchell et al., 2006; Ricciardi and Mottiar, 2006; Diez et al., 2008; Thuiller et al., 2010; Cadotte et al., 2018; Levin et al., 2020; Park et al., 2020). At fine spatial scales environmental conditions are essentially uniform and, therefore, competition for limited resources is largely responsible for structuring communities. On the other hand, at larger spatial scales such as regional scales, environmental heterogeneity is expected to strongly influence which species are likely to occur where. At these scales, exotic species are more likely to be preadapted to conditions in new ranges that harbour closely related species (Ricciardi and Mottiar, 2006; Cadotte et al., 2018).

Many hypotheses firmly embedded in evolutionary biology have emerged since Darwin's propositions to explain why some exotic species are more likely to become successful invaders than others (see Table 3.1, Catford et al., 2009; Jeschke and Heger, 2018; Enders et al., 2020). For

**Table 3.1** Main hypotheses in invasion ecology that invoke biotic interactions.

| Hypothesis | Description |
|---|---|
| Darwin's naturalisation hypothesis | Invasiveness is enhanced when recipient communities have few species that are closely to exotic species |
| Enemy release hypothesis | Invasiveness is enhanced when exotic species are liberated from their natural enemies (especially specialist enemies) |
| Evolution of increased competitive ability hypothesis | Following enemy release (see earlier), invasiveness is enhanced when exotic species reallocate resources used for defence against specialist enemies towards growth and reproduction |
| Increased susceptibility hypothesis | Founder events and genetic bottlenecks result in low genetic diversity that may lower enemy defence in exotic species, making them more susceptible to enemy attack in the new range |
| Limiting similarity hypothesis | Invasiveness is enhanced when exotic species have distinct functional traits, minimising competition and allowing exotics to fill empty niches |
| Missed mutualisms hypothesis | Establishment is impeded when exotic species lose beneficial mutualists during introduction |
| Naïve prey hypothesis | Exotic predators benefit from naïve resident prey that lack evolutionary history to them. Resident prey species suffer heavy predation because they exhibit ineffective antipredator responses to exotic predators |
| New associations hypothesis | Establishment is either hampered or enhanced when exotic species form novel associations with resident species |
| Novel weapons hypothesis | Establishment is enhanced when resident natives lack evolutionary experience to the growth-inhibiting chemicals (e.g. phytochemicals) produced by exotic plants |
| Preadaptation hypothesis | Establishment is enhanced when exotic species are preadapted to recipient communities and environments |
| Resource–enemy release hypothesis | Establishment is enhanced when exotic species lack specialist enemies (see enemy release hypothesis earlier) and when increased resource levels in the new range provide opportunities for invasion |

*Continued*

**Table 3.1** Main hypotheses in invasion ecology that invoke biotic interactions—cont'd

| Hypothesis | Description |
| --- | --- |
| Shifting defence hypothesis | Enemy release (see enemy release hypothesis earlier) allows exotic plants to evolve reduced resistance against specialist herbivores and increased resistance against generalist herbivores (also see evolution of increased competitive ability hypothesis earlier), facilitating establishment and invasion |
| Specialist-generalist hypothesis | Establishment is enhanced when resident enemies are specialists that are unable to utilise exotic species and when mutualists are generalists capable of forming effective associations with exotic species |

example, the 'enemy release hypothesis' states that a reduction or total absence of specialist enemies allows exotic species to attain high reproductive output (Colautti et al., 2004; Heger and Jeschke, 2018). It is intuitive why phylogenetic relatedness between exotic and native species influences the degree of enemy release experienced by exotics. All biotic interactions span a continuum of specificity. At one end of the spectrum, highly specialised interactions represent intimate co-evolved relationships (Le Roux et al., 2017, 2020; Le Roux, 2020) while, at the other end, organisms interact with a range of different partners. Many interactions fall somewhere in between these two extremes (Le Roux, 2020). While exotic species are expected to attract generalist enemies in their new ranges, they may also accumulate specialist enemies in areas where they co-occur with closely related native species (e.g. Crous et al., 2016). Exotic species may also be released from mutualists, which may negatively impact their establishment success (i.e. 'missed mutualisms hypothesis'; Mitchell et al., 2006). However, it appears that many exotic species that are introduced without specialised mutualists can form associations with generalist mutualists in the new range (so-called specialist-generalist hypothesis; Table 3.1, Callaway et al., 2004; Joshi and Vrieling, 2005). Competitive interactions are also influenced by past evolutionary history. For example, exotic plants may produce chemicals that inhibit resident species that lack evolutionary experience with them (i.e. 'novel weapons hypothesis'; Callaway and Ridenour, 2004).

It is clear from the examples before that past evolutionary history plays an important role in how exotic species deal with abiotic and biotic conditions in their new environments (Mitchell et al., 2006). Notwithstanding the role of phylogenetic relatedness between exotic and native species, it is also important to consider the historical context within which individual biotic interactions evolved to understand establishment success and invasion. This so-called eco-evolutionary experience (EEE) describes past ecological and evolutionary processes that shaped a species' experience with certain interactions or interaction types. Therefore, in the context of biological invasions, EEE emphasises the role of preadaptations to biotic interactions, or lack thereof, in driving the integration of exotic species into novel communities (Carroll, 2011; Saul et al., 2013; Saul and Jeschke, 2015; Heger et al., 2019). Exotic species are not only dissociated from interactions that shaped their own evolutionary trajectories and traits, but resident native species are also confronted by new species that they have never encountered before (Richardson et al., 2000). How native species respond to these novel interactions will therefore also be dependent on their EEE (Saul et al., 2013; Heger et al., 2019).

While we have gained many insights into the role of EEE with individual interaction types in the establishment success of exotic species (e.g. novel weapons, Callaway and Ridenour, 2004; enemy release, Colautti et al., 2004; etc.), virtually no studies have investigated EEE across various interaction types. For example, a resident soil microbial symbiont may lack EEE with the recruitment strategy of an exotic plant while, at the same time, the latter lack EEE to compete with resident plants for available soil symbionts. Saul et al. (2013) proposed metrics, barrowed from community ecology, to quantify EEE. Ecological interaction networks that incorporate phylogenetic data illustrate why quantitative estimates of EEE will help us to better understand how EEE influences invasion success. For example, Elias et al. (2013) reconstructed networks across four trophic levels of antagonistic interactions (plants, aphids, primary parasitoids, and secondary parasitoids) and found phylogenetic relatedness between resource species to be correlated with the overlap of their consumers. On the other hand, phylogenetic relatedness between consumers was not correlated, or negatively correlated, with the overlap of their resources (Elias et al., 2013). Taken together, this suggests that both phylogenetic constraints on resource breadth, and exploitative competition among phylogenetically closely related consumers, may shape biotic interactions and the composition of these communities. In this chapter we summarise various studies on EEE, with a focus on those that tested general hypotheses in invasion biology that invoke biotic interactions (Catford et al., 2009; Enders et al., 2018; Jeschke and Heger, 2018).

## 3.2 Eco-evolutionary experience and invasiveness

### 3.2.1 Exotic species x native herbivore/predator/pathogen interactions

Enemy release is a common feature of many invasive species. However, enemy release may only be temporary as resident enemies are expected to accumulate on invasive species over time (Crous et al., 2016; Stricker et al., 2016). Significant enemy release may persist when exotic species share high EEE with resident enemies that, in turn, share low EEE with them (Fig. 3.1a). As defence strategies against enemies are usually energetically costly, release from specialist enemies may cause a shift in the allocation of resources, away from defence and towards growth and reproduction (i.e. EICA; see Chapter 2, Section 2.2.4; Blossey and Nötzold, 1995). It follows that the evolution of increased competitive ability is more likely to occur when native enemies share little or no EEE with exotic species (Fig. 3.1a). While the assumption that native enemies will preferentially attack native species over exotic species is generally true, exotics may also be poorly adapted against them. So, while exotic species may evolve lower defence against specialist enemies (i.e. increased susceptibility hypotheses; Fig. 3.1, Colautti et al., 2004), they may also evolve higher resistance against generalist enemies (i.e. 'shifting defence hypothesis'; Fig. 3.1a; Table 3.1; Joshi and Vrieling, 2005; Doorduin and Vrieling, 2011, but see Wan et al., 2019).

Let us consider a situation where an exotic species lacks EEE with resident enemies. In this instance new interactions are likely to establish when native enemies have some EEE with the exotic species (i.e. 'new associations hypothesis'; Table 3.1; Colautti et al., 2004; Enders et al., 2018). Parker and Hay (2005) studied herbivory by two native freshwater crayfish species on aquatic plants in the United States. They presented crayfishes with various, and phylogenetically paired, native and exotic plant species in choice experiments. In all experiments the crustaceans preferentially consumed exotic species over native species, likely because of the exotic species' lack of EEE with, and thus defence against, crayfish herbivory (Parker and Hay, 2005). These authors also found this pattern to hold up in terrestrial ecosystems, by showing that native herbivores generally prefer invasive plants over native ones (Parker and Hay, 2005). Preferences for exotic resource species over native species may be particularly prevalent when resident enemies share high EEE with exotics. For example, in North America the milfoil weevil (*Euhrychiopsis lecontei*) prefers invasive Eurasian watermilfoil (*Myriophyllum spicatum*) over native congeneric watermilfoil species, again, possibly due to lower defence by the

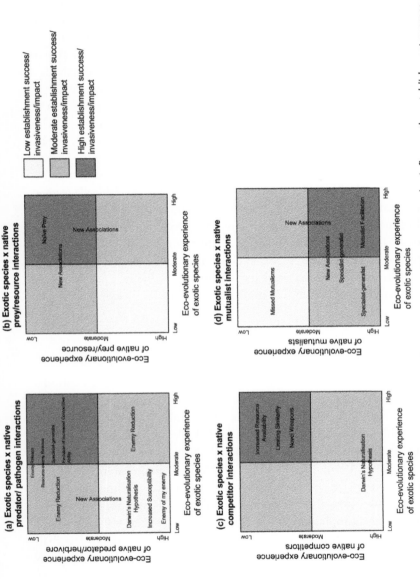

**Fig. 3.1** The eco-evolutionary experience (EEE) of introduced and native species is expected to influence the establishment success of exotic species and the severity of their ecological impacts. This involves EEE of exotic species with (a) native herbivores, predators, or pathogens, (b) prey or other resource species, (c) competitors, and (d) mutualists, and vice versa. Major hypotheses in invasion biology that invoke biotic interactions are indicated in relation to these predictions (also see Table 3.1; Catford et al., 2009; Enders et al., 2018).

exotic plant against the weevil (Solarz and Newman, 2001; Gross et al., 2001). When reared on Eurasian watermilfoil the weevil also preferentially oviposits on it (Solarz and Newman, 1996). This enemy spillover is thought to have significantly suppressed invasive Eurasian watermilfoil populations in North America (Creed, 2000). Limited EEE with resident enemies may not always hamper the performance of exotic species. For example, plants that are able to capitalise on abundant resources (i.e. being fast-growing) are generally also less defended against natural enemies (Fraser and Grime, 1999; Blumenthal, 2006; Blumenthal et al., 2009). When introduced into high-resource environments, such exotics may still outcompete resident species that still have to cope with their co-evolved specialist natural enemies (i.e. 'resource-enemy release hypothesis'; Fig. 3.1a; Table 3.1; Blumenthal, 2006).

Exotic species will obviously benefit from situations where resident enemies lack EEE with them. For example, invasive cane toads (*Rhinella marina*) produce toxins to which most predators in Australia lack EEE, which has resulted in the decimation of many native predator populations (Llewelyn et al., 2010; Shine, 2010). However, some Australian predators do show natural resistance against cane toads, such as the common keelback snake (*Tropidonophis mairii*). In this case, toad resistance is likely a result of this snake's toad-rich Southeast Asian heritage, and thus EEE with toad toxins (Llewelyn et al., 2010).

## 3.2.2 Exotic species x native prey/resource interactions

When native species lack EEE with experienced exotic predators, establishment success and ecological impacts are likely to be high (i.e. 'naïve prey hypothesis'; Fig. 3.1b; Table 3.1; Cox and Lima, 2006). For example, endemic cichlid populations plummeted in Lake Victoria in eastern Africa, following the introduction of predatory Nile perch (*Lates niloticus*) (Ogutu-Ohwayo, 1990; Witte et al., 1992). The disappearance of these native fishes took place over a remarkably short period of time, coinciding with the explosive growth of Nile perch populations (Kaufman, 1992; Witte et al., 1992). For example, in 1978 cichlids made up around 80% of biomass caught by fisherman in Lake Victoria, while Nile perch made up less than 2%. A mere 5 years later, catch records indicate that cichlid communities were virtually eliminated, while Nile perch made up more than 80% of all caught biomass (Kaufman, 1992). McGee et al. (2015) found that evolutionary constraints in the size and structure of cichlid jaws, associated with adaptations to cichlid diet, prohibited these fishes from ever occupying energy-rich top predator niches. As a consequence, they lack any EEE with top predators which, in turn, played a critical role in their demise following the introduction of Nile perch (McGee et al., 2015).

As illustrated by the example of cichlids in Lake Victoria, a lack of EEE to predation may reflect a lack of historical exposure to certain predator types, rather than predator species per se. Such unfamiliarity may be particularly prevalent in isolated habitats like lakes or islands. For example, the invasive brown tree snake (*Boiga irregularis*) has driven most native forest bird species on Guam to the brink of extinction (Fritts and Rodda, 1998), while exotic birds, such as the Chinese painted quail (*Coturnix chinensis*), are able to persist on the island (Engbring and Fritts, 1988). Although painted quails have no direct EEE with brown tree snake predation, their historical exposure to snakes equipped them with the necessary EEE to coexist with brown tree snakes on Guam (Cox and Lima, 2006).

### 3.2.3 Exotic species x native competitor interactions

As we discussed, a central tenet of Darwin's naturalisation hypothesis is that the relatedness of species corresponds to their ecological similarity and, therefore, the intensity of competition for available resources (Darwin, 1859). During invasion most interactions ultimately contribute to the competitive advantage that invasive species have over natives. In addition to direct competition, indirect competition (i.e. apparent and exploitative competition) must also play an important role in the success of exotic species. For example, Orrock et al. (2008) demonstrated that the highly competitive native grass, *Nassella pulchra*, cannot re-establish in the presence of invasive black mustard (*Brassica nigra*) plants as the latter increases herbivory pressure on the grass and, therefore, pre-empts direct competitive exclusion by it.

The influence of EEE on competition between exotic and native species is best demonstrated by the 'novel weapons hypothesis' (Callaway and Ridenour, 2004). As we discussed, novel weapons are phytochemicals produced by exotic plants that disrupt the metabolism and growth of naïve native plants (i.e. allelopathy; Fig. 3.1c; Callaway and Aschehoug, 2000; Callaway and Ridenour, 2004; Inderjit and Van der Putten, 2010; Novoa et al., 2012). Novel weapons may also indirectly suppress native plants, through their impacts on resident soil biota via antibiotic effects (Callaway et al., 2008), by protecting exotic species against herbivores (Schaffner et al., 2011) or by stimulating early germination of the invader's seeds (Yannelli et al., 2020).

Increased resource availability and enemy release often act in concert to enhance the competitive performance of invasive plants (see Section 3.2.1; Blumenthal, 2006; Blumenthal et al., 2009). This is especially true when fast-growing species are introduced into environments with high resource availability and where resident enemies lack appreciable EEE with them. For example, Blumenthal et al. (2009) studied 243 European plant species and found high resource-adapted species to have higher pathogen loads in their native ranges than less competitive slower-growing species—this pattern was the opposite in the invasive ranges of these species (Blumenthal et al., 2009). Conversely, the 'limiting similarity hypothesis' stems from MacArthur and Levins' (1967) idea that colonisation success is likely to be low when the functional overlap between resident and exotic species, and thus when competition between them for available resources, are high. Many authors, however, have found that an overlap in functional traits between native and exotic species rarely predicts establishment success (Price and Pärtel, 2013; Li et al., 2015; Yannelli et al., 2017). This suggests that any indirect effects on the competitive ability of exotic species stemming from their EEE with various different interaction types (e.g. degree of enemy release experienced) may benefit their direct competitive abilities, even when they are functionally similar to resident species.

## 3.2.4 Exotic species x native mutualist interactions

The generalist nature of many mutualists means that exotic species often form effective mutualisms with resident species, even when they share low EEE with them. This likely explains why plant–pollinator and –seed disperser networks are often easily infiltrated by invasive species (Olesen et al., 2002). For example, in South Africa's hyper diverse fynbos biome, native ants frequently disperse and bury the seeds of invasive plants like the Port Jackson willow (*Acacia saligna*; Holmes, 1990). These ants share high EEE with seeds that have nutritional rewards, in this case the elaiosomes of acacia seeds. On the other hand, invasive ants have negatively impacted seed dispersal of native fynbos plants. For example, the Argentine ant (*Linepithema humile*) often displaces native ants, impacting on the recruitment of some native plants (Bond and Slingsby, 1984). A lack of EEE with fynbos plants make Argentine ants slow to discover their seeds and to only move seeds over short distances without burying them. This leads to the majority of seeds being consumed by rodents, in some instances causing a 50-fold reduction in seedling emergence in plants like the Red-crested pagoda (*Mimetes cucullatus*; Bond and Slingsby, 1984). It is conceivable that these impacts may cause dramatic changes in the composition of fynbos plant communities over the medium to long term (Christian, 2001).

Some mutualist interactions are highly co-evolved and specialised. The 'missed mutualisms hypothesis' (Colautti et al., 2004; Mitchell et al., 2006) postulates that the establishment success of exotic species will be hampered when resident mutualists share no EEE with them (Fig. 3.1d). For example, pine trees in the genus *Pinus* have been extensively planted around the world, in many instances leading to widespread invasions (Richardson and Rejmánek, 2011). Despite the apparent invasiveness of the group, many initial pine introductions failed (Richardson, 2006). This was because of a lack of compatible ectomycorrhizal fungi in their new ranges (EMF; Pringle et al., 2009; Nuñez et al., 2017; Policelli et al., 2019). Ectomycorrhizal fungi are integral components of soil biodiversity and benefit plants through the absorption of nutrients and protection against diseases. These fungi are not free living, often show some degree of host specificity, and are patchily distributed across the globe (e.g. Gazol et al., 2016). It was only after the introduction of pine-specific EMF that many exotic pines successfully established and became widespread invaders (Vellinga et al., 2009). Interestingly, recent work suggests that more invasive pines are also more reliant on EMF mutualists than less invasive pines (Moyano et al., 2020, 2021). The tight-knit co-evolutionary history, and thus high EEE, between pines and their EMF illustrates how missed mutualisms may lead to reduced invasive performance or even establishment failure.

## 3.3 How does eco-evolutionary experience influence the pace of contemporary evolution and the ecological impacts caused by invasive species?

Selection will be strong when native species share high EEE with naïve exotic species, or vice versa (e.g. Mackin et al., 2021). When both partners share high EEE with one another, one may still expect moderate selection, given the expected high frequency of interactions. Interactions may be rare or absent, and selection therefore weak or non-existent, when both partners lack EEE with each other (Saul and Jeschke, 2015).

Balloon vines in the genus *Cardiospermum* provide a good example of how EEE impacts on rates of evolution in native species. As their common name suggests, these plants carry seeds in inflated capsules (Fig. 3.2), an adaptation to seed predators with piercing mouth parts. For example, soapberry

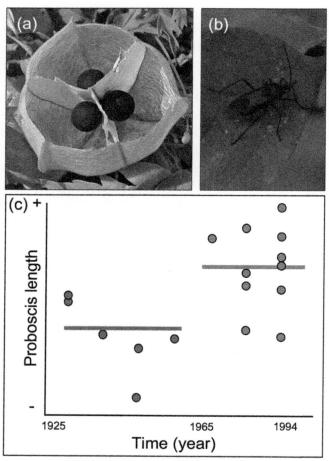

**Fig. 3.2** High eco-evolutionary experience (EEE) facilitates rapid evolution in native species: insights from invasive balloon vines and native soapberry bugs. (a) Balloon vines such as *Cardiospermum halicacabum* carry their seeds in inflated fruit (cut in half for illustration), protecting the seeds against predators with piercing mouthparts. (b) Invasive balloon vines have repeatedly been colonised by native soapberry bugs around the world (here shown *Leptocoris mutilatus* found in association with invasive *C. grandiflorum* in South Africa). Balloon vines and soapberry bugs share high levels of EEE and associations between them have often resulted in rapid evolution. For example, in Australia *Leptocoris tagalicus* evolved longer proboscides (beaks) following colonisation of invasive *C. grandiflorum*. (c) Beak lengths of *L. tagalicus* females prior to (i.e. pre-1965, *blue circles* (dark gray in the print version)) and after (i.e. post-1965, *red circles* (gray in the print version)) *C. grandiflorum* becoming an abundant and widespread invasive species along Australia's East coast suggest rapid evolution in this native insect to exploit its new host plant. Horizontal bars provide

*(Continued)*

bugs (various genera in the subfamily Serinethinae) use their long probosci-des (or 'beaks') to reach balloon vine seeds (Carroll et al., 2005b; Carroll and Loye, 2012). These predators of Sapindaceae plants are naturally found in many parts of the world (Carroll and Loye, 2012).

Invasive populations of balloon vines are found in many parts of the world (Gildenhuys et al., 2013). Three species in particular, *Cardiospermum corindum*, *C. grandiflorum*, and *C. halicacabum*, now have near-cosmopolitan distributions (Gildenhuys et al., 2013, 2015a, b). Given their high EEE with sapindacious plants, native soapberry bugs have colonised invasive balloon vines on several occasions (Carroll et al., 2005a; Carroll and Loye, 2012; Foster et al., 2019). Many of these new associations have led to rapid evo-lution in soapberry bugs, a direct consequence of the size of balloon vine fruits (Carroll et al., 2005a; Foster et al., 2019). For example, in Australia, the evolution of longer beaks in the bug, *Leptocoris tagalicus*, correlates strongly with its feeding efficiency on *Cardiospermum grandiflorum* (Fig. 3.2; Carroll et al., 2005a). This balloon vine has larger fruit than the bug's native host, the woolly rambutan (*Alectryon tomentosus*) (Fig. 3.2c; Car-roll et al., 2005a). In South Africa, host shifts by *Leptocoris mutilatus* onto *Cardiospermum halicacabum* and *C. grandiflorum* not only led to evolutionary responses in beak length, but also the formation of genetically distinct host races (Foster et al., 2019). We also know of soapberry bug host shifts in the opposite direction. For example, in Florida in the United States, *Jadera haematoloma* shifted from its *native* balloon vine host, *C. corindum*, onto exotic goldenrain trees (*Koelreuteria elegans*; Carroll and Boyd, 1992). In this instance the bug was presented with flatter seedpods by the exotic host plant leading to the rapid evolution of shorter beaks (Carroll and Boyd, 1992).

Strong selection is also expected when one interacting partner has high EEE and the other not. As we discussed earlier, invasive cane toads have had devastating impacts on some native predators in Australia. Australian anurans lack chemical defences similar to the toxins produced by cane toads, and therefore most native predators have no EEE with cane toad toxins. In situations like this, one would expect selection to be strong. Larger toads

---

**Fig. 3.2, cont'd**   average beak length for these two time periods. (For interpretation of the references to colour in this figure legend, the reader is referred to the web version of this article.) *Credit: Photograph (panel a), sourced from Wikimedia Commons under the Creative Commons Share-Alike Licence (CC-BY-SA 3.0). Original author: H. Zell. Photo panel b: Johannes Le Roux. Panel c: redrawn from Carroll, S.P., Loye, J.E., Dingle, H., Mathieson, M., Famula, T.R., Zalucki, M., 2005a. And the beak shall inherit—evolution in response to invasion. Ecol. Lett. 8, 944–951.*

are also more toxic than smaller ones, and therefore they pose the highest risk to predators. As we will discuss in Chapter 7, this has resulted in strong selection on predator gape size (i.e. mouth size), as predators with small mouths are more likely to survive the sublethal toxin doses of smaller toads (Phillips and Shine, 2004). As we already discussed, snakes such as keelbacks have resistance to cane toad poisoning given their EEE with toads (Llewelyn et al., 2010). As a result, keelbacks have not experienced significant changes in gape size since the arrival of cane toads in Australia (Phillips and Shine, 2004).

It is not only Australia's reptiles that have suffered severely from the impacts of cane toads, but also populations of other animals like the endangered carnivorous marsupial, the northern quoll *Dasyurus hallucatus* (Shine, 2010). Kelly and Phillips (2019) recently provided evidence for rapid evolution in response to these impacts in some northern quoll populations. These authors bred quolls that have been exposed to cane toads for over 70 years, as well as those from an offshore island that have never encountered cane toads. They also crossbred 'toad-exposed' and 'toad-naïve' quolls. Toad-naïve offspring was found to be more than twice as likely to consume cane toad meat than their toad-exposed counterparts. Interestingly, 'hybrid' offspring showed a similar aversion to cane toad meat as toad-exposed quolls, not only indicating that this behaviour is heritable, but also that it is dominant (Kelly and Phillips, 2019).

The ecological impacts of invasive species are generally linked to their range sizes and local abundances (e.g. Parker et al., 1999; Ricciardi, 2003; Vilà et al., 2015). It is also thought that invaders will cause more severe ecological impacts when they are functionally, or phylogenetically, distinct from dominant native species (Levine et al., 2003; Ma et al., 2016; Davis et al., 2019). For example, Ricciardi and Atkinson (2004) compared the taxonomic distinctiveness of invasive species that have been classified as high impact (having caused more than an 80% decline in one or more native species) or low impact (i.e. those not considered high impact) in various aquatic ecosystems. Their analysis found that high-impact invaders are more likely to come from genera that are absent from native communities. While these authors did not explicitly mention EEE they suggested that '... strong impacts are observed even in species-rich systems ... which have no evolutionary experience [with the invader]...'.

## 3.4 Conclusions and future directions

The context dependency of biological invasions limits our ability to predict which species will become established and invasive, and ultimately, which species will have the most severe ecological impacts (Novoa et al., 2020). It is clear that a consideration of past evolutionary history of biotic interactions will provide a more complete picture of the factors that determine the success of exotic species and their evolutionary and ecological impacts. Quantitative measures of EEE provide promising ways to achieve this (Saul et al., 2013).

A future research priority should be to examine the links between functional similarity, phylogenetic relatedness, and EEE across various interaction types, and how these influence the integration of exotic species into new communities. While some progress has been made by studies examining the impacts of phylogenetic relatedness on the structure of multitrophic ecological interaction networks (e.g. Elias et al., 2013), we need more information on how functional traits (Rezende et al., 2007) and phylogenetic constraints (Minoarivelo et al., 2014) influence the ecological integration of exotic species. For example, specialised plant–microbial interactions often show strong co-evolutionary phylogenetic signals (e.g. Le Roux et al., 2016, 2017; Le Roux, 2020), implying that integration by exotic plants will be contingent upon their phylogenetic relatedness to native plants. Lastly, metrics commonly used in community ecology allow quantification of EEE across different tropic levels (Saul et al., 2013), and, along with the vast amount of available phylogenetic data, makes the time ripe for studies to combine phylogeny and quantitative metrics of EEE to predict establishment success and invasion.

# References

Blossey, B., Nötzold, R., 1995. Evolution of increased competitive ability in invasive non-indigenous plants: a hypothesis. J. Ecol. 83, 887–889.

Blumenthal, D.M., 2006. Interactions between resource availability and enemy release in plant invasion. Ecol. Lett. 9, 887–895.

Blumenthal, D., Mitchell, C.E., Pyšek, P., Jarošik, V., 2009. Synergy between pathogen release and resource availability in plant invasion. Proc. Natl. Acad. Sci. U. S. A. 106, 7899–7904.

Bond, W., Slingsby, P., 1984. Collapse of an ant-plant mutualism: the Argentine ant (*Iridomyrmex humilis*) and myrmecochorous Proteaceae. Ecology 65, 1031–1037.

Cadotte, M.W., Campbell, S.E., Li, S.-.P., Sodhi, D.S., Mandrak, N.E., 2018. Preadaptation and naturalization of nonnative species: Darwin's two fundamental insights into species invasion. Annu. Rev. Plant Biol. 69, 661–684.

Callaway, R.M., Aschehoug, E.T., 2000. Invasive plants versus their new and old neighbors: a mechanism for exotic invasion. Science 290, 521–523.

Callaway, R.M., Ridenour, W.M., 2004. Novel weapons: invasive success and the evolution of increased competitive ability. Front. Ecol. Environ. 2, 436–443.

Callaway, R.M., Thelen, G.C., Rodriguez, A., Holben, W.E., 2004. Soil biota and exotic plant invasion. Nature 427, 731–733.

Callaway, R.M., Cipollini, D., Barto, K., Thelen, G.C., Hallett, S.G., Prati, D., Stinson, K., Klironomos, J., 2008. Novel weapons: invasive plant suppresses fungal mutualists in America but not in its native Europe. Ecology 89, 1043–1055.

Carroll, S.P., 2011. Conciliation biology: on the eco-evolutionary management of permanently invaded biotic systems. Evol. Appl. 4, 184–199.

Carroll, S.P., Boyd, C., 1992. Host race radiation in the soapberry bug: natural history with the history. Evolution 46, 1052–1069.

Carroll, S.P., Loye, J.E., 2012. Soapberry bug (Hemiptera: Rhopalidae: Serinethinae) native and introduced host plants: biogeographic background of anthropogenic evolution. Ann. Entomol. Soc. Am. 105, 671–684.

Carroll, S.P., Loye, J.E., Dingle, H., Mathieson, M., Famula, T.R., Zalucki, M., 2005a. And the beak shall inherit—evolution in response to invasion. Ecol. Lett. 8, 944–951.

Carroll, S.P., Loye, J.E., Dingle, H., Mathieson, M., Zalucki, M.P., 2005b. Ecology of *Leptocoris* Hahn (Hemiptera: Rhopalidae) soapberry bugs in Australia. Aust. J. Entomol. 44, 344–353.

Catford, J.A., Jansson, R., Nilsson, C., 2009. Reducing redundancy in invasion ecology by integrating hypotheses into a single theoretical framework. Divers. Distrib. 15, 22–40.

Christian, C.E., 2001. Consequences of a biological invasion reveal the importance of mutualism for plant communities. Nature 413, 635–639.

Colautti, R.I., Ricciardi, A., Grigorovich, I.A., MacIsaac, H.J., 2004. Is invasion success explained by the enemy release hypothesis? Ecol. Lett. 7, 721–733.

Cox, J.G., Lima, S.L., 2006. Naiveté and an aquatic-terrestrial dichotomy in the effects of introduced predators. Trends Ecol. Evol. 21, 674–680.

Creed, R.P., 2000. Is there a new keystone species in North American lakes and rivers? Oikos 91, 405–408.

Crous, C.J., Burgess, T.I., Le Roux, J.J., Richardson, D.M., Slippers, B., Wingfield, M.J., 2016. Ecological disequilibrium driving insect pest and pathogen accumulation in non-native trees in South Africa. AoB Plants 9, plw081.

Daehler, C.C., 2001. Darwin's naturalization hypothesis revisited. Am. Nat. 158, 324–330.

Darwin, C., 1859. On the Origin of Species by Means of Natural Selection. J. Murray, London, UK.

Davis, K.T., Callaway, R.M., Fajardo, A., Pauchard, A., Nuñez, M.A., Brooker, R.W., Maxwell, B.D., Dimarco, R.D., Peltzer, D.A., Mason, B., Ruotsalainen, S., McIntosh, A.C.S., Pakeman, R.J., Smith, A.L., Gundale, M.J., 2019. Severity of impacts of an introduced species corresponds with regional eco-evolutionary experience. Ecography 42, 12–22.

Diez, J.M., Sullivan, J.J., Hulme, P.E., Edwards, G., Duncan, R.P., 2008. Darwin's naturalization conundrum: dissecting taxonomic patterns of species invasions. Ecol. Lett. 11, 674–681.

Doorduin, L.J., Vrieling, K., 2011. A review of the phytochemical support for the shifting defence hypothesis. Phytochem. Rev. 10, 99–106.

Elias, M., Fontaine, C., van Veen, F.J.F., 2013. Evolutionary history and ecological processes shape a local multilevel antagonistic network. Curr. Biol. 23, 1355–1359.

Enders, M., Hutt, M.-.T., Jeschke, J.M., 2018. Drawing a map of invasion biology based on a network of hypotheses. Ecosphere 9, e02146.

Enders, M., Havemann, F., Ruland, F., Bernard-Verdier, M., Catford, J.A., Gómez-Aparicio, L., Haider, S., Heger, T., Kueffer, C., Kühn, I., Meyerson, L.A., Musseau, C., Novoa, A., Ricciardi, A., Sagouis, A., Schittko, C., Strayer, D.L., Vilà, M., Essl, F., Hulme, P.E., van Kleunen, M., Kumschick, S., Lockwood, J.L., Mabey, A.L., McGeoch, M.A., Palma, E., Pyšek, P., Saul, W.-C., Yannelli, F.A.,

Jeschke, J.M., 2020. A conceptual map of invasion biology: integrating hypotheses into a consensus network. Glob. Ecol. Biogeogr. 29, 978–991.

Engbring, J., Fritts, T.H., 1988. Demise of an Insular Avifauna: The Brown Tree Snake on Guam. Transactions of the Western Section of the Wildlife Society vol. 24, 31–37.

Foster, J.D., Ellis, A.G., Foxcroft, L.C., Carroll, S.P., Le Roux, J.J., 2019. The potential evolutionary impact of invasive balloon vines on native soapberry bugs in South Africa. NeoBiota 49, 19–35.

Fraser, L.H., Grime, J.P., 1999. Interacting effects of herbivory and fertility on a synthesized plant community. J. Ecol. 87, 514–525.

Fritts, T.H., Rodda, G.H., 1998. The role of introduced species in the degradation of island ecosystems: a case history of Guam. Annu. Rev. Ecol. Evol. Syst. 29, 113–140.

Gazol, A., Zobel, M., Cantero, J.J., Davison, J., Esler, K.J., Jairus, T., Opik, M., Vasar, M., Moora, M., 2016. Impact of alien pines on local arbuscular mycorrhizal fungal communities—evidence from two continents. FEMS Microbiol. Ecol. 92, fiw073.

Gildenhuys, E., Ellis, A.G., Carroll, S.P., Le Roux, J.J., 2013. The ecology, biogeography, history and future of two globally important weeds: *Cardiospermum halicacabum* Linn. and *C. grandiflorum* Sw. NeoBiota 19, 45–65.

Gildenhuys, E., Ellis, A.G., Carroll, S.P., Le Roux, J.J., 2015a. Combining natal range distributions and phylogeny to resolve biogeographic uncertainties in balloon vines (*Cardiospermum*, Sapindaceae). Divers. Distrib. 21, 163–174.

Gildenhuys, E., Ellis, A.G., Carroll, S.P., Le Roux, J.J., 2015b. From the neotropics to the Namib: evidence for rapid ecological divergence following extreme long-distance dispersal. Bot. J. Linn. Soc. 179, 477–486.

Gross, E.M., Johnson, R.L., Hairston, N.G., 2001. Experimental evidence for changes in submersed macrophyte species composition caused by the herbivore *Acentria ephemerella* (Lepidoptera). Oecologia 127, 105–114.

Heger, T., Jeschke, J.M., 2018. Enemy release hypothesis. In: Heger, T., Jeschke, J.M. (Eds.), Invasion Biology—Hypotheses and Evidence. CABI, pp. 92–102.

Heger, T., Bernard-Verdier, M., Gessler, A., Greenwood, A.D., Grossart, H.P., Hilker, M., Keinath, S., Kowarik, I., Kueffer, C., Marquard, E., Müller, J., Niemeier, S., Onandia, G., Petermann, J.S., Rillig, M.C., Rödel, M.-O., Saul, W.-C., Schittko, C., Tockner, K., Joshi, J., Jeschke, J.M., 2019. Towards an integrative, eco-evolutionary understanding of ecological novelty: studying and communicating interlinked effects of global change. BioScience 69, 888–899.

Holmes, P.M., 1990. Dispersal and predation of alien *Acacia* seeds: effects of season and invading stand density. S. Afr. J. Bot. 56, 428–434.

Inderjit, Van der Putten, W.H., 2010. Impacts of soil microbial communities on exotic plant invasions. Trends Ecol. Evol. 25, 512–519.

Jeschke, J.M., Heger, T. (Eds.), 2018. Invasion Biology: Hypotheses and Evidence. CABI, Wallingford, UK.

Joshi, J., Vrieling, K., 2005. The enemy release and EICA hypothesis revisited: incorporating the fundamental difference between specialist and generalist herbivores. Ecol. Lett. 8, 704–714.

Kaufman, L., 1992. Catastrophic change in species-rich freshwater ecosystems. BioScience 42, 846–858.

Kelly, E., Phillips, B.L., 2019. Targeted gene flow and rapid adaptation in an endangered marsupial. Conserv. Biol. 33, 112–121.

Le Roux, J.J., 2020. Molecular ecology of plant–microbial interactions during invasions: progress and challenges. In: Traveset, A., Richardson, D.M. (Eds.), Plant Invasions: The Role of Biotic Interactions. Invasive Species Series, CABI, UK, pp. 340–362.

Le Roux, J.J., Mavengere, N., Ellis, A.G., 2016. The structure of legume–rhizobium interaction networks and their response to tree invasions. AoB Plants 8, plw038.

Le Roux, J.J., Hui, C., Keet, J.-.H., Ellis, A.G., 2017. Co-introduction vs ecological fitting as pathways to the establishment of effective mutualisms during biological invasions. New Phytol. 119, 1319–1331.

Le Roux, J.J., Clusella-Trullas, S., Mokotjomela, T.M., Mairal, M., Richardson, D.M., Skein, L., Wilson, J.R., Weyl, O.L.F., Geerts, S., 2020. Biotic interactions as mediators of biological invasions: insights from South Africa. In: van Wilgen, B.W., Measey, J., Richardson, D.M., Wilson, J.R., Zengeya, T.A. (Eds.), Biological Invasions in South Africa. Springer, Berlin, pp. 387–427.

Levin, S.C., Crandall, R.M., Pokoski, T., Stein, C., Knight, T.M., 2020. Phylogenetic and functional distinctiveness explain alien plant population responses to competition. Proc. R. Soc. B Biol. Sci. 287, 20201070.

Levine, J.M., Vilà, M., D'Antonio, C.M., Dukes, J.S., Grigulis, K., Lavorel, S., 2003. Mechanisms underlying the impacts of exotic plant invasions. Proc. R. Soc. B Biol. Sci. 270, 775–781.

Li, S.P., Cadotte, M.W., Meiners, S.J., Hua, Z.S., Shu, H.Y., Li, J.T., Shu, W.-S., 2015. The effects of phylogenetic relatedness on invasion success and impact: deconstructing Darwin's naturalisation conundrum. Ecol. Lett. 18, 1285–1292.

Llewelyn, J., Phillips, B.L., Brown, G.P., Schwarzkopf, L., Alford, R.A., Shine, R., 2010. Adaptation or preadaptation: why are keelback snakes (*Tropidonophis mairii*) less vulnerable to invasive cane toads (*Bufo marinus*) than are other Australian snakes? Evol. Ecol. 25, 13–24.

Ma, C., Li, S., Pu, Z., Tan, J., Liu, M., Zhou, J., Li, H., Jiang, L., 2016. Different effects of invader–native phylogenetic relatedness on invasion success and impact: a meta-analysis of Darwin's naturalization hypothesis. Proc. Royal Soc. B Biol. Sci. 283, 1–8.

MacArthur, R., Levins, R., 1967. The limiting similarity, convergence, and divergence of coexisting species. Am. Nat. 101, 377–385.

Mackin, C.R., Peña, J.F., Blanco, M.A., Balfour, N.J., Castellanos, M.C., 2021. Rapid evolution of a floral trait following acquisition of novel pollinators. J. Ecol. 00, 1–13.

McGee, M.D., Borstein, S.R., Neches, R.Y., Buescher, H.H., Seehausen, O., Wainwright, P.C., 2015. A pharyngeal jaw evolutionary innovation facilitated extinction in Lake Victoria cichlids. Science 350, 1077–1079.

Minoarivelo, H.O., Hui, C., Terblanche, J.S., Kosakovsky Pond, S.L., Scheffler, K., 2014. Detecting phylogenetic signal in mutualistic interaction networks using a Markov process model. Oikos 123, 1250–1260.

Mitchell, C.E., Agrawal, A.A., Bever, J.D., Gilbert, G.S., Hufbauer, R.A., Klironomos, J.N., Maron, J.L., Morris, W.F., Parker, I.M., Power, A.G., Seabloom, E.W., Torchin, M.E., Vázquez, D.P., 2006. Biotic interactions and plant invasions. Ecol. Lett. 9, 726–740.

Moyano, J., Rodriguez-Cabal, M.A., Nuñez, M.A., 2020. Highly invasive tree species are more dependent on mutualisms. Ecology 101, e02997.

Moyano, J., Rodriguez-Cabal, M.A., Nuñez, M.A., 2021. Invasive trees rely more on mycorrhizas, countering the ideal-weed hypothesis. Ecology 102, e03330.

Novoa, A., González, L., Moravcová, L., Pyšek, P., 2012. Effects of soil characteristics, allelopathy and frugivory on establishment of the invasive plant *Carpobrotus edulis* and a co-occurring native, *Malcolmia littorea*. PLoS One 7, e53166.

Novoa, A., Richardson, D.M., Pyšek, P., Meyerson, L.A., Bacher, S., Canavan, S., Catford, J.A., Čuda, J., Essl, F., Foxcroft, L.C., Genovesi, P., Hirsch, H., Hui, C., Jackson, M.C., Kueffer, C., Le Roux, J.J., Measey, J., Mohanty, N.P., Moodley, D., Müller-Schärer, H., Packer, J.G., Pergl, J., Robinson, T.B., Saul, W.-C., Shackleton, R.T., Visser, V., Weyl, O.L.F., Yannelli, F.A., Wilson, J.R.U., 2020. Invasion syndromes: a systematic approach for predicting biological invasions and facilitating effective management. Biol. Invasions 22, 1801–1820.

Nuñez, M.A., Chiuffo, M.C., Torres, A., Paul, T., Dimarco, R.D., Raal, P., Policelli, N., Moyano, J., García, R.A., van Wilgen, B.W., Pauchard, A., Richardson, D.M., 2017.

Ecology and management of invasive Pinaceae around the world: progress and challenges. Biol. Invasions 19, 3099–3120.

Ogutu-Ohwayo, R., 1990. The decline of the native fishes of lakes Victoria and Kyoga (East Africa) and the impact of introduced species, especially the Nile perch, *Lates niloticus* and the Nile tilapia, *Oreochromis niloticus*. Environ. Biol. Fishes 27, 81–90.

Olesen, J.M., Eskildsen, L.I., Venkatasamy, S., 2002. Invasion of pollination networks on oceanic islands: importance of invader complexes and endemic super generalists. Divers. Distrib. 8, 181–192.

Orrock, J.L., Witter, M.S., Reichman, O.J., 2008. Apparent competition with an exotic plant reduces native plant establishment. Ecology 89, 1168–1174.

Park, D.S., Potter, D., 2013. A test of Darwin's naturalization hypothesis in the thistle tribe shows that close relatives make bad neighbors. Proc. Natl. Acad. Sci. U. S. A. 110, 17915–17920.

Park, D.S., Feng, X., Maitner, B.S., Ernst, K.C., Enquist, B.J., 2020. Darwin's naturalization conundrum can be explained by spatial scale. Proc. Natl. Acad. Sci. U. S. A. 117, 10904–10910.

Parker, J.D., Hay, M.E., 2005. Biotic resistance to plant invasions? Native herbivores prefer non-native plants. Ecol. Lett. 8, 959–967.

Parker, I.M., Simberloff, D., Lonsdale, W.M., Goodell, K., Wonham, M., Kareiva, P.M., Williamson, M.H., Von Holle, B., Moyle, P.B., Byers, J.E., Goldwasser, L., 1999. Impact: toward a framework for understanding the ecological effect of invaders. Biol. Invasions 1, 3–19.

Phillips, B.L., Shine, R., 2004. Adapting to an invasive species: toxic cane toads induce morphological change in Australian snakes. Proc. Natl. Acad. Sci. U. S. A. 101, 17150–17155.

Policelli, N., Bruns, T.D., Vilgalys, R., Nuñez, M.A., 2019. Suilloid fungi as global drivers of pine invasions. New Phytol. 222, 714–725.

Price, J.N., Pärtel, M., 2013. Can limiting similarity increase invasion resistance? A meta-analysis of experimental studies. Oikos 122, 649–656.

Pringle, A., Bever, J.D., Gardes, M., Parrent, J.L., Rillig, M.C., Klironomos, J.N., 2009. Mycorrhizal symbioses and plant invasions. Annu. Rev. Ecol. Evol. Syst. 40, 699–715.

Rezende, E.L., Jordano, P., Bascompte, J., 2007. Effects of phenotypic complementarity and phylogeny on the nested structure of mutualistic networks. Oikos 116, 1919–1929.

Ricciardi, A., 2003. Predicting the impacts of an introduced species from its invasion history: an empirical approach applied to zebra mussel invasions. Freshw. Biol. 48, 972–981.

Ricciardi, A., Atkinson, S.K., 2004. Distinctiveness magnifies the impact of biological invaders in aquatic ecosystems. Ecol. Lett. 7, 781–784.

Ricciardi, A., Mottiar, M., 2006. Does Darwin's naturalization hypothesis explain fish invasions? Biol. Invasions 8, 1403–1407.

Richardson, D.M., 2006. *Pinus*: a model group for unlocking the secrets of alien plant invasions? Preslia 78, 375–388.

Richardson, D.M., Rejmánek, M., 2011. Trees and shrubs as invasive alien species, a global review. Divers. Distrib. 17, 788–809.

Richardson, D.M., Allsopp, N., D'Antonio, C.M., Milton, S.J., Rejmánek, M., 2000. Plant invasions—the role of mutualisms. Biol. Rev. 75, 65–93.

Saul, W.-C., Jeschke, J.M., 2015. Eco-evolutionary experience in novel species interactions. Ecol. Lett. 18, 236–245.

Saul, W.-C., Jeschke, J.M., Heger, T., 2013. The role of eco-evolutionary experience in invasion success. NeoBiota 17, 57–74.

Schaffner, U., Ridenour, W.M., Wolf, V.C., Bassett, T., Müller, C., Müller-Schärer, H., Sutherland, S., Lortie, C.J., Callaway, R.M., 2011. Plant invasions, generalist herbivores, and novel defense weapons. Ecology 92, 829–835.

Shine, R., 2010. The ecological impact of invasive cane toads (*Bufo marinus*) in Australia. Q. Rev. Biol. 85, 253–291.

Solarz, S.L., Newman, R.M., 1996. Oviposition specificity and behavior of the watermilfoil specialist *Euhrychiopsis lecontei*. Oecologia 106, 337–344.

Solarz, S.L., Newman, R.M., 2001. Variation in hostplant preference and performance by the milfoil weevil, *Euhrychiopsis lecontei* Dietz, exposed to native and exotic watermilfoils. Oecologia 126, 66–75.

Strauss, S.Y., Webb, C.O., Salamin, N., 2006. Exotic taxa less related to native species are more invasive. Proc. Natl. Acad. Sci. U. S. A. 103, 5841–5845.

Stricker, K.B., Harmon, P.F., Goss, E.M., Clay, K., Flory, S.L., 2016. Emergence and accumulation of novel pathogens suppress an invasive species. Ecol. Lett. 19, 469–477.

Thuiller, W., Gallien, L., Boulangeat, I., De Bello, F., Münkemüller, T., Roquet, C., Lavergne, S., 2010. Resolving Darwin's naturalization conundrum: a quest for evidence. Divers. Distrib. 16, 461–475.

Vellinga, E.C., Wolfe, B.E., Pringle, A., 2009. Global patterns of ectomycorrhizal introductions. New Phytol. 181, 960–973.

Vilà, M., Rohr, R.P., Espinar, J.L., Hulme, P.E., Pergl, J., Le Roux, J.J., Schaffner, U., Pyšek, P., 2015. Explaining the variation in impacts of non-native plants on local-scale species richness: the role of phylogenetic relatedness. Glob. Ecol. Biogeogr. 24, 139–146.

Wan, J., Huang, B., Yu, H., Peng, S., 2019. Reassociation of an invasive plant with its specialist herbivore provides a test of the shifting defence hypothesis. J. Ecol. 107, 361–371.

Witte, F., Goldschmidt, T., Wanink, J.H., van Oijen, M.J.P., Goudswaard, P.C., Witte-Maas, E.L.M., Bouton, N., 1992. The destruction of an endemic species flock: quantitative data on the decline of the haplochromine species from the Mwanza Gulf of Lake Victoria. Environ. Biol. Fishes 34, 1–28.

Yannelli, F.A., Koch, C., Jeschke, J.M., Kollmann, J., 2017. Limiting similarity and Darwin's naturalization hypothesis: understanding the drivers of biotic resistance against invasive plant species. Oecologia 183, 775–784.

Yannelli, F.A., Novoa, A., Lorenzo, P., Rodríguez, J., Le Roux, J.J., 2020. No evidence for novel weapons: biochemical recognition modulates early ontogenetic processes in native species and invasive acacias. Biol. Invasions 22, 549–562.

# CHAPTER 4

# Phenotypic plasticity and the emerging field of 'invasion epigenetics'

## Chapter Outline

4.1 Introduction    56
4.2 Mechanisms of epigenetic variation    59
    4.2.1 DNA methylation    59
    4.2.2 Chromatin remodelling through histone modifications    61
    4.2.3 Non-coding RNAs    61
4.3 The ecological consequences of epigenetic variation    63
4.4 The evolutionary consequences of epigenetic variation    64
4.5 The role of epigenetic variation in establishment success    64
4.6 Epigenetic variation during range expansions    68
4.7 Epigenetics, hybridisation, and polyploidization: insights from invasive *Spartina anglica*    69
4.8 Conclusions    71
References    72

## Abstract

Epigenetic variation provides an important, but complex, layer of information involving gene regulation. Epigenetic variation represents various biochemical modifications of the genetic code (DNA) that influence gene expression, best known for its role in governing plastic responses to changing environmental conditions. This chapter summarises the main epigenetic mechanisms (DNA methylation, histone modification, non-coding RNAs) and remarks on how these may influence the ecological and evolutionary dynamics of organisms. The chapter then takes a closer look at examples involving invasive species to synthesise current knowledge on the role of epigenetic variation during different stages of invasion. The compensatory role of epigenetic variation in the performance of invasive species with no or low genetic diversity is discussed, in particular in relation to stress conditions. The chapter also briefly discusses the link between hybridisation and genome doubling (i.e. polyploidization) and epigenetic remodelling.

**Keywords:** Adaptive plasticity, DNA methylation, Epigenetic, Histone modification, Invasion epigenetics, Non-coding RNAs, Phenotypic plasticity, Polyploidization

*The Evolutionary Ecology of Invasive Species*
https://doi.org/10.1016/B978-0-12-818378-6.00011-5

*But DNA isn't really like that. It's more like a script. Think of Romeo and Juliet, for example. In 1936 George Cukor directed Leslie Howard and Norma Shearer in a film version. Sixty years later Baz Luhrmann directed Leonardo DiCaprio and Claire Danes in another movie version of this play. Both productions used Shakespeare's script, yet the two movies are entirely different. Identical starting points, different outcomes.*

**Carey (2012)**

## 4.1 Introduction

In 1849 Philip Von Siebold imported a single Japanese knotweed (*Reynoutria japonica* aka *Fallopia japonica*) plant into the Netherlands (Bailey and Conolly, 2000). He later sent clones of this plant to the Royal Botanical Gardens at Kew, from where it was introduced to the Royal Botanical Gardens in Edinburgh (Conolly, 1977). The first documented escape of Japanese knotweed in the United Kingdom dates back to 1886 (Storrie, 1886) and by 1996 the species was recorded in over 50% of sites surveyed by the Biological Records Centre (Hollingsworth and Bailey, 2000). Remarkably, almost all exotic Japanese knotweed populations around the world are related to the single plant Von Siebold originally brought to Europe (Del Tredici, 2017). Numerous examples, similar to Japanese knotweed, of highly successful invasive species that have no, or very low, genetic diversity, exist (e.g. Le Roux et al., 2007; Verhoeven et al., 2010; Vogt, 2017). These species often have high levels of phenotypic plasticity to help them survive in a wide variety of habitats, suggesting that epigenetics must play an important role in their ecological success.

It is only recently that we have begun to grasp of the molecular mechanisms that govern phenotypic plasticity (Richards, 2006). Epigenetic variation describes various biochemical modifications to the genetic code, often induced by environmental conditions, that affect gene expression and thus phenotypic variation. In its most elementary way, one can think of epigenetic variation as 'dials' that provide differential access to the same underlying genetic code (Richards, 2006). The ability of these modifications to quickly generate phenotypic variation, even in the absence of appreciable levels of genetic variation, is now widely appreciated (Fig. 4.1). Moreover, phenotypes resulting from epigenetic variation can in some instances be (semi) heritable (Holeski et al., 2012), with important consequences for the establishment and spread of exotic species. For example, invasive populations with low genetic diversity often harbour high epigenetic diversity

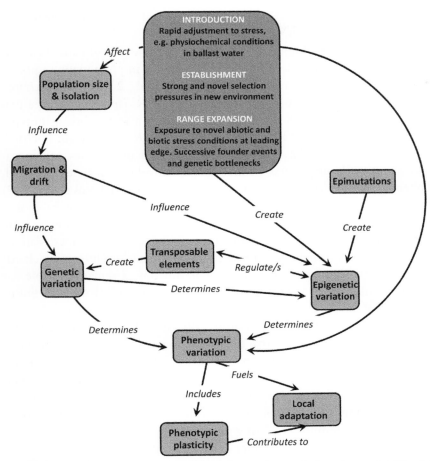

**Fig. 4.1** During the introduction, establishment, and spread of exotic species (*blue box* (gray in the print version)), genetic variation (influenced by demographic processes), epigenetic variation, and novel abiotic and biotic conditions interact to generate novel phenotypic variation that is either transient (i.e. phenotypic plasticity) or transgenerational (i.e. heritable and possibly adaptive). (For interpretation of the reference to colour in this figure legend, the reader is referred to the web version of this article.) *Adapted and modified from Richards, C.L., Alonso, C., Becker, C., Bossdorf, O., Bucher, E., Colomé-Tatché, M., Durka, W., Engelhardt, J., Gaspar, B., Gogol-Döring, A., Grosse, I., van Gurp, T.P., Heer, K., Kronholm, I., Lampei, C., Latzel, V., Mirouze, M., Opgenoorth, L., Paun, O., Prohaska, S. J., ... Verhoeven, K., 2017. Ecological plant epigenetics: evidence from model and non-model species, and the way forward. Ecol. Lett. 20, 1576–1590.*

(e.g. Schrey et al., 2012; Hawes et al., 2018a), possibly to compensate for low genetic diversity (Hawes et al., 2018a).

Invasive Japanese knotweed populations represent ideal natural experiments to explore the role of epigenetic variation in phenotypic responses to novel environmental conditions (Richards et al., 2012). In the United States, invasive populations consist of the single *R. japonica* genotype that was imported by Von Siebold into Europe and a few hybrid genotypes (*R. × bohemica*; a hybrid between *R. japonica* and *R. sachalinensis*; Richards et al., 2008, 2012). These populations are found on beaches and in salt marshes in the United States, habitats that are novel for Japanese knotweed. Given the low genetic diversity in these populations, Richards et al. (2012) wanted to know whether trait variation within and between habitats is linked to epigenetic diversity. They found significant levels of epigenetic differentiation (measured as differences in genome-wide methylation) between populations to be linked to local habitat conditions (Richards et al., 2012). Moreover, these epigenetic differences remained stable for up to 6 months in plants grown under common garden conditions (Richards et al., 2012). Similar observations have been made in invasive Japanese knotweed populations in Europe. Zhang et al. (2017) sampled knotweed clones from habitats across central Europe and found them to harbour high levels of epigenetic variation that was linked, in some instances, to variation in functional traits. Some of these epigenotypes persisted for several years under common garden conditions, again illustrating that phenotypic responses stemming from epigenetic modifications may be heritable over the short to medium term (Zhang et al., 2017).

The Japanese knotweed studies suggest that phenotypic plasticity, modulated by epigenetic responses, may be crucial in facilitating the survival and success of exotic species. More importantly, it also illustrates that rapid phenotypic responses can occur when genetic variation is limited or absent (Hawes et al., 2018a). Furthermore, within an individual's lifespan, stress-induced epigenetic variation may lead to more rapid and stronger responses to future exposure to the same stressors, in what has been termed 'molecular memory' (Jaskiewicz et al., 2011; Alvarez et al., 2020), suggesting that organisms can make predictions about their future environment based on cues received in earlier life (Gluckman et al., 2005; Varley et al., 2009).

This chapter provides a cursory overview of the main mechanisms that govern epigenetic variation (for detailed reviews, see Goldberg et al., 2007; Duncan et al., 2014; Frias-Lasserre and Villagra, 2017; Ashe et al., 2021) and briefly describes the ecological and evolutionary consequences of such variation. We then summarise recent insights into the roles of epigenetic variation in facilitating the establishment and spread of exotic species.

## 4.2 Mechanisms of epigenetic variation

We are now beginning to understand some of the molecular mechanisms underlying epigenetic variation in great detail (e.g. see reviews by Rapp and Wendel, 2005; Richards, 2006; Slotkin and Martienssen, 2007; Costa, 2008; Pimpinelli and Piacentini, 2020). The best known of these are (i) methylation of DNA bases in regulatory gene regions that can silence or downregulate gene activity; (ii) chromatin remodelling through biochemical modifications of histone proteins that can cause up- or downregulation, or silencing, of genes; and (iii) gene regulation through processes governed by non-coding RNAs (ncRNAs). Importantly, and as we will discuss later, these mechanisms do not necessarily operate in isolation, but often regulate gene expression in interactive and complex ways (Grant-Downton and Dickinson, 2006; Ashe et al., 2021).

Epigenetic diversity can be classified into three broad categories according to its dependence on the underlying genetic code: obligate, facilitated, and pure (Richards, 2006). Obligate epigenetic variation is 100% reliant on genetic variation. For example, a gene may have to undergo mutation to condition its alleles to regulation via epigenetic mechanisms. The contribution of obligate epigenetic variation to evolutionary responses therefore largely reflects that of genetic variation (Richards et al., 2017). Facilitated epigenetic variation is when an individuals' genotype probabilistically determines its epigenotype, e.g. genetic mutations condition alleles to fall stochastically under epigenetic control (Angers et al., 2020). Epigenetic variation is considered pure when it is completely independent of the underlying DNA code. Irrespective of the mechanisms or nature of epigenetic variation, the phenotypes that it creates can be permanent (i.e. developmental plasticity) or transient (i.e. activational plasticity, sensu Snell-Rood, 2015; Pimpinelli and Piacentini, 2020). Because epigenetically controlled phenotypes can sometimes be (semi) heritable they may provide additional variation for natural selection to act on (Jablonka et al., 1995; Lachman and Jablonka, 1996).

### 4.2.1 DNA methylation

DNA methylation is currently the best understood epigenetic mechanism. It modifies DNA sequences through the addition/removal of methyl groups at cytosine bases which affects the packaging of DNA (coiling around histone proteins), causing changes in the binding of transcriptional factors. Although DNA methylation is ubiquitous in all organisms, the extent of genome methylation can vary dramatically between species (Feng et al., 2010). Where in the genome DNA methylation occurs determines how gene regulation is affected, but generally elevated methylation, especially in pro- moter regions, leads to down-regulation or inactivation of genes (Jones, 2012; Angers et al., 2020). DNA methylation is also the most-studied epigenetic mechanism in invasive species (for a review, see Hawes et al., 2018a). These studies have generally found higher variation in DNA methylation in invasive populations with low standing genetic diversity, compared with genetically diverse native range populations (Hawes et al., 2018a).

DNA methylation can also influence phenotypic variation in more permanent ways such as through its impacts on the activity of transposable elements (or transposons, Fig. 4.1). Informally referred to as 'jumping genes', transposons are DNA segments that can move (translocate) from one region in the genome to another (Biémont, 2010). These genetic elements are abundant in most genomes (SanMiguel et al., 1996), e.g. making up around 45% of the human genome (Lander et al., 2001) and up to 85% of the maize genome (Schnable et al., 2009). This raises the important question of what the evolutionary significance of these mobile genetic elements is. Much of what a transposon does depends on where it inserts itself in the genome. Transposons have the ability to generate mutations of great variety and mag- nitude in phenotypic effects, ranging from subtle effects on gene regulation to large genomic rearrangements with complex effects (Feschotte, 2008; Pimpinelli and Piacentini, 2020). It is generally believed that transposons insert themselves at random, and, therefore, that their effects are often del- eterious. It is thus expected that their activity will be suppressed in most genomes and that deleterious insertions will be quickly eliminated by purifying selection. However, a growing body of research suggests that transposon insertions may not always be deleterious or random and may, in fact, often hold significant adaptive value (Casacuberta and González, 2013; Oliver et al., 2013; Pimpinelli and Piacentini, 2020). Only a few examples of transposon diversity and its ecological and evolutionary effects in invasive species are known. For example, transposons likely played an important role in mediating rapid evolution in two independent invasions by the tramp ant (*Cardiocondyla obscurior*) (Schrader et al., 2014). Schrader et al. (2014) found 7% of this ant's genome to have distinct and rapidly evolving accumulations of transposons in specific gene regions or so-called transposon 'islands'. Transposon islands were found to have higher rates of molecular evolution and sequence diversity, including in functional genes, compared with regions with low transposon densities (Schrader et al., 2014). These findings suggest that, following the introduction of the tramp ant, novel stress conditions in new environments may have caused changes in DNA methylation which triggered increased transposon activity, in turn creating novel genetic variation over just a few generations. This variation may have fuelled rapid evolutionary responses to adverse conditions, like pesticide resistance (Schrader et al., 2014). This example also supports the general belief that transposons activity in response to stress conditions often facilitates rapid adaptive responses (Stapley et al., 2015; Maggert, 2019).

## 4.2.2 Chromatin remodelling through histone modifications

Chromatin is organised into nucleosomes, i.e. segments of DNA wound around histone proteins. Histone proteins are frequently modified through various epigenetic mechanisms (Peterson and Laniel, 2004; Duncan et al., 2014; Angers et al., 2020). These modifications usually involve the addition of biochemical groups that change the physical properties of chromatin fibre and alter the accessibility of DNA to transcription and, therefore, gene expression (Bell et al., 2010; Duncan et al., 2014). Some histone modifications cause DNA to unwind, which will activate or enhance gene expression, while others condense chromatin which suppresses or inactivates gene expression (Turner, 2000; Duncan et al., 2014). While the role of histone modifications in phenotypic variation is well understood in model organisms like *Arabidopsis thaliana* (e.g. Li et al., 2011; Alvarez et al., 2020), little is known about their role during biological invasions. Limited insights have been gained from research on plant–pathogen interactions. For example, a review on plant pathogenic fungi found histone modifications to have important effects on the interaction between fungi and their host plants, i.e. pathogenesis (Jeon et al., 2014).

## 4.2.3 Non-coding RNAs

Most eukaryotes only have a few protein-coding genes relative to the overall size of their genomes. This 'redundant' genetic information has historically been referred to as 'junk' DNA (Biémont, 2010). We now know that much of this DNA is transcribed into non-coding RNAs: gene products that are not translated into proteins. The abundance of ncRNA-coding DNA in all genomes suggests that they have high evolutionary significance, recently recognised as a hidden layer of internal signalling that profoundly affects gene expression (Couzin, 2002; Yu et al., 2019). For example, the genomes of humans and mice share 85% similarity in their protein-coding genes (Mattick, 2001), and therefore the principal mechanisms governing pheno-typic divergence between us and mice reside in the non-protein-coding parts of our genomes. Various types of ncRNAs have been described, of which small ncRNAs (>25 nucleotides in length) are the best known. Their discovery was hailed as one of the most significant in recent history (Couzin, 2002), and since then, our appreciation of the importance and evolutionary roles of these molecules has grown from negligible to far reaching. Being generated in response to (a)biotic conditions, ncRNAs are capable of transmitting information from the environment, which is stored in the epigenome (Mattick, 2001; Glastad et al., 2019). The sequences of small ncRNAs show high evolutionary conservatism across taxonomic groups and their roles in gene regulation lie in their ability to control epigenetic mechanisms like DNA methylation and histone modifi-cations (Mattick and Makunin, 2006; Duncan et al., 2014). So how exactly do they do this? Only a handful of enzymes (e.g. DNA methyltransferases, histone methyltransferases, etc.) are known to control major epigenetic modifications, very few of which have any affinity for particular DNA sequences. However, epigenetic modifications must be purposefully directed to specific regions in the genome, in different cells, and in associ-ation with different environmental cues, which implies that there must be a communication system in place to guide these epigenetic modifications along the genome, a role often played by small ncRNAs. Small ncRNAs have high gene specificity, e.g., they may be highly complementarity to transposons, where they guide levels of DNA methylation and thus of trans-poson activity (Rey et al., 2016).

In addition to guiding epigenetic changes in the genome, ncRNAs also process and store information from the environment, especially cues in stress conditions, and therefore play key roles in the regulation of plasticity, molecular memory, and rapid responses to stress (Wang et al., 2017; Glastad et al., 2019; Alvarez et al., 2020). They can further influence gene expression through targeted degradation of mRNA (Angers et al., 2020). Despite their apparent role in epigenetic gene regulation, few studies have directly tested the diversity and role of ncRNAs in phenotypic responses in invasive spe-cies. For example, Qin et al. (2015) wanted to know whether differences in gene expression in response to salinity stress in invasive smooth cordgrass (*Spartina alterniflora*) are related to ncRNA activities. They found ncRNAs to be more active under stress conditions, providing some evidence for their possible involvement in mediating salt tolerance in this invasive grass (Qin et al., 2015).

## 4.3 The ecological consequences of epigenetic variation

Many researchers have inferred population epigenetic structure of species (see review by Richards et al., 2017 and references therein). From these studies it has emerged that most taxa, whether native or invasive, have strong population epigenetic structure, likely because of differences in the environmental conditions experienced by different populations (Richards et al., 2017). Despite this wealth of information, we know surprisingly little about how epigenetic variation impacts ecologically relevant traits. The poster child for how epigenetic variation impacts traits is the yellow toadflax (*Linaria vulgaris*). In natural populations of this plant, bilateral or radial flower morphologies are modulated by DNA methylation at a single locus (Cubas et al., 1999). Model organisms also provide examples of significant trait variation that is modulated by one or a few genes via epigenetic mechanisms. For instance, variation in flowering time and development in *A. thaliana* are strongly influenced by methylation of a few genes (e.g. Bastow et al., 2004; Peragine et al., 2004; Yaish et al., 2009). Despite a paucity in data, it is clear that morphological responses modulated by epigenetic mechanisms, even when involving one or only a few loci, may hold significant implications for ecological processes, e.g., plant–pollinator interactions in response to different flower morphologies.

Examples of intriguing and complex epigenetic effects also exist. For example, colonies of honey bees (*Apis mellifera*) typically consist of three castes: workers, drones, and queens. Worker bees are undeveloped females, and a colony will have several thousand of them cooperating in nest building, food collection, and brood rearing. Workers and the fertile queen (usually one per colony) develop from genetically identical larvae, but only those fed an exclusive diet of royal jelly develop into queens. The underlying mechanisms behind this transformation are not entirely understood but evidently involve epigenetic mechanisms. For example, DNA in the brain tissue of workers and queens and the genomes of their larvae are differentially methylated (Lyko et al., 2010; He et al., 2017). Royal jelly also contains compounds that inhibit histone deacetylation (Spannhoff et al., 2011). Interestingly, methylation differences between worker and queen bees, across hundreds of genes, do not appear to completely silence genes in either caste.

Instead, genes are differently expressed at low to moderate levels, illustrating that subtle differences in gene regulation across many genes have profound phenotypic effects (Lyko et al., 2010).

## 4.4 The evolutionary consequences of epigenetic variation

The potential influence of epigenetic diversity on ecologically relevant traits suggests that the Modern Evolutionary Synthesis, i.e. heritable phenotypic variation and its origin through random mutation, is no longer sufficient to explain evolution via natural selection (Mounger et al., 2020). Although genetic variation provides the most faithful mode of trait inheritance, less stable epigenetic responses may be favoured by selection if the phenotypes they produce have high fitness (Richards, 2006; Laland et al., 2015). The evolutionary significance of epigenetic variation has enjoyed a lot of attention (Richards et al., 2017; Pimpinelli and Piacentini, 2020; Mounger et al., 2020), with theoretical models suggesting that the dynamic, and often reversible, nature of epigenetic variation may add 'adaptive flexibility' to populations (Jablonka et al., 1995; Lachman and Jablonka, 1996). While some epigenetic variants can persist over multiple generations (Feng et al., 2010; Verhoeven et al., 2010), data from model organisms suggest that this is rarely the case for more than a few generations (Hagmann et al., 2015). It is therefore conceivable that epigenetic variation may fuel evolutionary responses by providing sufficient plasticity for organisms to cope with rapid environmental change (Jablonka et al., 1995; Lachman and Jablonka, 1996; Angers et al., 2020; Pimpinelli and Piacentini, 2020; Vogt, 2021), allowing for phenotypes with potentially high fitness to be 'held' for long enough for more permanent genetically based traits to accrue and stabilise them (West-Eberhard, 2005). To the best of my knowledge, no studies have demonstrated local adaptation in invasive species as a consequence of epigenetic mechanisms (also see review by Mounger et al., 2020). However, intriguing examples exist in other species. For example, Paun et al. (2010) found epigenetic differentiation between three recently diverged *Dactylorhiza* orchid species to be linked to their morphology, geographic distributions, and levels of habitat specialisation.

## 4.5 The role of epigenetic variation in establishment success

Invasive house sparrows (*Passer domesticus*) provide an interesting example of how interactions between novel environmental conditions, genetic diversity, and epigenetic variation can influence phenotypic responses during invasion. The house sparrow has been deliberately introduced to many other parts of the world and is the most widespread and abundant bird species in the world today (Callaghan et al., 2021). Population genetic studies have provided insights into the global introduction history of house sparrows. For example, in Australia and New Zealand, populations have similar levels of genetic diversity than native populations in Europe (Andrew et al., 2017), while in Kenya, invasive populations have experienced a population genetic bottleneck (Schrey et al., 2011). These differences in introduction effort provide opportunities to study the role of epigenetic variation during invasion in relation to genetic diversity, while differences in residence times allow inferences of temporal changes in epigenetic variation to be made. For example, comparisons between long-established and genetically diverse house sparrow populations in the United States, and more recently introduced and bottlenecked populations in Kenya, found similar levels of genome-wide DNA methylation in populations on both continents (Schrey et al., 2012). This suggests that epigenetic diversity may compensate for the low genetic diversity present in Kenyan populations (Schrey et al., 2012). In Kenyan populations epigenetic diversity is negatively correlated with genetic diversity, and positively with inbreeding (Liebl et al., 2013), and may act to boost phenotypic variation in these populations (Liebl et al., 2013). An inverse relationship between genetic and epigenetic diversity is frequently observed in invasive species, e.g. the common reed (*Phragmites australis*; Liu et al., 2018), the blue grass (*Poa annua*; Chwedorzewska and Bednarek, 2012), alligator weed (*Alternanthera philoxeroides*; Gao et al., 2010; Shi et al., 2019), Asian tiger mosquitoes (*Aedes albopictus*; Oppold et al., 2015), pygmy mussels (*Xenostrobus securis*; Ardura et al., 2017), and tubeworms (*Ficopomatus enigmaticus*; Ardura et al., 2017). These examples give credence to the perceived role of epigenetic variation in facilitating population persistence and/or aiding evolutionary responses in populations that have limited genetic diversity (Pérez et al., 2006; Angers et al., 2010; Kilvitis et al., 2014; Mounger et al., 2020; Vogt, 2021).

Invasive species with no genetic variation (i.e. clones) may best inform us on the ecological and evolutionary significance of epigenetic variation. For example, invasive marbled crayfish (*Procambarus virginalis*; see Chapter 2, Section 2.2.2) has a massive genome, comprising mare than 21,000 genes (Gutekunst et al., 2018). DNA methylation in this clonal triploid species is particularly prevalent in regions that harbour housekeeping genes. A large number of marbled crayfish genes also appear to have lower methylation levels compared to the same genes in its diploid ancestral species, the slough crayfish (*P. fallax*; Gatzmann et al., 2018). Variability in gene expression patterns also appears to be higher in marbled crayfish than in slough crayfish, which may buffer it against fluctuating environmental conditions (Gatzmann et al., 2018). Similarly, methylation changes contributed significantly to phenotypic variation in a single clone of the water flea, *Daphnia pulex*, that has invaded diverse freshwater habitats across eastern and southern Africa (Mergeay et al., 2006; Asselman et al., 2016; Vogt, 2017).

Epigenetic responses to stress conditions must have high ecological significance. Verhoeven et al. (2010) studied a single clone of the invasive dandelion (*Taraxacum officinale*) to determine how exposure to low nutrients, low salinity, herbivory, and pathogens impacted genome-wide DNA methylation in the species. While these authors found DNA methylation patterns to change under all stress conditions, they also found these changes to be most prominent in clones that were exposed to biotic stress conditions (i.e. herbivory and pathogens). Moreover, most of these epigenetic changes were heritable and persisted in second-generation plants grown under common garden conditions (Verhoeven et al., 2010). However, as we discussed before, epigenetic variation often only persists for a few generations or may reset entirely (Hagmann et al., 2015). Such phenotypic flexibility may have important implications for exotic species during transport or immediately after their release into new environments. For example, Huang et al. (2017) tested the effects of temperature and salinity on genome-wide methylation in the invasive ascidian, *Ciona savignyi*. They found significant and rapid changes in DNA methylation in response to each stress condition. Interestingly, however, stress-induced DNA methylation patterns reverted back to pre-stress levels under prolonged periods of stress exposure, indicating that epigenetic responses to these conditions are highly transient and reversible in this species (Fig. 4.2). Hawes et al. (2018b) assessed DNA methylation responses to salinity and temperature in another invasive ascidian, *Didemnum vexillum*. These authors also found genome-wide methylation patterns to change in response to temperature, but unlike in *Ciona savignyi*, not in response to salinity, possibly because *D. vexillum* is preadapted to saline conditions (Hawes et al., 2018b). In this instance, epigenetic responses to temperature stress were also correlated with decreased growth (Hawes et al., 2018b). Flexibility in phenotypic responses, such as those observed in ascidians, may benefit exotic species shortly after they are introduced into new environments, when conditions often change rapidly and unpredictably. Rapid epigenetic responses within an individual's lifespan may buffer them against adverse conditions, by increasing chances of survival or by

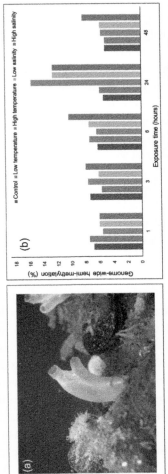

**Fig. 4.2** (a) The ascidian, *Ciona savignyi*, is a globally invasive species. (b) Using DNA methylation fingerprinting, Huang et al. (2017) found the genome of this species to be significantly methylated under various salinity and temperature stress conditions following 24 h of exposure. However, methylation levels reverted back to pre-stress levels after 48 h of stress exposure, indicating that these epigenetic responses are highly transient and reversible. *Figure drawn from data in Huang, X., Li, S., Ni, P., Gao, Y., Jiang, B., Zhou, Z., Zhan, A., 2017. Rapid response to changing environments during biological invasions: DNA methylation perspectives. Mol. Ecol. 26, 6621–6633. Credit: Photograph (panel A): Janna Nichols.*

providing time for them to disperse to areas where conditions are more favourable.

## 4.6 Epigenetic variation during range expansions

Rapid range expansions will create novel genotype x environment interactions and may, therefore, strongly impact epigenetic variation in invasive populations (for reviews, see Kilvitis et al., 2017; Mounger et al., 2020). Moreover, rapid range expansions may also result in successive founder events and genetic bottlenecks (Dlugosch et al., 2015), leading to low genetic diversity in leading-edge populations (Kilvitis et al., 2017). Despite this, outlying populations often establish successfully and it is thought that phenotypic plasticity, modulated by epigenetic mechanisms, plays important roles in facilitating this (Kilvitis et al., 2017). For example, epigenetic regulation of cold tolerance contributed to the invasive spread of crofton weed (*Ageratina adenophora*) into northern regions of China (Xie et al., 2015). Xie et al. (2015) found levels of DNA methylation of a specific cold response regulator gene (which forms part of the main physiological pathway that underlie cold tolerance in plants) to correlate with the range of cold tolerance levels measured among a wide geographic sample of crofton weed in China. These authors also found methylation patterns associated with increased cold tolerance to be heritable.

Despite the perceived benefits of epigenetic variation and its associated plasticity in leading-edge populations, there is also increasing evidence to suggest that fitness costs associated with plasticity could lead to spatial sorting of genotypes with only modest epigenetic potential (see Chapter 2, Section 2.2.2, for a detailed discussion on spatial sorting; Ghalambor et al., 2007; Huang et al., 2015). Huang et al. (2015) proposed that the dynamics of stress conditions in novel environments (i.e. the presence or absence of stress conditions at the leading edge) will greatly impact on the costs and benefits of plasticity, and therefore whether plasticity is adaptive or not. For example, the exposure of a range-expanding plant species to harsh conditions, such as nutrient-poor soils at the leading edge, may increase the costs, and reduce the benefits, of plasticity (via the reallocation of resources away from reproduction and towards nutrient acquisition), resulting in plasticity that is maladaptive (Huang et al., 2015). Alternatively, relaxed stress conditions at the leading edge, such as nutrient-rich soils, may reduce the costs of plasticity and increase its benefits and adaptive value (Huang et al., 2015).

House sparrow invasions again provide important insights into how epigenetic variation may facilitate rapid phenotypic responses during range expansions. In Kenya, birds at the leading edge secrete more corticosterone in response to stress than birds at the core (Liebl and Martin, 2012; Martin and Liebl, 2014). Corticosterone is one of the main steroid hormones involved in energy regulation, immunity, and stress responses (Hau et al., 2010). In addition, compared with birds at the core, leading-edge birds also express higher levels of glucocorticoid stress hormone receptors (Liebl and Martin, 2013). While the direct link between epigenetic diversity and stress hormone levels in these populations has not been definitively established, the low genetic diversity and high epigenetic diversity in these populations (see Section 4.5) point to a potential causal mechanism (Fig. 4.3; Schrey et al., 2012; Martin and Liebl, 2014; Hanson et al., 2021). Similar to stress hormone levels, expression of the Toll-like receptor 4 gene (*TLR4*) is higher in leading-edge house sparrow populations compared with core populations in Kenya (Martin et al., 2015). *TLR4* is an important pathogen recognition receptor, in particular of Gram-negative bacteria, and therefore a strong stimulator of inflammation. Armed with this information, Kilvitis et al. (2019) investigated whether differential *TLR4* expression in Kenyan house sparrow populations is directed by DNA methylation, genetic variation, or both. They found DNA methylation levels, but not genetic variation, at one particular *TLR4* promotor region to explain differences in *TLR4* expression (Fig. 4.3). Moreover, levels of DNA methylation were independent of genetic variation at the promotor region.

## 4.7 Epigenetics, hybridisation, and polyploidization: Insights from invasive *Spartina anglica*

Whole-genome duplication (i.e. polyploidization) has drastic and immediate effects on the genetic make-up of organisms, and thus their morphological, physiological, and ecological traits. Organisms can undergo polyploidization through intraspecific genome duplication (i.e. autopolyploidization) or following hybridisation and genome duplication (i.e. allopolyploidization). The link between polyploidy and invasiveness has been repeatedly shown in various taxonomic groups (e.g. te Beest et al., 2011).

The structural and functional reorganisation of polyploid genomes is frequently accompanied by major epigenetic remodelling, the degree of which is influenced by the mode of polyploidization (i.e. auto- vs allopolyploidization; Parisod et al., 2009). Our understanding of how epigenetic

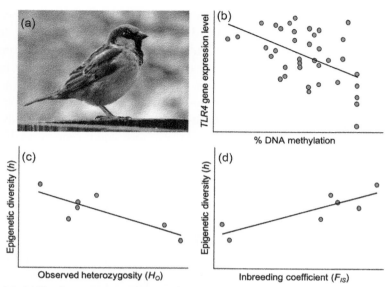

**Fig. 4.3** (a) The house sparrow (*Passer domesticus*) is the world's most widespread and abundant bird. (b) In invasive house sparrow populations in Kenya, expression of the Toll-like receptor 4 (*TLR4*) gene, an important pathogen recognition receptor, is higher in leading-edge populations compared with populations at the core. *TLR4* expression levels are also directly linked with levels of DNA methylation. (c) Genome-wide methylation diversity (*h*) was also found to be inversely correlated with genetic diversity ($H_O$; Observed heterozygosity) and (d) molecular estimates of inbreeding ($F_{IS}$; inbreeding coefficient) in Kenyan populations. *Panel b: redrawn from Kilvitis, H.J., Schrey, A.W., Ragsdale, A.K., Berrio, A., Phelps, S.M., Martin, L.B., 2019. DNA methylation predicts immune gene expression in introduced house sparrows Passer domesticus. J. Avian Biol. 50, 1–10; panels c and d: redrawn from Liebl, A.L., Schrey, A.W., Richards, C.L., Martin L.B., 2013. Patterns of DNA methylation throughout a range expansion of an introduced songbird. Integr. Comp. Biol. 53, 351–358. Credit: Photograph (panel a): sourced from Wikimedia Commons under the Creative Commons Share-Alike Licence (CC-BY-SA 3.0). Original author: Joe Ravi.*

remodelling in polyploids effects gene expression mostly comes from allo-polyploid crop species (e.g. Paun et al., 2011; te Beest et al., 2011; Li et al., 2019). Invasive smooth cordgrass (*Spartina anglica*) provides insights into how such remodelling may unfold following hybridisation and allopolyploid speciation, and its consequences for invasiveness. The North American salt-marsh cordgrass (*Spartina alterniflora*) was accidently introduced into southern England and western France where it hybridised with native European cordgrass (*S. maritima*) at least twice in the last century, resulting in two distinct hybrids: *Spartina × townsendii* and *Spartina × neyrautii* (Thompson,

1991). Genome doubling in *Spartina × townsendii* gave rise to the fertile and vigorous allododecaploid, *S. anglica* (Mounger et al., 2020). This species has rapidly spread across Europe and other parts of the world, whereas *Spartina × townsendii* and the two ancestral species have remained localised (Thompson, 1991; Baumel et al., 2003).

*Spartina anglica* has low levels of genetic diversity (Baumel et al., 2001; Ainouche et al., 2004) and hybridisation appears to have triggered both genetic and epigenetic changes in the species' ancestor, *Spartina × townsendii*, demonstrated by losses of genomic regions (Salmon et al., 2005), various transposon insertions (Baumel et al., 2002), and changes in DNA methylation (Salmon et al., 2005). Using genome-wide methylation analyses, Salmon et al. (2005) found both *Spartina × townsendii* and *Spartina × neyrautii* hybrids to have experienced substantial methylation changes (up to 35%), but little or no genetic changes, compared with the parental species that gave rise to them. A large proportion of methylated loci were also shared between these two hybrids, suggesting that epigenetic remodelling was somewhat 'directed' and not completely random. Moreover, a substantial amount of epigenetic variation in invasive *S. anglica* was already present in its ancestral hybrid lineage, *Spartina × townsendii*, i.e. prior to genome duplication. Taken together, these findings suggest that the most important genetic and epigenetic changes in *S. anglica*'s genome were triggered by incompatibilities between the genomes of *S. alterniflora* and *S. maritima*, rather than by genome doubling following hybridisation between these two species (Ainouche et al., 2009; Parisod et al., 2009). Genome doubling, however, does appear to influence gene expression in *S. anglica*, by increasing the number of overexpressed genes (Chelaifa et al., 2010) and through post-transcriptional gene regulation via small ncRNAs (Cavé-Radet et al., 2020). This species therefore appears capable of maintaining high intra-genomic diversity, which may explain its high levels of phenotypic plasticity in response to salinity and other stress conditions (Cavé-Radet et al., 2019).

## 4.8 Conclusions

The role of epigenetic variation in gene regulation and phenotypic plasticity is now widely acknowledged. Aside from model organisms, our understanding of the evolutionary significance of epigenetic variation in natural populations remains scant, including in invasive species. Available information, however, suggests that epigenetic variation plays a critical role in the establishment and spread of exotic species.

Quantifying the contributions of epigenetic and genetic variation to phenotypic variation and plasticity remains challenging. While field-based studies have provided us with examples of how local habitat conditions cause strong epigenetic structure, in some instances linked to trait variation, it remains difficult to determine the functional importance of such variation. An important future consideration for epigenetic studies should be to include measures of fitness. For example, the stability of differential *TLR4* expression in invasive house sparrow populations in Kenya we discussed can be determined over multiple generations under common garden conditions. If epigenetic variation linked to *TLR4* expression persists, then birds can be challenged with different stresses associated with these epigenetic responses, such as exposure to pathogenic bacteria. Differences in performance and fitness (e.g. survival, clutch size, etc.) can then be correlated with measures of epigenetic and functional divergence.

Lastly, isolating environmentally induced epigenetic variation from other forms of epigenetic diversity requires situations where both the genetic background of organisms and environmental conditions they experience are controlled. Here, clonal organisms that are easy to propagate, such as Japanese knotweed, provide particularly useful study systems. It is now also possible to chemically demethylate genomes (e.g. see Latzel et al., 2020). This means that field-collected individuals can be epi-genotyped to assess DNA methylation variation and then be chemically demethylated and reintroduced into their respective environments. Re-genotyping these reintroduced individuals will allow researchers to assess how predictable DNA methylation is at the genome level under specific environmental conditions. The study of invasion epigenetics opens the door to understanding important and exciting, but hitherto largely overlooked, aspects of the evolutionary ecology of invasive species.

# References

Ainouche, M.L., Baumel, A., Salmon, A., 2004. *Spartina anglica* C. E. Hubbard: a natural model system for analysing early evolutionary changes that affect allopolyploid genomes. Biol. J. Linn. Soc. 82, 475–484.

Ainouche, M.L., Fortune, M., Salmon, A., Parisod, C., Grandbastien, M.-A., Fukunaga, K., Ricou, M., Misset, M.T., 2009. Hybridization, polyploidy and invasion: lessons from *Spartina* (Poaceae). Biol. Invasions 11, 1159–1173.

Alvarez, M., Bleich, A., Donohue, K., 2020. Genotypic variation in the persistence of transgenerational responses to seasonal cues. Evolution 74, 2265–2280.

Andrew, S.C., Awasthy, M., Bolton, P.E., Rollins, L.A., Nakagawa, S., Griffith, S.C., 2017. The genetic structure of the introduced house sparrow populations in Australia and New

Zealand is consistent with historical descriptions of multiple introductions to each country. Biol. Invasions 20, 1507–1522.

Angers, B., Castonguay, E., Massicotte, R., 2010. Environmentally induced phenotypes and DNA methylation: how to deal with unpredictable conditions until the next generation and after. Mol. Ecol. 19, 1283–1295.

Angers, B., Perez, M., Menicucci, T., Leung, C., 2020. Sources of epigenetic variation and their applications in natural populations. Evol. Appl. 13, 1262–1278.

Ardura, A., Zaiko, A., Moran, P., Planes, S., Garcia-Vazquez, E., 2017. Epigenetic signatures of invasive status in populations of marine invertebrates. Sci. Rep. 7, 42193.

Ashe, A., Colot, V., Oldroyd, B.P., 2021. How does epigenetics influence the course of evolution? Philos. Trans. R. Soc. B Biol. Sci. 376, 20200111.

Asselman, J., De Coninck, D.I.M., Pfrender, M.E., De Schamphelaere, K.A., 2016. Gene body methylation patterns in *Daphnia* are associated with gene family size. Genome Biol. Evol. 8, 1185–1196.

Bailey, J.P., Conolly, A.P., 2000. Prize-winners to pariahs—a history of Japanese knotweed s. l. (Polygonaceae) in the British Isles. Watsonia 23, 93–110.

Bastow, R., Mylne, J., Lister, C., Lippman, Z., Martiennsen, R., Dean, C., 2004. Vernalization requires epigenetic silencing of FLC by histone methylation. Nature 427, 164–167.

Baumel, A., Ainouche, M.L., Levasseur, J.E., 2001. Molecular investigations in populations of *Spartina anglica* C.E. Hubbard (Poaceae) invading coastal Brittany (France). Mol. Ecol. 10, 1689–1701.

Baumel, A., Ainouche, M.L., Kalendar, R., Schulman, A.H., 2002. Retrotransposons and genomic stability in populations of the young allopolyploid species *Spartina anglica* C.E. Hubbard (Poaceae). Mol. Biol. Evol. 19, 1218–1227.

Baumel, A., Ainouche, M.L., Misset, M.-T., Gourret, J.P., Bayer, R.J., 2003. Genetic evidence for hybridization between the native *Spartina maritima* and the introduced *Spartina alterniflora* (Poaceae) in south-west France: *Spartina* × *neyrautii* re-examined. Plant Syst. Evol. 237, 87–97.

Bell, O., Schwaiger, M., Oakeley, E.J., Lienert, F., Beisel, C., Stadler, M.B., Schübeler, D., 2010. Accessibility of the *Drosophila* genome discriminates PcG repression, H4K16 acetylation and replication timing. Nat. Struct. Mol. Biol. 17, 894–900.

Biémont, C., 2010. A brief history of the status of transposable elements: from junk DNA to major players in evolution. Genetics 186, 1085–1093.

Callaghan, C.T., Nakagawa, S., Cornwell, W.K., 2021. Global abundance estimates for 9,700 bird species. Proc. Natl. Acad. Sci. U. S. A. 118, e2023170118.

Carey, N., 2012. The Epigenetics Revolution: How Modern Biology Is Rewriting our Understanding of Genetics, Disease and Inheritance. Icon Books Ltd., London.

Casacuberta, E., González, J., 2013. The impact of transposable elements in environmental adaptation. Mol. Ecol. 22, 1503–1517.

Cavé-Radet, A., Salmon, A., Lima, O., El Aïnouche, M., Amrani, A., 2019. Increased tolerance to organic xenobiotics following recent allopolyploidy in *Spartina* (Poaceae). Plant Sci. 280, 143–154.

Cavé-Radet, A., Giraud, D., Lima, O., El Amrani, A., Aïnouche, M., Salmon, A., 2020. Evolution of small RNA expression following hybridization and allopolyploidization: insights from *Spartina* species (Poaceae, Chloridoideae). Plant Mol. Biol. 102, 55–72.

Chelaifa, H., Monnier, A., Ainouche, M.L., 2010. Transcriptomic changes following recent natural hybridization and allopolyploidy in the salt marsh species *Spartina* × *townsendii* and *Spartina anglica* (Poaceae). New Phytol. 186, 161–174.

Chwedorzewska, K., Bednarek, P., 2012. Genetic and epigenetic variation in a cosmopolitan grass *Poa annua* from Antarctic and Polish populations. Pol. Polar Res. 33, 63–80.

Conolly, A.P., 1977. The distribution and history in the British Isles of some alien species of *Polygonum* and *Reynoutria*. Watsonia 11, 291–311.

Costa, F.F., 2008. Non-coding RNAs, epigenetics and complexity. Gene 410, 9–17.

Couzin, J., 2002. Breakthrough of the year: small RNAs make a big splash. Science 298, 2296–2297.

Cubas, P., Vincent, C., Coen, E., 1999. An epigenetic mutation responsible for natural variation in floral symmetry. Nature 401, 157–161.

Del Tredici, P., 2017. The introduction of Japanese knotweed, *Reynoutria japonica*, into North America. J. Torrey Bot. Soc. 144, 406–416.

Dlugosch, K.M., Anderson, S.R., Braasch, J., Cang, F.A., Gillette, H.D., 2015. The devil is in the details: genetic variation introduced populations and its contributions to invasion. Mol. Ecol. 24, 2095–2111.

Duncan, E.J., Gluckman, P.D., Dearden, P.K., 2014. Epigenetics, plasticity, and evolution: how do we link epigenetic change to phenotype? J. Exp. Zool. B Mol. Dev. Evol. 322, 208–220.

Feng, S., Cokus, S.J., Zhang, X., Chen, P.Y., Bostick, M., Goll, M.G., Hetzel, J., Jain, J., Strauss, S.H., Halpern, M.E., Ukomadu, C., Sadler, K.C., Pradhan, S., Pellegrini, M., Jacobsen, S.E., 2010. Conservation and divergence of methylation patterning in plants and animals. Proc. Natl. Acad. Sci. U. S. A. 107, 8689–8694.

Feschotte, C., 2008. Transposable elements and the evolution of regulatory networks. Nat. Rev. Genet. 9, 397–405.

Frias-Lasserre, D., Villagra, C.A., 2017. The importance of ncRNAs as epigenetic mechanisms in phenotypic variation and organic evolution. Front. Microbiol. 8, 2483.

Gao, L., Geng, Y., Li, B.O., Chen, J., Yang, J.I., 2010. Genome-wide DNA methylation alterations of *Alternanthera philoxeroides* in natural and manipulated habitats: implications for epigenetic regulation of rapid responses to environmental fluctuation and phenotypic variation. Plant Cell Environ. 33, 1820–1827.

Gatzmann, F., Falckenhayn, C., Gutekunst, J., Hanna, K., Raddatz, G., Carneiro, V.C., Lyko, F.F., 2018. The methylome of the marbled crayfish links gene body methylation to stable expression of poorly accessible genes. Epigenetics Chromatin 11, 57.

Ghalambor, C.K., McKay, J.K., Carroll, S.P., Reznick, D.N., 2007. Adaptive versus non-adaptive phenotypic plasticity and the potential for contemporary adaptation in new environments. Funct. Ecol. 21, 394–407.

Glastad, K.M., Hunt, B.G., Goodisman, M.A.D., 2019. Epigenetics in insects: genome regulation and the generation of phenotypic diversity. Annu. Rev. Entomol. 64, 185–203.

Gluckman, P.D., Hanson, M.A., Spencer, H.G., Bateson, P., 2005. Environmental influences during development and their later consequences for health and disease: implications for the interpretation of empirical studies. Proc. R. Soc. B Biol. Sci. 272, 671–677.

Goldberg, A.D., Allis, C.D., Bernstein, E., 2007. Epigenetics: a landscape takes shape. Cell 128, 635–638.

Grant-Downton, R.T., Dickinson, H.G., 2006. Epigenetics and its implications for plant biology: 2. The epigenetic epiphany: epigenetics, evolution and beyond. Ann. Bot. 97, 11–27.

Gutekunst, J., Andriantsoa, R., Falckenhayn, C., Hanna, K., Stein, W., Rasamy, J., Lyko, F., 2018. Clonal genome evolution and rapid invasive spread of the marbled crayfish. Nat. Ecol. Evol. 2, 567–573.

Hagmann, J., Becker, C., Müller, J., Stegle, O., Meyer, R.C., Wang, G., Schneeberger, K., Fitz, J., Altmann, T., Bergelson, J., Borgwardt, K., Weigel, D., 2015. Century-scale methylome stability in a recently diverged *Arabidopsis thaliana* lineage. PLoS Genet. 11, e1004920.

Hanson, H.E., Wang, C., Schrey, A.W., Liebl, A.K., Jiang, R.H.Y., Martin, L.B., 2021. Epigenetic potential and DNA methylation in an ongoing House Sparrow (*Passer domesticus*) range expansion. bioRxiv. 2020.03.07.981886.

Hau, M., Ricklefs, R.E., Wikelski, M., Lee, K.A., Brawn, J.D., 2010. Corticosterone, testosterone and life-history strategies of birds. Proc. R. Soc. B Biol. Sci. 277, 3203–3212.

Hawes, N.A., Fidler, A.E., Tremblay, L.A., Pochon, X., Dunphy, B.J., Smith, K.F., 2018a. Understanding the role of DNA methylation in successful biological invasions: a review. Biol. Invasions 20, 2285–2300.

Hawes, N.A., Tremblay, L.A., Pochon, X., Dunphy, B., Fidler, A.E., Smith, K.F., 2018b. Effects of temperature and salinity stress on DNA methylation in a highly invasive marine invertebrate, the colonial ascidian *Didemnum vexillum*. Peer J 6, e5003.

He, X.J., Zhou, L.B., Pan, Q.Z., Barron, A.B., Yan, W.Y., Zeng, Z.J., 2017. Making a queen: an epigenetic analysis of the robustness of the honeybee (*Apis mellifera*) queen developmental pathway. Mol. Ecol. 26, 1598–1607.

Holeski, L.M., Jander, G., Agrawal, A.A., 2012. Transgenerational defense induction and epigenetic inheritance in plants. Trends Ecol. Evol. 27, 618–626.

Hollingsworth, M.L., Bailey, J.P., 2000. Evidence for massive clonal growth in the invasive weed *Fallopia japonica* (Japanese knotweed). Bot. J. Linn. Soc 133, 463–472.

Huang, Q.Q., Pan, X.Y., Fan, Z.W., Peng, S.L., 2015. Stress relief may promote the evolution of greater phenotypic plasticity in exotic invasive species: a hypothesis. Ecol. Evol. 5, 1169–1177.

Huang, X., Li, S., Ni, P., Gao, Y., Jiang, B., Zhou, Z., Zhan, A., 2017. Rapid response to changing environments during biological invasions: DNA methylation perspectives. Mol. Ecol. 26, 6621–6633.

Jablonka, E., Oborny, B., Molnár, E., Kisdi, E., Hofbauer, J., Czárán, T., 1995. The adaptive advantage of phenotypic memory. Philos. Trans. R. Soc. B Biol. Sci. 350, 133–141.

Jaskiewicz, M., Conrath, U., Peterhansel, C., 2011. Chromatin modification acts as a memory for systemic acquired resistance in the plant stress response. EMBO Rep. 12, 50–55.

Jeon, J., Kwon, S., Lee, Y.H., 2014. Histone acetylation in fungal pathogens of plants. Plant Pathol. J 30, 1–9.

Jones, P.A., 2012. Functions of DNA methylation: islands, start sites, gene bodies and beyond. Nat. Rev. Genet. 13, 484–492.

Kilvitis, H., Alvarez, M., Foust, C., Schrey, A.W., Robertson, M., Richards, C.L., 2014. Ecological epigenetics. In: Landry, C.R., Aubin-Horth, N. (Eds.), Ecological Genomics. Springer, The Netherlands.

Kilvitis, H.J., Hanson, H., Schrey, A.W., Martin, L.B., 2017. Epigenetic potential as a mechanism of phenotypic plasticity in vertebrate range expansions. Integr. Comp. Biol. 57, 385–395.

Kilvitis, H.J., Schrey, A.W., Ragsdale, A.K., Berrio, A., Phelps, S.M., Martin, L.B., 2019. DNA methylation predicts immune gene expression in introduced house sparrows *Passer domesticus*. J. Avian Biol. 50, 1–10.

Lachman, M., Jablonka, E., 1996. The inheritance of phenotypes: an adaptation to fluctuating environment. J. Theor. Biol. 181, 1–9.

Laland, K.N., Uller, T., Feldman, M.W., Sterelny, K., Müller, G.B., Moczek, A., Jablonka, E., Odling-Smee, J., 2015. The extended evolutionary synthesis: its structure, assumptions and predictions. Proc. R. Soc. B Biol. Sci. 282, 1813.

Lander, E.S., Linton, L.M., Birren, B., Nusbaum, C., Zody, M.C., Baldwin, J., Devon, K., Dewar, K., Doyle, M., FitzHugh, W., Linton, L.M., Birren, B., Nusbaum, C., International Human Genome Sequencing Consortium, 2001. Initial sequencing and analysis of the human genome. Nature 409, 860–921. https://doi.org/10.1038/35057062.

Latzel, V., Münzbergová, Z., Skuhrovec, J., Novák, O., Strnad, M., 2020. Effect of experimental DNA demethylation on phytohormones production and palatability of a clonal plant after induction via jasmonic acid. Oikos 129, 1867–1876.

Le Roux, J.J., Wieczorek, A.M., Wright, M.G., Tran, C.T., 2007. Super-genotype: global monoclonality defies the odds of nature. PLoS One 2, e590.

Li, W., Liu, H., Cheng, Z.J., Su, Y.H., Han, H.N., Zhang, Y., Zhang, X.S., 2011. DNA methylation and histone modifications regulate de novo shoot regeneration in *Arabidopsis* by modulating WUSCHEL expression and auxin signaling. PLoS Genet. 7, e1002243.

Li, N., Xu, C., Zhang, A., Lv, R., Meng, X., Lin, X., Gong, L., Wendel, J.F., Liu, B., 2019. DNA methylation repatterning accompanying hybridization, whole genome doubling and homoeolog exchange in nascent segmental rice allotetraploids. New Phytol. 223, 979–992.

Liebl, A.L., Martin, L.B., 2012. Exploratory behaviour and stressor hyper-responsiveness facilitate range expansion of an introduced songbird. Proc. R. Soc. B Biol. Sci. 279, 4375–4381.

Liebl, A.L., Martin, L.B., 2013. Stress hormone receptors change as range expansion progresses in house sparrows. Biol. Lett. 9, 20130181.

Liebl, A.L., Schrey, A.W., Richards, C.L., Martin, L.B., 2013. Patterns of DNA methylation throughout a range expansion of an introduced songbird. Integr. Comp. Biol. 53, 351–358.

Liu, L., Pei, C., Liu, S., Guo, X., Du, N., Guo, W., 2018. Genetic and epigenetic changes during the invasion of a cosmopolitan species (*Phragmites australis*). Ecol. Evol. 8, 6615–6624.

Lyko, F., Foret, S., Kucharski, R., Wolf, S., Falckenhayn, C., Maleszka, R., 2010. The honey bee epigenomes: differential methylation of brain DNA in queens and workers. PLoS Biol. 8, e1000506.

Maggert, K.A., 2019. Stress: an evolutionary mutagen. Proc. Natl. Acad. Sci. U. S. A. 116, 17616–17618.

Martin, L.B., Liebl, A.L., 2014. Physiological flexibility in an avian range expansion. Gen. Comp. Endocrinol. 206, 227–234.

Martin, L.B., Liebl, A.L., Kilvitis, H.J., 2015. Covariation in stress and immune gene expression in a range expanding bird. Gen. Comp. Endocrinol. 211, 14–19.

Mattick, J.S., 2001. Non-coding RNAs: the architects of eukaryotic complexity. EMBO Rep. 2, 986–991.

Mattick, J.S., Makunin, I.V., 2006. Non-coding RNA. Hum. Mol. Genet. 15, R17–R29.

Mergeay, J., Verschuren, D., De Meester, L., 2006. Invasion of an asexual American water flea clone throughout Africa and rapid displacement of a native sibling species. Proc. R. Soc. B Biol. Sci. 273, 2839–2844.

Mounger, J., Ainouche, M., Bossdorf, O., CavéRadet, A., Li, B., Parepa, M., Salmon, A., Yang, J., Richards, C.L., 2020. Epigenetics and the success of invasive plants. Philos. Trans. R. Soc. B, Biol. Sci. 376, 20200117.

Oliver, K.R., McComb, J.A., Greene, W.K., 2013. Transposable elements: powerful contributors to angiosperm evolution and diversity. Genome Biol. Evol. 5, 1886–1901.

Oppold, A., Kreß, A., Bussche, J.V., Diogo, J.B., Kuch, U., Oehlmann, J., Vandegehuchte, M.B., Müller, R., 2015. Epigenetic alterations and decreasing insecticide sensitivity of the Asian tiger mosquito *Aedes albopictus*. Ecotoxicol. Environ. Saf. 122, 45–53.

Parisod, C., Salmon, A., Zerjal, T., Tenaillon, M., Grandbastien, M.A., Ainouche, M.L., 2009. Rapid structural and epigenetic reorganization near transposable elements in hybrid and allopolyploid genomes in Spartina. New Phytol. 184, 1003–1015.

Paun, O., Bateman, R.M., Fay, M.F., Hedren, M., Civeyrel, L., Chase, M.W., 2010. Stable epigenetic effects impact adaptation in allopolyploid orchids (*Dactylorhiza*: Orchidaceae). Mol. Biol. Evol. 27, 2465–2473.

Paun, O., Bateman, R.M., Fay, M.F., Luna, J.A., Moat, J., Hedren, M., Chase, M.W., 2011. Altered gene expression and ecological divergence in sibling allopolyploids of *Dactylorhiza* (Orchidaceae). BMC Evol. Biol. 11, 113.

Peragine, A., Yoshikawa, M., Wu, G., Albrecht, H., Poethig, R., 2004. SGS3 and SGS2/ SDE1/RDR6 are required for juvenile development and the production of transacting siRNAs in *Arabidopsis*. Genes Dev. 18, 2368–2379.

Pérez, J., Nirchio, M., Alfonsi, C., Muñoz, C., 2006. The biology of invasions: the genetic adaptation paradox. Biol. Invasions 8, 1115–1121.

Peterson, C.L., Laniel, M., 2004. Histones and histone modifications. Curr. Biol. 14, R546–R551.

Pimpinelli, S., Piacentini, L., 2020. Environmental change and the evolution of genomes: transposable elements as translators of phenotypic plasticity into genotypic variability. Funct. Ecol. 34, 428–441.

Qin, Z., Chen, J., Jin, L., Duns, G.J., Ouyang, P., 2015. Differential expression of miRNAs under salt stress in *Spartina alterniflora* leaf tissues. J. Nanosci. Nanotechnol. 15, 1554–1561.

Rapp, R.A., Wendel, J.F., 2005. Epigenetics and plant evolution. New Phytol. 168, 81–91.

Rey, O., Danchin, E., Mirouze, M., Loot, C., Blanchet, S., 2016. Adaptation to global change: a transposable element–epigenetics perspective. Trends Ecol. Evol. 31, 514–526.

Richards, E.J., 2006. Inherited epigenetic variation—revisiting soft inheritance. Nat. Rev. Genet. 7, 395–401.

Richards, C.L., Walls, R., Bailey, J.P., Parameswaran, R., George, T., Pigliucci, M., 2008. Plasticity in salt tolerance traits allows for invasion of salt marshes by Japanese knotweed s. l. (*Fallopia japonica* and *F. X bohemica*, Polygonaceae). Am. J. Bot. 95, 931–942.

Richards, C.L., Schrey, A., Pigliucci, M., 2012. Invasion of diverse habitats by few Japanese knotweed genotypes is correlated with epigenetic differentiation. Ecol. Lett. 15, 1016–1025.

Richards, C.L., Alonso, C., Becker, C., Bossdorf, O., Bucher, E., Colomé-Tatché, M., Durka, W., Engelhardt, J., Gaspar, B., Gogol-Döring, A., Grosse, I., van Gurp, T.P., Heer, K., Kronholm, I., Lampei, C., Latzel, V., Mirouze, M., Opgenoorth, L., Paun, O., Prohaska, S.J., Verhoeven, K., 2017. Ecological plant epigenetics: evidence from model and non-model species, and the way forward. Ecol. Lett. 20, 1576–1590. https://doi.org/10.1111/ele.12858.

Salmon, A., Ainouche, M.L., Wendel, J.F., 2005. Genetic and epigenetic consequences of recent hybridization and polyploidy in *Spartina* (Poaceae). Mol. Ecol. 14, 1163–1175.

SanMiguel, P., Tikhonov, A., Jin, Y.K., Motchoulskaia, N., Zakharov, D., Melake-Berhan, A., Springer, P.S., Edwards, K.J., Lee, M., Avramova, Z., Bennetzen, J.L., 1996. Nested retrotransposons in the intergenic regions of the maize genome. Science 274, 765–768.

Schnable, P.S., Ware, D., Fulton, R.S., Stein, J.C., Wei, F., Pasternak, S., Liang, C., Zhang, J., Fulton, L., Graves, T.A., Minx, P., Reily, A.D., Courtney, L., Kruchowski, S.S., Tomlinson, C., Strong, C., Delehaunty, K., Fronick, C., Courtney, B., Rock, S.M., Wilson, R.K., 2009. The B73 maize genome: complexity, diversity, and dynamics. Science 326, 1112–1115. https://doi.org/10.1126/science.1178534.

Schrader, L., Kim, J.W., Ence, D., Zimin, A., Klein, A., Wyschetzki, K., Weichselgartner, T., Kemena, C., Stökl, J., Schultner, E., Wurm, Y., Smith, C.D., Yandell, M., Heinze, J., Gadau, J., Oettler, J., 2014. Transposable element islands facilitate adaptation to novel environments in an invasive species. Nat. Commun. 5, 5495.

Schrey, A.W., Grispo, M., Awad, M., Cook, M.B., 2011. Broad-scale latitudinal patterns of genetic diversity among native European and introduced house sparrow (*Passer domesticus*) populations. Mol. Ecol. 20, 1133–1143.

Schrey, A.W., Coon, C.A.C., Grispo, M.T., Awad, M., Imboma, T., McCoy, E.D., Mushinsky, H.R., Richards, C.L., Martin, L.B., 2012. Epigenetic variation may compensate for decreased genetic variation with introductions: a case study using house sparrows (*Passer domesticus*) on two continents. Genet. Res. Int. 2012, 979751.

Shi, W., Chen, X., Gao, L., Xu, C.-Y., Ou, X., Bossdorf, O., Yang, J., Geng, Y., 2019. Transient stability of epigenetic population differentiation in a clonal invader. Front. Plant Sci. 9, 1851.

Slotkin, R.K., Martienssen, R., 2007. Transposable elements and the epigenetic regulation of the genome. Nat. Rev. Genet. 8, 272–285.

Snell-Rood, E.C., 2015. An overview of the evolutionary causes and consequences of behavioural plasticity. Anim. Behav. 85, 1004–1011.

Spannhoff, A., Kim, Y.K., Raynal, N.J., Gharibyan, V., Su, M.B., Zhou, Y.Y., Li, J., Castellano, S., Sbardella, G., Issa, J.P., Bedford, M.T., 2011. Histone deacetylase inhibitor activity in royal jelly might facilitate caste switching in bees. EMBO Rep. 12, 238–243.

Stapley, J., Santure, A.W., Dennis, S.R., 2015. Transposable elements as agents of rapid adaptation may explain the genetic paradox of invasive species. Mol. Ecol. 24, 2241–2252.

Storrie, J., 1886. The Flora of Cardiff. The Cardiff Naturalists' Society, Cardiff.

te Beest, M., Le Roux, J.J., Richardson, D.M., Brysting, A.K., Suda, J., Kubesová, M., Pysek, P., 2011. The more the better? The role of polyploidy in facilitating plant invasions. Ann. Bot. 109, 19–45.

Thompson, J.D., 1991. The biology of an invasive plant. What makes *Spartina anglica* so successful? BioScience 41, 393–401.

Turner, B.M., 2000. Histone acetylation and an epigenetic code. BioEssays 22, 836–845.

Varley, K.E., Mutch, D.G., Edmonston, T.B., Goodfellow, P.J., Mitra, R.D., 2009. Intratumor heterogeneity of MLH1 promoter methylation revealed by deep single molecule bisulfite sequencing. Nucleic Acids Res. 37, 4603–4612.

Verhoeven, K.J.F., Jansen, J.J., van Dijk, P.J., Biere, A., 2010. Stress-induced DNA methylation changes and their heritability in asexual dandelions. New Phytol. 185, 1108–1118.

Vogt, G., 2017. Facilitation of environmental adaptation and evolution by epigenetic phenotype variation: insights from clonal, invasive, polyploid, and domesticated animals. Environ. Epigenet. 3, 1–17.

Vogt, G., 2021. Epigenetic variation in animal populations: sources, extent, phenotypic implications, and ecological and evolutionary relevance. J. Biosci. 46, 24.

West-Eberhard, M.J., 2005. Developmental plasticity and the origin of species differences. Proc. Natl. Acad. Sci. U. S. A. 102, 6543–6549.

Xie, H.J., Li, A.H., Liu, A.D., Dai, W.M., He, J.Y., Lin, S., Duan, H., Liu, L.L., Chen, S.G., Song, X.L., Valverde, B.E., Qiang, S., 2015. ICE1 demethylation drives the range expansion of a plant invader through cold tolerance divergence. Mol. Ecol. 24, 835–850.

Yaish, M.W., Peng, M., Rothstein, S.J., 2009. AtMBD9 modulates *Arabidopsis* development through the dual epigenetic pathways of DNA methylation and histone acetylation. Plant J. 59, 123–135.

Yu, Y., Zhang, Y., Chen, X., Chen, Y., 2019. Plant noncoding RNAs: hidden players in development and stress responses. Annu. Rev. Cell Dev. Biol. 35, 407–431.

Zhang, Y.Y., Parepa, M., Fischer, M., Bossdorf, O., 2017. Epigenetics of colonizing species? A study of Japanese knotweed in Central Europe. In: Barrett, S.C.H., Colautti, R.I., Dlugosch, K.M., Rieseberg, L.H. (Eds.), Invasion Genetics: The Baker and Stebbins Legacy. John Wiley and Sons, Oxford, UK, pp. 328–340.

# CHAPTER 5

# Drivers of rapid evolution during biological invasions

## Contents

5.1 Introduction                                          80
5.2 Changes in abiotic conditions                         82
    5.2.1 Habitat availability and heterogeneity      82
    5.2.2 Climate and physicochemical conditions     85
5.3 Changes in biotic conditions                          88
    5.3.1 Mutualisms                                 88
    5.3.2 Competition                                90
    5.3.3 Food resources                             91
5.4 Conclusions                                           92
References                                                93

## Abstract

A paradigm shift occurred in the field of evolutionary biology during the 1990s. Biologists, who traditionally viewed evolution as a slow process, came to the realisation that evolution can often be rapid, sometimes occurring over timescales as short as a few generations. Much of the empirical evidence underlying this paradigm shift came from studies involving invasive species. It is now commonly accepted that abrupt changes in (a)biotic conditions in the new range can cause strong selection on, and rapid evolutionary responses in, invasive populations. This chapter summarises key abiotic (e.g. habitat availability and heterogeneity, climate, soil physicochemical properties) and biotic (e.g. mutualisms, competition, novel food resources) drivers that frequently underlie rapid evolution during invasion.

**Keywords:** Climate, Clinal variation, Competition, Local adaptation, Mutualism, Phenotypic plasticity, Rapid evolution, Spatial sorting

*It may be said that natural selection is daily and hourly scrutinising, throughout the world, every variation, even the slightest; rejecting that which is bad, preserving and adding up all that is good; silently and insensibly working, whenever and wherever opportunity offers, at the improvement of each organic being in relation to its organic and inorganic conditions of life. We see nothing of these slow changes in progress, until the hand of time has marked the long lapse of ages, and then*

*The Evolutionary Ecology of Invasive Species*
https://doi.org/10.1016/B978-0-12-818378-6.00008-5

*so imperfect is our view into long past geological ages, that we only see that the forms of life are now different from what they formerly were.*

**Darwin (1859)**

## 5.1 Introduction

Biologists have traditionally viewed the ecology and evolution of organisms as separate processes that operate on distinct timescales. Many still perceive evolution via natural selection to be slow, its biological outcomes manifesting over timescales far beyond anything that could be considered 'ecological'. Like many high school students, I was taught the classical example of evolved industrial melanism in peppered moths (*Biston betularia*), a rapid evolutionary response in camouflage coloration to predation in habitats that had recently become coated with industrial soot. Back in the day, observations like these were often not considered 'natural enough', and that anthropic selection was simply too strong to mimic the conditions needed for evolution via natural selection (Reznick et al., 2019). In his book, *Natural Selection in the Wild*, John Endler (1986) was one of the first to point out that selection under natural conditions is indeed often very strong, and therefore, that it should be possible to observe adaptive evolutionary change over short periods of 'ecological' time. Since this seminal publication, an appreciation that rapid evolution is more commonplace than general wisdom would predict has emerged (e.g. Lescak et al., 2015; Reznick et al., 2019).

Early evolutionary hypotheses in invasion biology were also firmly embedded in the traditional view of evolution as a slow process that is out of sync with the ecological dynamics of populations. For example, as we discussed earlier in this book, Darwin's naturalisation hypothesis postulates that evolutionary relatedness between exotic and native species corresponds to their ecological similarity, and therefore inversely with the ease by which exotics integrate into novel communities. This is based on Darwin's idea that contemporary ecological strategies are the products of past evolution and that closely related species will experience intense competition for shared resources. An upsurge from the mid-1990s in studies investigating the evolutionary ecology and genetics of organisms, including invasive species, have changed our perception of the supposed temporal mismatch between ecology and evolution, to one that recognises a dynamic two-way interaction between them over short contemporary timescales (Schoener, 2011; Reznick et al., 2019). This has had tremendous implications for all disciplines in biology. From an ecological viewpoint, it was quickly realised that rapid evolution can change the way in which species interact, and thus ecological dynamics and even ecosystem ecology, often over timescales as short as a few generations (Thompson, 1998; Schoener, 2011; Reznick et al., 2019). Indeed, empirical data from different taxa and ecosystems illustrate that intraspecific genetic variation can affect ecosystem processes such as resource fluctuation, whether due to differences between individual genotypes (e.g. Compson et al., 2018) or locally adapted populations (e.g. Simon et al., 2017; Gillis and Walsh, 2017). Of course, the recognition that evolution can occur over short timescales does not dismiss the fact that ecological change itself can drive evolutionary processes through natural selection over long periods of time, but rather, emphasises that we should never have ignored the contemporary interplay between ecology and evolution (Sultan, 2015; Garnas, 2018).

Abrupt changes in an organism's environmental conditions, like habitats blackened by industrial soot making light-coloured peppered moths easy pickings for predators, promote rapid adaptive evolution (Reznick et al., 2019; Moran and Alexander, 2014). In their influential review, Reznick and Ghalambor (2001) identified changes in host plant or food resources, the biophysical environment, mortality rates, and competitive interactions as the main drivers of rapid evolution. Most of these drivers act strongly on invasive populations of many types of organisms (Moran and Alexander, 2014; van Kleunen et al., 2018). Reznick and Ghalambor (2001) found studies documenting rapid evolution to often involve colonising species (both native and exotic species). This is hardly surprising, as introductions into new environments create unique associations between species that often share no co-evolutionary history, and thus biotic interactions (i.e. parasite-host, mutualist interactions, competition) and food resources (e.g. predator-prey interactions), simultaneously exposing colonists to novel abiotic environmental conditions (Reznick et al., 2019; Le Roux et al., 2020). Rapid population expansion that is typical during invasion also continuously exposes individuals to novel environmental conditions, creating further opportunities for rapid adaptive evolution. Of course, different ecological conditions that promote rapid evolution almost never act in isolation, and as a result, heterogenous abiotic habitat conditions are usually mirrored by high biotic turnover. The fast population growth rates and spread of invasive species may further promote rapid evolution. Rapid population declines in response to abrupt changes in the environment are expected to be less severe, and the odds of local extinction low, in these populations. Large and dense populations are also better buffered against demographic stochasticity than small populations (Lande, 1988).

As we discussed in Chapter 2, the demographic processes associated with the introduction, establishment, and spread of exotic species may also lead to non-adaptive evolutionary changes, e.g. shifts in trait values due to genetic drift. These demographic processes may also cause the strength of different drivers of adaptive evolution to vary over time. For example, broad-scale habitat heterogeneity is expected to impose strong selection on widespread invasive species, while immediate habitat availability will strongly affect exotic populations shortly after their arrival in the new environment and/ or directly after dispersal from initial introduction sites. Differences in the sizes of introduced, naturalised, and invasive populations also suggest that the impacts of stochastic processes on trait variation will be dependent on residence time. That is, small founder populations are expected to be more sensitive to stochastic events than widespread, and usually abundant, invasive populations. It follows that neutral (e.g. genetic drift) and deterministic (i.e. natural selection) drivers of microevolution will impact exotic populations in diverse ways, depending on their progression from initial introduction, to becoming established and widespread.

In this chapter we discuss examples of rapid (and often adaptive) evolution in invasive species in response to abrupt changes in key ecological conditions. These examples are by no means exhaustive, but rather illustrative of the main environmental conditions thought to often promote rapid evolution (Table 5.1).

## 5.2 Changes in abiotic conditions

### 5.2.1 Habitat availability and heterogeneity

Populations in suboptimal or changing habitats can respond in numerous ways. They can either tolerate adverse conditions via phenotypic plasticity, adapt, disperse, or become extinct. If dispersal allows species to find suitable habitat elsewhere, then directional selection on dispersal-related traits may occur. Theory predicts that, under heterogenous environmental conditions, decreased dispersal will evolve when recipient habitats are less favourable than the original habitat (Shigesada et al., 1986; Lindenmayer and Fischer, 2013). On the other hand, enhanced dispersal is expected to evolve when large and continuous patches of unoccupied and suitable habitat are available (Williams et al., 2016a). Under these circumstances, directional selection or spatial sorting can lead to the evolution of enhanced dispersal (Phillips et al., 2010; Shine et al., 2011). Yet, recent experimental work, tracking changes in dispersal over six generations of the model species, *Arabidopsis thaliana*, found increased dispersal to evolve more rapidly, and to be higher, under heterogenous than homogenous *and* suitable habitat conditions (Williams et al., 2016b). This suggests that increased dispersal may benefit individuals

**Table 5.1** Selected examples of rapid evolutionary responses in invasive species to different ecological drivers.

| Species | Common name | Ecological driver | Traits(s) | Residence time (years) | No. of generations | Reference |
|---|---|---|---|---|---|---|
| *Ambrosia artemisiifolia* | Ragweed | Abiotic (climate) | Growth rate, flowering onset, dichogamy, seed weight, total reproductive biomass, specific leaf area | c. 110–130 | c. 110–130 | van Boheemen et al. (2019) |
| *Lythrum salicaria* | Purple loosestrife | Abiotic (climate) | Onset of flowering, plant size at flowering | c. 200 | c. 100–200 | Colautti and Barrett (2013) |
| *Coccinella septempunctata* | Seven-spot ladybird | Abiotic (climate) | Melanism | 61 | c. 61 | O'Neill et al. (2017) |
| *Nicotiana glauca* | Tree tobacco | Biotic (mutualist availability) | Stigma-to-anther distances | Unknown | Unknown | Ollerton et al. (2012) |
| *Drosophila subobscura* | Fruit fly | Abiotic (climate) | Wing size | 20 | c. 100 | Huey et al. (2000) |
| *Helicoverpa zea* | Corn earworm | Biotic (host plant) | Bt toxin resistance | 6[a] | c. 30–42 | Tabashnik et al. (2013) |
| *Hypericum perforatum* | St. John's wort | Biotic (mutualist availability) | Responsiveness to arbuscular mycorrhization | 240 | c. 240 | Seifert et al. (2009) |
| *Rhinella marina* | Cane toad | Habitat availability and spatial sorting | Leg length | 70 | c. 70–140 | Phillips et al. (2006) |

[a] *Note*: Here residence time refers to the time since introduction of the novel food source and not time since introduction of the invasive species.

under heterogenous environmental conditions by allowing their escape from local competition or suboptimal habitat conditions (Hui et al., 2012).

The demographic dynamics of rapidly expanding invasive populations make them ideal systems to investigate the link between habitat availability, habitat heterogeneity, and evolution in dispersal–related traits. Limited data suggest that the conditions (i.e. continuous vs. patchy suitable habitat) under which enhanced or reduced dispersal may evolve during invasion are not easy to predict. For instance, Phillips et al. (2006) famously showed the evolution of increased dispersal ability in invasive cane toads (*Rhinella marina*) in Australia. Compared to individuals at the core of the geographic range, toads at the leading edge have longer legs, higher endurance, and can travel longer distances. These evolutionary changes, however, did not result from natural selection but rather from spatial sorting, whereby highly dispersive toads interbreed more frequently with each other at the leading edge than at the core (Shine et al., 2011). This evolution happened over the last 70 years, or between 70 and 140 toad generations (Rick Shine, personal communication). Others have since identified similar instances of increased dispersal ability in invasive populations due to spatial sorting (e.g. the Indian myna *Acridotheres tristis* in South Africa, Berthouly-Salazar et al., 2012; the ladybird *Harmonia axyridis* in Western Europe, Lombaert et al., 2014; the beach gladiolus *Gladiolus gueinzii* in Australia, Tabassum and Leishman, 2018). In the case of cane toads, increased dispersal in leading-edge populations is likely a response to continuously available habitat unoccupied by competitors (i.e. lack of niche competition). More recently, Berthouly-Salazar et al. (2013) used population genetic approaches to indirectly track dispersal in invasive European starlings (*Sturnus vulgaris*) in South Africa. Not only did they infer dispersal to be higher in leading edge than in core populations, but also that dispersal accelerated along the species' route of expansion. Unlike the cane toad example from Australia, increased dispersal in these populations coincided with birds encountering unsuitable habitats (related to rainfall and primary production), suggesting that evolution of higher dispersal abilities may facilitate avoidance of these habitats (Berthouly-Salazar et al., 2013).

Habitat heterogeneity may not always cause changes in dispersal–related traits. Distinct populations, those with high population genetic structure, as a consequence of local adaptation to different environmental conditions, are a common feature of most widespread species, including some invasive species. For example, in California, the invasive Canary Islands St. John's wort (*Hypericum canariense*) rapidly evolved latitudinal clines in flowering time and growth rates in response to varying climate conditions (Dlugosch and Parker, 2008). In the following section we discuss examples of such evolutionary responses in invasive species to broad-scale variation in abiotic environmental conditions in the new range.

## 5.2.2 Climate and physicochemical conditions

Climate often imposes strong selection on exotic species, illustrated by the usually strong relationship between establishment success and the climatic match between the native and non-native ranges of introduced species (e.g. Petitpierre et al., 2012; Hill et al., 2017; Liu et al., 2020). This may be especially true for sessile organisms like most plants, where establishment failure is almost certain when climate conditions are unsuitable in the new area of introduction. Ectotherms like insects, particularly those with limited dispersal capabilities, may suffer similar fates.

Introduced populations may adapt to novel climate conditions when these are only marginally unsuitable. For example, van Boheemen et al. (2019) found common ragweed (*Ambrosia artemisiifolia*) to have rapidly evolved clinal trait variation in response to climate conditions in the species' invasive ranges. These authors collected ragweed seeds across multiple, but similar, latitudinal gradients in the species' native range in North America and invasive ranges in Europe and Australia and planted these in a common garden. Trait data, such as maximum growth rates and onset of flowering, not only showed that similar clines evolved independently in the two invasive ranges, but also that these resemble clines from the native range. A follow-up population genomics study further found that clines on all three continents have similar genetic architectures, sharing between 17% and 26% of genes identified to be under climate selection (van Boheemen and Hodgins, 2020). This is remarkable as it suggests that, despite the very different demographic histories of these ragweed populations, similar and independent evolutionary responses were, to some degree, also modulated by the same genes. The independent evolution of these clines in Australia and Europe also strongly suggests that these responses are adaptive. Similar clinal responses have been documented for other widespread invasive plants (Fig. 5.1; Colautti et al., 2010; Colautti and Barrett, 2013; Li et al., 2014). Clinal responses often lead to trade-offs. For example, shorter seasons at higher latitudes generally lead to evolved earlier onset of flowering (Colautti and Barrett, 2013), while the opposite is true at low latitudes. The former often translates into trade-offs between plant size and

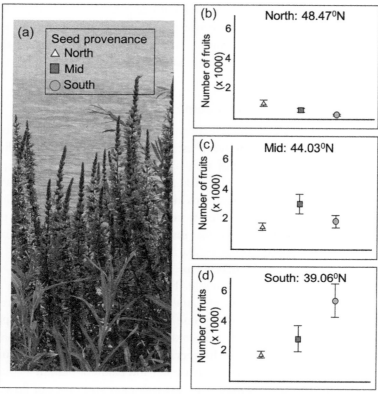

**Fig. 5.1** Evidence for rapid adaptation by an invasive species in response to climate conditions. (a) Purple loosestrife (*Lythrum salicaria*) was introduced into North America in the early 1800s. Colautti and Barrett (2013) found evidence for rapid local adaptation in invasive populations along a 1000-km latitudinal gradient. These authors conducted reciprocal transplant experiments between three sites (indicated here as North, Mid, and South) and found 'local' genotypes at each site to have higher fitness than 'foreign' genotypes (b–d). Note that symbols and colours in graphs (i.e. mean number of fruits produced ±95% confidence intervals) indicate the seed provenances used in each reciprocal transplant site (see inset frame in panel a). *(Photograph (panel a): Sourced from Wikimedia Commons under the Creative Commons Share-Alike License (CC-BY-SA 3.0). Original author: Manfred Heyde. Graphs were redrawn from Colautti and Barrett (2013).)*

reproductive maturity, as smaller earlier–flowering individuals will have the highest fitness when growing seasons are short (Colautti et al., 2010; Colautti and Barrett, 2013).

Sessile organisms will also experience strong selection due to non-climatic factors such as the physical and chemical properties of soils. Another study involving ragweed tested whether invasive populations in China have undergone rapid evolution in response to soil nutrient availability (Sun and Roderick, 2019). This study quantified population genetic structure within and between native and Chinese populations of ragweed using data from both neutral molecular markers and quantitative traits, allowing the researchers to disentangle the contributions stochastic processes (i.e. from neutral genetic marker data) and selection (i.e. from quantitative trait data) to population differentiation. As we will discuss in Chapter 8 (Section 8.3.1), these so-called $Q_{ST}$-$F_{ST}$ comparisons are useful to distinguish between trait differentiation that is the result of drift or (mostly past) selection. To infer levels of phenotypic plasticity, ragweed seeds from both ranges were grown under common garden conditions with treatments that mimicked low and high nutrient availability. Plasticity in height, total biomass, and stem biomass were significantly higher in Chinese than in North American ragweed populations. Moreover, $Q_{ST}$-$F_{ST}$ comparisons suggested that these differences could not be explained by genetic drift alone and that selection must have played a significant role. For example, heritability in plasticity was lower in North American populations than in Chinese populations, suggesting that historical selection may have purged most genetic variation for trait plasticity in the native range. This is plausible since ragweed probably occupies most available suitable habitat in North America. In China, on the other hand, ongoing and rapid range expansion is continuously exposing genotypes to novel soil conditions, creating opportunities for the evolution and maintenance of high phenotypic plasticity (Sun and Roderick, 2019).

Examples of invasive insects that have undergone rapid evolution in response to climate conditions also exist. For instance, the seven-spot ladybird (*Coccinella septempunctata*) was first introduced into North America from Europe as a biological control agent for aphids in 1956. Additional introductions from various native sources followed suit, resulting in a genetically diverse North American population (Kajita et al., 2012). O'Neill and co-workers (2017) wanted to know whether clinal patterns in melanism in North America in this ladybird were the result of rapid post-introduction evolution or phenotypic plasticity (they observed a negative correlation between melanism levels and temperature). Their hypothesis was based on the general idea that darker individuals have more efficient thermoregulation in low-temperature environments (Clusella-Trullas et al., 2007). Common garden experiments confirmed heritable genetic variation, and not plasticity, to underlie melanism in these populations (O'Neill et al., 2017), suggesting that the North American cline is indeed a product of rapid evolution. In another example, Huey et al. (2000) compared variation in wing size (as a proxy for body size) between invasive North American populations of the fruit fly, *Drosophila subobscura*, and native populations from Europe. Flies were sampled in both continents over roughly the same latitudinal range. Female flies showed a 4% increase in wing size from southern to northern latitudes in both ranges. In Europe, increased wing size resulted from the lengthening of wing parts closest to the body of flies, while in North America it was due to lengthening of outer segments of the wing. Thus, unlike the ragweed example discussed before, similarities in these clines probably did not involve selection on the same genomic regions.

Many invasive species attain spectacular geographic distributions, with populations often found over wide latitudinal ranges. As illustrated by the examples we discussed in this section, such widespread species often provide fascinating examples of rapid adaptive evolution. Moreover, these species may be ideal systems to test the predictability of rapid adaptive responses to climate conditions when their populations are found across similar latitudinal ranges in different locations.

## 5.3 Changes in biotic conditions

### 5.3.1 Mutualisms

Many organisms rely on intimate interactions with others to complete their life cycles, e.g. for reproduction (e.g. pollination), dispersal (e.g. myrmecochory), acquisition of essential nutrients (e.g. mycorrhization), immune system function (e.g. aphids with *Buchnera* bacteria). Upon introduction into new environments, exotic species leave most, if not all, of their symbionts behind in their native ranges (Le Roux et al., 2020). The loss of highly specialised mutualists may hamper establishment success (Catford et al., 2009). For example, many pine trees (family Pinaceae) initially failed to establish in the southern Hemisphere despite concerted introduction efforts to the region for commercial exploitation. It was only after the introduction of pine-specific ectomycorrhizal fungi that these species established, eventually taking off to become widespread invaders in many cases (Policelli et al., 2019). Intuitively, one would expect the loss of generalist mutualist interactions to pose less constraints on the establishment success of introduced exotic species. For example, many invasive plant species have generalist mutualist requirements and therefore often replace lost mutualisms by forming new associations with (usually also generalist) resident mutualists (Le Roux et al., 2020). Changes in mutualist interactions are expected to exert selection on exotic species if they are able to interact successfully with mutualists from the introduced range, especially those exotics with specialised requirements (Le Roux et al., 2017). Tree tobacco (*Nicotiana glauca*) provides us with a relevant example. This tree is native to South America and has been widely introduced around the world and is now considered invasive in many regions. In the native range various species of hovering hummingbirds pollinate this plant, an ecological specialisation has led to the evolution of tubular flowers that exclude most other pollinators (Nattero and Cocucci, 2007; Ollerton et al., 2012). In South Africa, invasive populations are pollinated by three different species of native hovering sunbirds (Geerts and Pauw, 2009). This is unusual, as sunbirds have a perching flower visiting behaviour rather than hovering around the native plants they visit (Anderson et al., 2005), and suggests that a behavioural adjustment towards hovering has occurred in these birds to exploit (and pollinate) tree tobacco (Geerts and Pauw, 2009). In regions like Greece and the Canary Islands, tree tobacco rapidly evolved shorter stigma-to-anther distances to facilitate self-fertilisation, a response to the lack of bird pollinators in these regions (Ollerton et al., 2012). Examples like this, where primarily self-incompatible or outcrossing plants evolved high levels of self-fertilisation in their new ranges, whether due to pollinator or mate limitations, are numerous (e.g. Ward et al., 2012; Pannell et al., 2015).

Evidence for rapid evolutionary responses in other types of mutualisms also exists. For example, Seifert et al. (2009) found invasive St. John's wort (*Hypericum perforatum*) in North America to be less responsive to arbuscular mycorrhizal fungi (AMF) than native European plants, an apparent adaptation to low AMF availability in North America. These authors inoculated plants from both ranges with *Glomus mosseae*, an AM fungus that naturally occurs in soils on both continents. So, what can explain the lower dependence of North American plants on AMF if *G. mosseae* is naturally found on the continent? As St. John's wort primarily invades highly disturbed agricultural ecosystems in North America, Seifert et al. (2009) speculated that degradation, and thus availability, of AMF communities led to the evolution of reduced AMF dependence. This was supported by trait data showing that North American plants had finer, thinner, and more branched roots than European plants. This trade-off between fine-root architecture and mycorrhizal responsiveness is well established in numerous plant species as a generalised response to varied AMF prevalence (Hetrick et al., 1992).

The examples discussed before illustrate several important points. First, exotic species with specialised mutualist requirements may experience particularly strong selection in the new range which may lead to establishment failure, unless these species can rapidly adapt to new mutualists (e.g. Mackin et al., 2021), or adopt survival strategies that are not dependent on mutualists (e.g. Ollerton et al., 2012). Second, even exotic species with generalist mutualisms may be subject to significant selection pressures when the availability or effectiveness of resident mutualists is low in the introduced range.

## 5.3.2 Competition

Significant changes in interspecific competition often lead to rapid evolution (Reznick and Ghalambor, 2001). During biological invasions, such responses are likely to be strongest in *native* species that have to compete with their new, and usually abundant and highly competitive, invasive neighbours. The high competitiveness of invasive species stems, in part, from a lack of many of the key biotic interactions, such as with specialist enemies, that kept their numbers in check in their historical ranges. On the other hand, resident native species not only have to compete with co-evolved residents, but also with invasive species that they typically share no eco-evolutionary experience with (Catford et al., 2009). These naïve native species are therefore expected to experience strong selection from competitive interactions with invasive species (e.g. Stuart et al., 2014). Invasive species, however, often do undergo rapid evolution in response to relaxed competitive conditions. As we discussed in Chapter 2 (Section 2.2.4) one such response is the evolution of increased competitive ability (EICA). The EICA hypothesis was originally formulated by Blossey and Nötzold (1995) when they observed invasive purple loosestrife in North America to have significantly higher reproductive output and vegetative growth than native European plants. They also observed a concurrent decline in plant investment in defence mechanisms in invasive North American populations, against two specialist European herbivores that was the absent in the invaded range (both of which have been subsequently introduced as biocontrol agents into North America). These observations led them to conclude that "in absence of [specialist] herbivores [in the invasive range], selection will favour [purple loosestrife] genotypes with improved competitive abilities and reduced resource allocation to herbivore defence." How common EICA is remains an open question and, while empirical support has been mixed (reviewed in Bossdorf et al., 2005 and Felker-Quinn et al., 2013), some have suggested it may occur frequently (for a recent meta-analysis, see Rotter and Holeski, 2018).

### 5.3.3 Food resources

Many specialised organisms are restricted to utilise one, or a limited number of, food resource(s). Strong directional selection may act on traits involved in food acquisition and processing when these specialist feeders switch to novel food resources. With the exception of some agricultural pests, we generally have few examples of rapid evolution in invasive species in response to novel food resources. This is probably because most invasive species are highly generalist feeders, negating the need to adapt to food resources in the new range. Specialist exotic feeders are often unable to find host or prey species in the new range with which they share any eco-evolutionary experience, which may lead to establishment failure. We have examples of biological control agents that have expanded their prey or host ranges. However, none of these host shifts have been demonstrated to be adaptive, and largely reflect the generalist diets of these species, and are simply expanded diet ranges in already polyphagous species (Mark Wright, personal communication; Wright and Bennett, 2018; but see Hill et al., 2017).

Invasive pests of agricultural crops, especially herbivorous insects, provide us with examples of rapid evolutionary responses to novel food resources. Here we will discuss two examples: one involving adaptive host shifts and the other resistance evolution against host defence. The codling moth (*Cydia pomonella*) is native to the Middle East. This moth is arguably one of the most damaging agricultural pests in the world, especially of apple and pear orchards. The species was introduced to the USA in 1750, quickly becoming widespread throughout the country's apple and pear production regions. However, within 150 years of its arrival, the species was reported on various new agricultural plants, including plums and walnuts (Phillips and Barnes, 1975). These shifts were accompanied by genetically based preferences for oviposition sites by moths on their newly adopted hosts, in some instances leading to the rapid development of reproductive isolation between moth populations on different host plants, i.e. the formation of host races (Bush, 1969). In the case of codling moths, temporal isolation likely led to host race formation. The moth's seasonal emergence closely follows the development of its host plants and the latter is asynchronous between apple, plum, and walnut orchards (Cox, 2004). The formation of host races has been documented in various other agricultural pests (e.g. Craig et al., 1993; Emelianov et al., 2004).

Agricultural pests also often evolve resistance against management practices like the build-up of resistance to pesticides, and more recently, changes in their food resources. The diamondback moth (*Plutella xylostella*) is a serious pest of cruciferous crops (family Brassicaceae), found today wherever these are grown (Talekar and Shelton, 1993). Field-evolved resistance against pesticides containing Bt toxins (insecticidal proteins isolated from the bacterium *Bacillus thuringiensis* or whole *B. thuringiensis* bacteria) was first discovered in diamondback moth in the 1980s in the USA (Tabashnik et al., 1990).

The effectiveness of Bt toxins caught the attention of biotechnologists, leading to the creation, and widespread planting, of numerous crops engineered to express Bt toxins. For instance, by 2014, millions of hectares of Bt corn and cotton were grown globally (Tabashnik et al., 2013; Aldemita et al., 2015; also see Chapter 9, Section 9.4). Widespread planting of these transgenic crops generated one of the strongest selection pressures for insect resistance ever known (Tabashnik et al., 2013). Laboratory experiments confirmed that many insect pests have the capacity to evolve resistance against Bt crops (Tabashnik, 1994; Ferré and Van Rie, 2002). Despite resistant management strategies like refugia, whereby a proportion of crops planted must be non-Bt to avoid resistance build-up, cases of field-evolved resistance have been identified in the corn earworm (*Helicoverpa zea*) and pink bollworm (*Pectinophora gossypiella*) (Tabashnik et al., 2013).

Invasive species are also frequently encountered and utilised by native species as new food resources. This, in turn, may impose strong selection on native species. Indeed, many examples of rapid evolutionary responses in native herbivorous insects (e.g. Carroll et al., 2005) and parasitoids (e.g. Knapp et al., 2019) utilising invasive species as food resources exist. Strong selection on native species is also expected when they become the hosts or prey items of widespread invasive herbivores and predators (e.g. Freeman and Byers, 2006). These and other evolutionary impacts of invasive species on native species are discussed in detail in Chapter 7.

## 5.4 Conclusions

A recent explosion in studies documenting rapid evolution has caused a paradigm shift in how we perceive evolution. An appreciation for interactions between the ecology and evolution of species, often resulting in rapid evolutionary change, is now firmly cemented in the minds of biologists. Some of the most intriguing examples of rapid adaptive evolution come from studies on invasive species. These have shown that evolution can occur over timescales that are sometimes as short as a few generations. Adaptive evolutionary responses in invasive species result from the abrupt changes they experience in most, if not all, abiotic and biotic conditions in their new ranges. These responses are further aided by their usually high abundances, short generation times, high reproductive capacities, and high dispersal capabilities (Hoffmann et al., 2008; Moran and Alexander, 2014). Out of the environmental drivers of rapid evolution, abrupt changes in climate conditions are most frequently invoked as responsible for rapid evolution in invasive species, especially in sessile organisms. Changes in biotic interactions may also impose selection on invasive species, e.g., mutualist requirements and availability, novel food resources, and conditions of relaxed competition. Changes in mortality rates, such as via increased predation, have often been linked to rapid evolution. For invasive species, lower mortality rates, owing to the absence of natural enemies, may lead to the development of larger populations for selection to act on.

In summary, invasive species represent ideal systems to study rapid evolution, not only over a range of timescales, but also in a variety of organisms. How pervasive rapid evolution is, under which conditions it occurs, and how predicable its outcomes are, will remain priorities on the research agendas of biologists. Since interactions between species are instrumental in the way communities are organised, rapid evolution may have far-reaching implications for ecosystem processes. Yet, how such evolution scales up to community interactions and ecosystem ecology remains largely unexplored. Biological invasions are ideal opportune natural experiments to fill this research gap.

# References

Aldemita, R.R., Reaño, I.M.E., Solis, R.O., Hautea, R.A., 2015. Trends in global approvals of biotech crops (1992–2014). GM Crops Food 6, 150–166.

Anderson, B.W., Cole, W.W., Barrett, S.C.H., 2005. Specialized bird perch aids cross-pollination: a plant scores by providing an access point for visiting sunbirds to feed on its nectar. Nature 435, 41–42.

Berthouly-Salazar, C., van Rensburg, B., Le Roux, J.J., van Vuuren, B.J., Hui, C., 2012. Spatial sorting drives morphological variation in the invasive bird, Acridotheris tristis. PLoS One 7, e38145.

Berthouly-Salazar, C., Hui, C., Blackburn, T.M., Gaboriaud, C., van Rensburg, B.J., van Vuuren, B.J., Le Roux, J.J., 2013. Long-distance dispersal maximizes evolutionary potential during rapid geographic range expansion. Mol. Ecol. 22, 5793–5804.

Blossey, B., Nötzold, R., 1995. Evolution of increased competitive ability in plants: a hypothesis. J. Ecol. 83, 887–889.

Bossdorf, O., Auge, H., Lafuma, L., Rogers, W.E., Siemann, E., Prati, D., 2005. Phenotypic and genetic differentiation between native and introduced plant populations. Oecologia 144, 1–11.

Bush, G.L., 1969. Sympatric host race formation and speciation in frugivorous flies of the genus *Rhagoletis* (Diptera, Tephritidae). Evolution 23, 237–251.

Carroll, S.P., Loye, J.E., Dingle, H., Mathieson, M., Famula, T.R., Zalucki, M.P., 2005. And the beak shall inherit—evolution in response to invasion. Ecol. Lett. 8, 944–951.

Catford, J.A., Jansson, R., Nilsson, C., 2009. Reducing redundancy in invasion ecology by integrating hypotheses into a single theoretical framework. Divers. Distrib. 15, 22–40.

Clusella-Trullas, S., van Wyk, J.H., Spotila, J.R., 2007. Thermal melanism in ectotherms. J. Therm. Biol. 32, 235–245.

Colautti, R.I., Barrett, S.C.H., 2013. Rapid adaptation to climate facilitates range expansion of an invasive plant. Science 342, 364–366.

Colautti, R.I., Eckert, C.G., Barrett, S.C., 2010. Evolutionary constraints on adaptive evolution during range expansion in an invasive plant. Proc. R. Soc. B Biol. Sci. 277, 1799–1806.

Compson, Z.G., Hungate, B.A., Whitham, T.G., Koch, G.W., Dijkstra, P., Siders, A.C., Wojtowicz, T., Jacobs, R., Rakestraw, D.N., Allred, K.E., Sayer, C.K., Marks, J.C., 2018. Linking tree genetics and stream consumers: isotopic tracers elucidate controls on carbon and nitrogen assimilation. Ecology 99, 1759–1770.

Cox, G.W., 2004. Alien Species and Evolution: The Evolutionary Ecology of Exotic Plants, Animals, Microbes, and Interacting Native Species. Island Press, Washington, DC.

Craig, T.P., Itami, J.K., Abrahamson, W.G., Horner, J.D., 1993. Behavioral evidence for host-race formation in *Eurosta solidaginis*. Evolution 47, 1696–1710.

Darwin, C.R., 1859. The Origin of Species. John Murray, London.

Dlugosch, K.M., Parker, I.M., 2008. Founding events in species invasions: genetic variation, adaptive evolution, and the role of multiple introductions. Mol. Ecol. 17, 431–449.

Emelianov, I., Marec, F., Mallet, J., 2004. Genomic evidence for divergence with gene flow in host races of the Larch Budmoth. Proc. R. Soc. B Biol. Sci. 271, 97–105.

Endler, J.A., 1986. Natural Selection in the Wild. Princeton University Press, Princeton, NJ.

Felker-Quinn, E., Schweitzer, J.A., Bailey, J.K., 2013. Meta-analysis reveals evolution in invasive plant species but little support for evolution of increased competitive ability (EICA). Ecol. Evol. 3, 739–751.

Ferré, J., Van Rie, J., 2002. Biochemistry and genetics of insect resistance to *Bacillus thuringiensis*. Annu. Rev. Entomol. 47, 501–533.

Freeman, A.S., Byers, J.E., 2006. Divergent induced responses to an invasive predator in marine mussel populations. Science 313, 831–833.

Garnas, J.R., 2018. Rapid evolution of insects to global environmental change: conceptual issues and empirical gaps. Curr. Opin. Insect Sci. 29, 93–101.

Geerts, S., Pauw, A., 2009. African sunbirds hover to pollinate an invasive hummingbird-pollinated plant. Oikos 118, 573–579.

Gillis, M.K., Walsh, M.R., 2017. Rapid evolution mitigates the ecological consequences of an invasive species (*Bythotrephes longimanus*) in lakes in Wisconsin. Proc. R. Soc. B Biol. Sci. 284, 20170814.

Hetrick, B.A.D., Wilson, G.W.T., Todd, T.C., 1992. Relationships of mycorrhizal symbiosis, rooting strategy, and phenology among tallgrass prairie forbs. Can. J. Bot. 70, 1521–1528.

Hill, M.P., Gallardo, B., Terblanche, J.S., 2017. A global assessment of climatic niche shifts and human influence in insect invasions. Glob. Ecol. Biogeogr. 26, 679–689.

Hoffmann, A.A., Reynolds, K.T., Michael, N.A., Weeks, A.R., 2008. A high incidence of parthenogenesis in agricultural pests. Proc. R. Soc. B Biol. Sci. 275, 2473–2481.

Huey, R.B., Gilchrist, G.W., Carlson, M.L., Berrigan, D., Serra, L., 2000. Rapid evolution of a geographic cline in size in an introduced fly. Science 287, 308–309.

Hui, C., Roura-Pascual, N., Brotons, L., Robinson, R.A., Evans, K.L., 2012. Flexible dispersal strategies in native and non-native ranges: environmental quality and the "good–stay, bad–disperse" rule. Ecography 35, 1024–1032.

Kajita, Y., O'Neill, E.M., Zheng, Y., Obrycki, J.J., Weisrock, D.W., 2012. A population genetic signature of human releases in an invasive ladybeetle. Mol. Ecol. 21, 5473–5483.

Knapp, M., Řeřicha, M., Maršíková, S., Harabiš, F., Kadlec, T., Nedvěd, O., Teder, T., 2019. Invasive host caught up with a native parasitoid: field data reveal high parasitism of *Harmonia axyridis* by *Dinocampus coccinellae* in Central Europe. Biol. Invasions 21, 2795–2802.

Lande, R., 1988. Genetics and demography in biological conservation. Science 241, 1455–1460.

Le Roux, J.J., Hui, C., Keet, J.H., Ellis, A.G., 2017. Co-introduction vs ecological fitting as pathways to the establishment of effective mutualisms during biological invasions. New Phytol. 215, 1354–1360.

Le Roux, J.J., Clusella-Trullas, S., Mokotjomela, T.M., Mairal, M., Richardson, D.M., Skein, L., Wilson, J.R., Weyl, O.L.F., Geerts, S., 2020. Biotic interactions as mediators of biological invasions: insights from South Africa. In: van Wilgen, B.W., Measey, J., Richardson, D.M., Wilson, J.R., Zengeya, T.A. (Eds.), Biological Invasions in South Africa. Springer, Berlin, pp. 387–427.

Lescak, E.A., Bassham, S.L., Catchen, J., Gelmond, O., Sherbick, M.L., von Hippel, F.A., Cresko, W.A., 2015. Evolution of stickleback in 50 years on earthquake-uplifted islands. Proc. Natl. Acad. Sci. U. S. A. 112, E7204–E7212.

Li, X.-M., She, D.-Y., Zhang, D.-Y., Liao, W.-J., 2014. Life history trait differentiation and local adaptation in invasive populations of *Ambrosia artemisiifolia* in China. Oecologia 177, 669–677.

Lindenmayer, D.B., Fischer, J., 2013. Habitat Fragmentation and Landscape Change: An Ecological and Conservation Synthesis. Island Press, Washington, DC.

Liu, C., Wolter, C., Xian, W., Jeschke, J.M., 2020. Most invasive species largely conserve their climatic niche. Proc. Natl. Acad. Sci. U. S. A. 117, 23643–23651.

Lombaert, E., Estoup, A., Facon, B., Joubard, B., Grégoire, J.-.C., Jannin, A., Blin, A., Guillemaud, T., 2014. Rapid increase in dispersal during range expansion in the invasive ladybird *Harmonia axyridis*. J. Evol. Biol. 27, 508–517.

Mackin, C.R., Peña, J.F., Blanco, M.A., Balfour, N.J., Castellanos, M.C., 2021. Rapid evolution of a floral trait following acquisition of novel pollinators. J. Ecol. 109, 2234–2246.

Moran, E.V., Alexander, J.M., 2014. Evolutionary responses to global change: lessons from invasive species. Ecol. Lett. 17, 637–649.

Nattero, J., Cocucci, A.A., 2007. Geographical variation in floral traits of the tree tobacco in relation to its hummingbird pollinator fauna. Biol. J. Linn. Soc. 90, 657–667.

Ollerton, J., Watts, S., Connerty, S., Lock, J., Parker, L., Wilson, I., Schueller, S., Nattero, J., Cocucci, J.A., Izhaki, I., Geerts, S., Pauw, A., Stout, J.C., 2012. Pollination ecology of the invasive tree tobacco *Nicotiana glauca*: comparisons across native and non-native ranges. J. Pollinat. Ecol. 9, 85–95.

O'Neill, E.M., Hearn, E.J., Cogbill, J.M., Kajita, Y., 2017. Rapid evolution of a divergent ecogeographic cline in introduced lady beetles. Evol. Ecol. 31, 695–705.

Pannell, J.R., Auld, J.R., Brandvain, Y., Burd, M., Busch, J.W., Cheptou, P.-.O., Conner, J.K., Goldberg, E.E., Grant, A.-.G., Grossenbacher, D.L., Hovick, S.M., Igic, B., Kalisz, S., Petanidou, T., Randle, A.M., de Casas, R.R., Pauw, A., Vamosi, J.C., Winn, A.A., 2015. The scope of Baker's law. New Phytol. 208, 656–667.

Petitpierre, B., Kueffer, C., Broennimann, O., Randin, C., Daehler, C., Guisan, A., 2012. Climatic niche shifts are rare among terrestrial plant invaders. Science 335, 1344–1348.

Phillips, P.A., Barnes, M.M., 1975. Host race formation among sympatric apple, walnut, and plum populations of the codling moth, *Laspeyresiapo monella*. Ann. Entomol. Soc. Am. 68, 1053–1060.

Phillips, B.L., Brown, G.P., Webb, J.K., Shine, R., 2006. Invasion and the evolution of speed in toads. Nature 439, 803.

Phillips, B.L., Brown, G.P., Shine, R., 2010. Life-history evolution in range-shifting populations. Ecology 91, 1617–1627.

Policelli, N., Bruns, T.D., Vilgalys, R., Nuñez, M.A., 2019. Suilloid fungi as global drivers of pine invasions. New Phytol. 222, 714–725.

Reznick, D.N., Ghalambor, C.K., 2001. The population ecology of contemporary adaptations: what empirical studies reveal about the conditions that promote adaptive evolution. Genetica 112, 183–198.

Reznick, D.N., Losos, J., Travis, J., 2019. From low to high gear: there has been a paradigm shift in our understanding of evolution. Ecol. Lett. 22, 233–244.

Rotter, M.C., Holeski, L.M., 2018. A meta-analysis of the evolution of increased competitive ability hypothesis: genetic-based trait variation and herbivory resistance trade-offs. Biol. Invasions 20, 2647–2660.

Schoener, T.W., 2011. The newest synthesis: understanding the interplay of evolutionary and ecological dynamics. Science 331, 426–429.

Seifert, E.K., Bever, J.D., Maron, J.L., 2009. Evidence for the evolution of reduced mycorrhizal dependence during plant invasion. Ecology 90, 1055–1062.

Shigesada, N., Kawasaki, K., Teramoto, E., 1986. Traveling periodic waves in heterogeneous environments. Theor. Popul. Biol. 30, 143–160.

Shine, R., Brown, G.P., Phillips, B.L., 2011. An evolutionary process that assembles phenotypes through space rather than through time. Proc. Natl. Acad. Sci. U. S. A. 108, 5708–5711.

Simon, T.N., Bassar, R.D., Binderup, A.J., Flecker, A.S., Freeman, M.C., Gilliam, J.F., Marshall, M.C., Thomas, S.A., Travis, J., Reznick, D.N., Pringle, C.M., 2017. Local adaptation in Trinidadian guppies alters stream ecosystem structure at landscape scales despite high environmental variability. Copeia 105, 504–513.

Stuart, Y., Campbell, T., Hohenlohe, P., Reynolds, R., Revell, L., Losos, J., 2014. Rapid evolution of a native species following invasion by a congener. Science 346, 463–466.

Sultan, S., 2015. Organism and Environment, first ed. Oxford University Press, New York.

Sun, Y., Roderick, G.K., 2019. Rapid evolution of invasive traits facilitates the invasion of common ragweed, *Ambrosia artemisiifolia*. J. Ecol. 107, 2673–2687.

Tabashnik, B.E., 1994. Evolution of resistance to *Bacillus thuringiensis*. Annu. Rev. Entomol. 39, 47–79.

Tabashnik, B.E., Cushing, N.L., Finson, N., Johnson, M.W., 1990. Field development of resistance to *Bacillus thuringiensis* in diamondback moth (Lepidoptera: Plutellidae). J. Econ. Entomol. 83, 1671–1676.

Tabashnik, B., Brévault, T., Carrière, Y., 2013. Insect resistance to Bt crops: lessons from the first billion acres. Nat. Biotechnol. 31, 510–521.

Tabassum, S., Leishman, M.R., 2018. Have your cake and eat it too: greater dispersal ability and faster germination towards range edges of an invasive plant species in eastern Australia. Biol. Invasions 20, 1199–1210.

Talekar, N.S., Shelton, A.M., 1993. Biology, ecology and management of diamondback moth. Annu. Rev. Entomol. 38, 275–301.

Thompson, J.N., 1998. Rapid evolution as an ecological process. Trends Ecol. Evol. 13, 329–332.

van Boheemen, L.A., Hodgins, K.A., 2020. Rapid repeatable phenotypic and genomic adaptation following multiple introductions. Mol. Ecol. 29, 4102–4117.

van Boheemen, L.A., Atwater, D.Z., Hodgins, K.A., 2019. Rapid and repeated local adaptation to climate in an invasive plant. New Phytol. 222, 614–627.

van Kleunen, M., Bossdorf, O., Dawson, W., 2018. The ecology and evolution of alien plants. Annu. Rev. Ecol. Evol. Syst. 49, 25–47.

Ward, M., Johnson, S.D., Zalucki, M.P., 2012. Modes of reproduction in three invasive milkweeds are consistent with Baker's rule. Biol. Invasions 14, 1237–1250.

Williams, J.L., Snyder, R.E., Levine, J.M., 2016a. The influence of evolution on population spread through patchy landscapes. Am. Nat. 188, 15–26.

Williams, J.L., Kendall, B.E., Levine, J.M., 2016b. Rapid evolution accelerates plant population spread in fragmented experimental landscapes. Science 353, 482–485.

Wright, M.G., Bennett, G.M., 2018. Evolution of biological control agents following introduction to new environments. BioControl 63, 105–116.

# CHAPTER 6

# The current state of research on the evolutionary ecology of invasive species

## Chapter Outline

6.1 Introduction                                                                100
6.2 Causes of rapid evolution                                                   102
    6.2.1 Abiotic factors                                   102
    6.2.2 Biotic factors                                    114
    6.2.3 Stochastic factors                                117
6.3 Traits that frequently undergo rapid evolution during invasion             118
6.4 Introduction history                                                       119
6.5 Experimental design                                                        122
6.6 Conclusions                                                                123
References                                                                     125

## Abstract

This chapter reviews the mechanisms underlying, and dynamics of, rapid evolution in invasive species. By synthesising recent research, the relative roles of adaptive (abiotic and biotic) and non-adaptive (founder events, bottlenecks, drift, spatial sorting) processes in causing rapid evolution in invasive species are discussed. Traits that often undergo rapid evolution in invasive species, such as clinal variation in phenology, increased competitive ability, changes in reproductive biology, etc., are examined in detail. The need to account for geographic structure in traits, as well as stochastic processes associated with the introduction history and spread of exotic species, when studying rapid evolution, is discussed. The chapter concludes by suggesting future avenues for the study of rapid evolution in invasive species.

**Keywords:** Common garden experiments, Clines, Founder events, Genetic drift, Genetic paradox, Local adaptation, Rapid evolution, Reciprocal transplant experiments, Spatial sorting

*...throughout nature the forms of life which are now dominant tend to become still more dominant by leaving many modified and dominant descendants.*
***Darwin (1859)***

*The Evolutionary Ecology of Invasive Species*
https://doi.org/10.1016/B978-0-12-818378-6.00006-1

## 6.1 Introduction

As we discussed throughout this book, invasive species may adapt in response to novel interactions with herbivores, predators, mutualists, and abiotic conditions in their new environments (e.g. Barrett et al., 2008; Prentis et al., 2008; Zenni et al., 2014; Oduor et al., 2016), or they may experience non-adaptive shifts in trait values due to introduction history or other stochastic processes (Shine et al., 2011; Colautti and Lau, 2015; Schrieber et al., 2017; Hodgins et al., 2018). As a result, invasive species afford multiple opportunities to study evolution over short ecological timescales, and under 'natural laboratory' conditions, and have led to a recent surge in research.

The Indian dwarf mongoose (*Urva auropunctata*) is one of the worst invasive animals on tropical islands (Lowe et al., 2004; Louppe et al., 2021). From my days as a graduate student in Hawaii, I remember this species, along with rats, to be one of the most abundant mammals on the islands. This is somewhat ironic, as mongooses were originally introduced to the islands by well-meaning farmers to control these rodents in sugarcane plantations! This 'biocontrol' introduction was wholly misguided, partly because of the different ecologies of mongooses and their intended prey (rats are mostly nocturnal and mongooses are diurnal), and partly because of the mongoose's generalist diet and subsequent negative impacts on native species (Hays and Conant, 2007). In the absence of predators and competitors in places like Hawaii, mongoose populations quickly exploded, reaching densities up to 100 times that in the species' native range (Owen and Lahti, 2020).

The Indian dwarf mongoose is promiscuous, solitary, and non-territorial, and males use scent to advertise themselves to females. Scent is placed in the environment by fleshy protrusions around the anus, known as anal pads (Owen and Lahti, 2020). Anal pad size shows variation typical of strong sexual selection, being more than twice as large in males as in females (Owen and Lahti, 2015, 2020). High population densities on invaded islands likely affect the ways in which mongooses communicate and, therefore, how males maximise their own fitness. Compared with males in native range populations, invasive males are expected to encounter females more frequently and, therefore, costly long-distance scent advertising is likely to be less critical. High encounter rates between males and females in invasive populations may also lead to situations where sperm of different males compete for fertilisation of the same egg, a phenomenon known as sperm competition (Parker, 2020). Owen and Lahti (2020) tested these two hypotheses by comparing the size of anal pads and testes between mongooses from invasive populations on the Hawaiian Islands, Jamaica, Mauritius, and St. Croix and those from native populations in India. They found sexual dimorphism in anal pad size to be between 63% and 88% lower in island populations, largely because of less investment by males in these structures. This supported their first hypothesis that males are experiencing relaxed sexual selection in dense invasive populations. They also found invasive males to have larger testes than those from native range populations, supporting their second hypothesis of increased sperm competition. Higher sperm-producing males are likely selected in dense invasive populations, where the likelihood of females mating with multiple partners, and thus of sperm competition occurring, is higher than in the native range (Owen and Lahti, 2020).

Despite a recent surge in studies such as those on invasive Indian dwarf mongooses, few authors have attempted to synthesise these data (but see Colautti and Lau, 2015; Moran and Alexander, 2014; Hodgins et al., 2018; Reznick et al., 2019). In this chapter we take a closer look at the contribution of recent work to our understanding of rapid evolution during invasion. To examine the main drivers, dynamics, and consequences of rapid evolution in invasive species I conducted a literature search in the ISI Web of Knowledge database (Thomson Reuters) by searching combinations of the *topic* (TS) keywords 'invasive species' and 'rapid evolution', and restricting my search to the subject categories 'Evolutionary biology' and 'Ecology' (February 2021) and papers published between 2010 and 2021. Papers were then screened to identify those that made use of quantitative trait data to infer rapid evolution. The native and invasive ranges of study species and their broad taxonomic or functional classification (e.g. annual vs perennial plant, invertebrate vs vertebrate, etc.) were noted. Functional trait(s) that underwent rapid evolution and whether inferences of rapid evolution were based on trait comparisons between invasive and native populations *or* trait comparisons between invasive populations were also recorded. Where possible, I also determined whether selection was due to abiotic or biotic factors, or both; the residence time of the species (i.e. years since introduction); whether invasive populations originated from multiple or single introduction events; and the number of generations that have elapsed since introduction. Most of these data were obtained from additional literature sources. Studies reporting rapid evolution in the same species, but from multiple and independent invaded regions (e.g. invasive Mozambican tilapia *Oreochromis mossambicus* in New Caledonia and Guadeloupe; Firmat et al., 2012) were treated as independent case studies. The same applied to different studies reporting rapid evolution in the same invasive species in the same invaded range but measuring different functional traits (e.g. ragweed *Ambrosia artemisiifolia*; Chun et al., 2011; Hodgins and Rieseberg, 2011; van Boheemen et al., 2019).

In total, 73 papers involving 62 species and 87 independent case studies were identified (Table 6.1). Fifty-six of all case studies involved plants, 30 animals, and 1 a fungus (Robin et al., 2017). For plants, 45 cases involved annual and 11 perennial species. Insects were the most studied animal taxon (27% of all animal studies) and birds the least (1 study, Le Gros et al., 2016). The earliest known introduction date was 1520 (i.e. Jimsonweed *Datura stramonium*, introduced to Spain from Mexico; Castillo et al., 2019) and the most recent 1990 (i.e. the Pacific oyster *Crassostrea gigas*, introduced to Europe from East Asia; Wendling and Wegner, 2015). On average, 210 generations have lapsed since the introduction of species, ranging between 47 (i.e. the guttural toad *Sclerophrys gutturalis*, on the Mascarene Islands; Baxter-Gilbert et al., 2020) and 1280 (i.e. the Asian tiger mosquito *Aedes albopictus*, in the United States; Urbanski et al., 2012) generations. After accounting for pseudo replication (i.e. different studies reporting on the same species in the same country), 26 instances were identified where invasive populations have originated from multiple introductions (19 plant, 7 animal), and 3 instances where invasive populations have originated from single introduction events (1 plant, 2 animal; Table 6.1). In the remainder of this chapter we expand on these findings.

## 6.2 Causes of rapid evolution

As we discussed in Chapter 5, rapid evolution generally occurs when species experience abrupt changes in host plant or food resources, the biophysical environment, mortality rates, and competitive interactions, all of which are typical features of biological invasions (Reznick and Ghalambor, 2001; Reznick et al., 2019). Here, we briefly discuss the most common adaptive (i.e. natural selection) and non-adaptive (e.g. genetic drift) drivers of rapid evolution identified in recent studies (also see Table 6.1 for more details).

### 6.2.1 Abiotic factors

The climatic match between the native and non-native ranges of species has been consistently linked to their establishment success and invasiveness (Liu et al., 2020a). Climate is therefore expected to impose strong selection on

**Table 6.1** Summary of selected studies published between 2010 and 2021 documenting rapid evolution in 62 invasive species from 87 independent invasive ranges.

| Taxon | Native range | Invaded range | Intro. date | No. gen. | Intro. history | Evolutionary driver(s) | Trait(s) | Reference |
|---|---|---|---|---|---|---|---|---|
| **Animals** | | | | | | | | |
| *Aedes aegypti* | Africa | USA | N/A | N/A | Multiple | Biotic, competition | Resistance to satyrisation (females being rendered infertile) | Bargielowski and Lounibos (2014) |
| *Aedes albopictus* | East and South–east Asia | USA | 1985 | 1280 | Multiple | Abiotic, photoperiod | Reduced diapause | Urbanski et al. (2012) |
| | | Europe | N/A | N/A | Multiple | Abiotic | Clinal variation in wing size | Sherpa et al. (2019) |
| | | USA | 1985 | 1280 | Multiple | Abiotic, temperature | Increased survival of diapause-induced eggs | Medley et al. (2019) |
| *Alosa pseudoharengus* | Eastern USA | Laurentian Great Lakes | 1873 | 147 | Unknown | Biotic, novel prey/host | Variable gill raker spacing | Smith et al. (2020) |
| *Anolis cristatellus* | Puerto Rico | USA | 1975 | N/A | Multiple | Abiotic, temperature | Increased low-temperature acclimation (plasticity) | Kolbe et al. (2012) |
| *Bythotrephes longimanus* | Europe, Asia | Canada | 1990 | N/A | Unknown | Biotic, predator/herbivore | Longer distal spines | Miehls et al. (2014) |

*Continued*

**Table 6.1** Summary of selected studies published between 2010 and 2021 documenting rapid evolution in 62 invasive species from 87 independent invasive ranges—cont'd

| Taxon | Native range | Invaded range | Intro. date | No. gen. | Intro. history | Evolutionary driver(s) | Trait(s) | Reference |
|---|---|---|---|---|---|---|---|---|
| *Ciona robusta* | Northwest Pacific Ocean | Red Sea | 2015 | N/A | Unknown | Abiotic, temperature and salinity | Increased tolerance of high temperatures and salinity | Chen et al. (2018) |
| *Cocinella septempunctata* | Europe, Asia | North America | 1956 | 61 | Unknown | Abiotic, temperature | Clinal variation in melanin pigmentation | O'Neill et al. (2017) |
| *Crassostrea gigas* | East Asia | Europe | 1990 | N/A | Multiple | Biotic, parasitism | Increased pathogen resistance | Wendling and Wegner (2015) |
| *Eurytemora affinis* | Atlantic Ocean | USA | N/A | N/A | Unknown | Abiotic, salinity | Increased tolerance to variable salinity | Lee et al. (2011) |
| *Mus musculus* | India | Kerguelen Archipelago | 1850 | N/A | Unknown | Biotic, novel prey/host | Modified jaw morphology | Renaud et al. (2015) |
| *Oreochromis mossambicus* | Mozambique | New Caledonia | 1950 | N/A | Unknown | Unknown | More streamlined body shape | Firmat et al. (2012) |
| | | Guadeloupe | 1950 | N/A | Unknown | Unknown | More streamlined body shape | Firmat et al. (2012) |
| *Pheidole megacephala* | Africa | Australia, USA, South Africa, Mauritius | N/A | N/A | Unknown | Biotic, competition | Larger body and head size | Wills et al. (2014) |

| Species | Native range | Introduced range | Year | Generations | Driver | Introductions | Trait change | Reference |
|---|---|---|---|---|---|---|---|---|
| *Pieris rapae* | Europe | North America | 1860 | N/A | Abiotic, temperature | Unknown | Clinal variation in development | Seiter et al. (2013) |
| | | Japan | 1700 | N/A | Abiotic, temperature | Unknown | Clinal variation in development | Seiter et al. (2013) |
| *Pycnonotus jocosus* | South–east Asia | La Reunion Island | 1965 | N/A | Environmental heterogeneity | Single | Morphological divergence (body size, wing, tail, beak) | Le Gros et al. (2016) |
| *Python bivittatus* | South–east Asia | USA | 1980 | N/A | Abiotic, temperature | Multiple | Increased cold resistance | Card et al. (2018) |
| *Rhinella marina* | South and central America | Australia | 1935 | 70–140 | Abiotic, temperature | Unknown | Increased heat tolerance | Kosmala et al. (2018) |
| | | USA | 1930 | 90–180 | Abiotic, temperature | Multiple | Increased cold tolerance | Mittan and Zamudio (2019) |
| | | Australia | 1935 | 70–140 | Unknown | Unknown | Increased dispersal | Hudson et al. (2020) |
| *Salmo trutta* | Europe, Asia | Canada | 1883 | 137 | Unknown | Unknown | Evidence for local adaptation based on survival rates | Westley et al. (2013) |
| *Sclerophrys gutturalis* | South Africa | Mauritius | 1925 | 47 | Unknown | Multiple | Decreased body size and hind limb length | Baxter-Gilbert et al. (2020) |
| *Sclerophrys gutturalis* | South Africa | La Reunion | 1925 | 47 | Unknown | Unknown | Decreased body size and hind limb length | Baxter-Gilbert et al. (2020) |

*Continued*

**Table 6.1** Summary of selected studies published between 2010 and 2021 documenting rapid evolution in 62 invasive species from 87 independent invasive ranges—cont'd

| Taxon | Native range | Invaded range | Intro. date | No. gen. | Intro. history | Evolutionary driver(s) | Trait(s) | Reference |
|---|---|---|---|---|---|---|---|---|
| *Urva auropunctata* | South Asia | Hawaii | 1890 | N/A | Unknown | Biotic, density-dependent sexual selection | Decreased anal pad size; increased testicle size | Owen and Lahti (2020) |
| | | Jamaica | 1890 | N/A | Unknown | Biotic, density-dependent sexual selection | Decreased anal pad size; increased testicle size | Owen and Lahti (2020) |
| | South Asia | St. Croix | 1890 | N/A | Unknown | Biotic, density-dependent sexual selection | Decreased anal pad size; increased testicle size | Owen and Lahti (2020) |
| | | Mauritius | 1890 | N/A | Unknown | Biotic, density-dependent sexual selection | Decreased anal pad size; increased testicle size | Owen and Lahti (2020) |
| *Xenopus laevis* | sub-Saharan Africa | Europe | 1985 | 136 | Single | Spatial sorting | Increased dispersal | Courant et al. (2019) |
| **Plants** | | | | | | | | |
| *Aegilops triuncialis* | Europe, Asia | USA | 1914 | 106 | Multiple | Abiotic, nutrients | Increased seed mass and seed number, earlier flowering time | Meimberg et al. (2010) |
| *Alliaria petiolata* | Europe | USA | 1868 | 150 | Multiple | Abiotic, temperature | Strong divergence in seedling emergence in invaded range | Blossey et al. (2017) |

| Species | Native range | Introduced range | Year | Number | Multiple/Unknown | Factors | Observed change | Reference |
|---|---|---|---|---|---|---|---|---|
| *Ambrosia artemisiifolia* | North America | Europe | 1850 | 160 | Multiple | Abiotic, photoperiod | Clinal variation in flowering time | McGoey et al. (2020) |
| | | China | 1935 | 100 | Unknown | Abiotic, nutrients | Increased growth and plasticity | Sun and Roderick (2019) |
| | | Europe | 1850 | 160 | Multiple | Abiotic, temperature and precipitation | Increased growth and reproduction, later flowering | Hodgins and Rieseberg (2011) |
| | | Europe | 1850 | 160 | Multiple | Abiotic, temperature and precipitation | Geographic variation in size, reproduction and flowering time | Chun et al. (2011) |
| | | Australia | 1900 | 120 | Unknown | Abiotic, temperature and precipitation | Clinal variation in defensive phenolics | van Boheemen et al., 2019 |
| | | Europe | 1850 | 160 | Multiple | Abiotic, temperature and precipitation | Clinal variation in defensive phenolics | van Boheemen et al. (2019) |
| *Arabidopsis thaliana* | Europe, Asia | North America | 1850 | 170 | Unknown | Abiotic, temperature | Clinal variation in flowering time | Samis et al. (2012) |
| *Beta vulgaris* | Europe | Europe | N/A | N/A | Multiple | Biotic, hybridisation | Earlier flowering | Arnaud et al. (2010) |
| *Bidens pilosa* | Central and South America | China | 1857 | 648 | Unknown | Abiotic, salinity | Increased salinity tolerance | Liu et al. (2019) |
| *Brachypodium sylvaticum* | Europe, Asia, North Africa | USA | 1920 | N/A | Multiple | Abiotic, precipitation | Increased drought resistance | Marchini et al. (2018) |

*Continued*

**Table 6.1** Summary of selected studies published between 2010 and 2021 documenting rapid evolution in 62 invasive species from 87 independent invasive ranges—cont'd

| Taxon | Native range | Invaded range | Intro. date | No. gen. | Intro. history | Evolutionary driver(s) | Trait(s) | Reference |
|---|---|---|---|---|---|---|---|---|
| *Bromus inermis* | Europe | USA | 1888 | 124 | Unknown | Biotic, mutualism | Reduced mycorrhization; local adaption (based on biomass production) | Sherrard and Maherali (2012) |
| *Centaurea diffusa* | Europe, Asia | North America | 1900 | 120 | Unknown | Unknown | Increased size and later maturation | Turner et al. (2015) |
| *Centaurea solstitialis* | Europe, Asia | USA | 1850 | 170 | Multiple | Unknown | Increased seed size, growth rate and size; earlier flowering | Barker et al. (2017) |
| | | USA | 1850 | 170 | Multiple | Abiotic, geographic isolation | Reproductive isolation | Montesinos et al. (2012) |
| *Centaurea sulphurea* | Spain | USA | 1923 | 170 | Unknown | Abiotic, geographic isolation | Reproductive isolation | Montesinos et al. (2012) |
| *Cynara cardunculus* | Southern Europe, Northern Africa | USA | 1850 | 170 | Unknown | Unknown | Increased reproduction, delayed and extended flowering | Ellstrand et al. (2010) |
| *Datura stramonium* | Mexico | Spain | 1520 | 500 | Multiple | Biotic, predator/ herbivore | Reduced defence (alkaloids) | Castillo et al. (2019) |

| Species | Native range | Invasive range | | | | | | Reference |
|---|---|---|---|---|---|---|---|---|
| *Erodium cicutarium* | Europe | USA | 1700 | 320 | Unknown | Abiotic, precipitation | Earlier flowering and increased size | Latimer et al. (2019) |
| *Erodium cicutarium* | Europe | Chile | 1850 | 170 | Unknown | Abiotic, precipitation | Earlier flowering and increased size | Latimer et al. (2019) |
| *Reynoutria x bohemica* | East Asia | Europe, North America | 1850 | N/A | Unknown | Biotic, hybridisation | Increased competitiveness | Parepa et al. (2014) |
| *Impatiens glandulifera* | India, Pakistan | Europe | 1800 | 220 | Multiple | Biotic, predator/herbivore | Loss and subsequent gain of enemy resistance | Gruntman et al. (2017) |
| *Lonicera japonica* | East Asia | North America | 1900 | N/A | Multiple | Unknown | Increased size and growth rates | Kilkenny and Galloway (2013) |
| *Lupinus polyphyllus* | United States | Finland | 1890 | 130 | Multiple | unknown | Increased size, less frequent flowering | Ramula and Kalske (2020) |
| *Lythrum salicaria* | Europe, Asia | North America | 1800 | 220 | Multiple | Abiotic, likely photoperiod | Varying flowering time and fecundity | Colautti et al. (2010) |
| *Lythrum salicaria* | Europe, Asia | North America | 1800 | 220 | Multiple | Abiotic, likely photoperiod | Flowering time and size clines | Colautti and Barrett (2013) |
| *Medicago polymorpha* | Europe | North America | 1750 | 270 | Unknown | Unknown | Increased size | Getman–Pickering et al. (2018) |
| *Medicago polymorpha* | Europe | North America | 1750 | 270 | Unknown | Abiotic, photoperiod | Cline in flowering time | Helliwell et al. (2018) |
| *Microstegium vimineum* | Asia | North America | 1919 | 101 | Unknown | Abiotic, photoperiod | Clinal variation in flowering time and size | Novy et al. (2013) |
| *Microstegium vimineum* | Asia | North America | 1919 | 101 | Unknown | Unknown | Increased size | Flory et al. (2011) |

*Continued*

**Table 6.1** Summary of selected studies published between 2010 and 2021 documenting rapid evolution in 62 invasive species from 87 independent invasive ranges—cont'd

| Taxon | Native range | Invaded range | Intro. date | No. gen. | Intro. history | Evolutionary driver(s) | Trait(s) | Reference |
|---|---|---|---|---|---|---|---|---|
| *Mikania micrantha* | Central and South America | China | 1910 | N/A | Unknown | Abiotic, salinity | Increased salinity tolerance | Liu et al. (2019) |
| *Oryza sativa japonica* | Unknown | China | 1910 | N/A | Unknown | Unknown | Increased dispersal | Huang et al. (2015) |
| | | China | N/A | N/A | Unknown | Unknown | Increased dispersal | Ellstrand et al. (2010) |
| *Phragmites australis* | Europe, Canary Islands | North America | 1800 | N/A | Unknown | Unknown | Increased growth and size; higher salinity tolerance; more efficient photosynthesis | Guo et al. (2014) |
| *Plantago virginica* | North America | China | 1951 | 70 | Single | Unknown | Increased and faster germination | Xu et al. (2019) |
| | | China | 1951 | 70 | Single | Abiotic, nutrients | Increased size and seed production | Luo et al. (2019) |
| *Prosopis juliflora* | Central and South America | Kenya | 1980 | 26 | Multiple | Unknown | Higher seed mass and production, germination, stem numbers, and trait plasticity | Castillo et al. (2021) |

| Species | Native range | Introduced range | | | | | | Reference |
|---|---|---|---|---|---|---|---|---|
| *Polygonum cespitosum* | Asia | North America | 1900 | 120 | Multiple | Abiotic, likely photoperiod | Increased reproduction and higher plasticity | Sultan et al. (2013) |
| | | USA | 1900 | 120 | Unknown | Environmental heterogeneity | Increased phenotypic plasticity (biomass allocation and photosynthetic rate) | Sultan and Matesanz (2015) |
| *Raphanus raphanistrum* | Eurasia | USA | 1850 | 170 | Multiple | Environmental heterogeneity | Increased reproduction | Ridley and Ellstrand (2010) |
| *Raphanus sativus x raphanistrum* | Eurasia | USA | 1850 | 170 | Multiple | Biotic, hybridisation | Increased reproduction | Ellstrand et al. (2010) |
| *Secale cereale* | Eurasia | USA | N/A | N/A | Unknown | Unknown | Increased dispersal, delayed flowering | Ellstrand et al. (2010) |
| *Senecio pterophorus* | Eastern South Africa | Western South Africa | 1915 | 100 | Unknown | Abiotic, likely drought | Smaller size | Colomer-Ventura et al. (2015) |
| *Senecio pterophorus* | Eastern South Africa | Australia | 1915 | 100 | Unknown | Abiotic, likely drought | Smaller size | Colomer-Ventura et al. (2015) |
| *Senecio pterophorus* | Eastern South Africa | Europe | 1985 | 30 | Unknown | Abiotic, likely drought | Smaller size | Colomer-Ventura et al. (2015) |

*Continued*

**Table 6.1** Summary of selected studies published between 2010 and 2021 documenting rapid evolution in 62 invasive species from 87 independent invasive ranges—cont'd

| Taxon | Native range | Invaded range | Intro. date | No. gen. | Intro. history | Evolutionary driver(s) | Trait(s) | Reference |
|---|---|---|---|---|---|---|---|---|
| *Senecio squalidus* | Italy | United Kingdom | 1720 | 300 | Unknown | Unknown | Faster germination, increased height, higher leaf chlorophyll, increased dispersal | Brennan et al. (2012) |
| *Solidago altissima* | United States | Japan | 1900 | 120 | Unknown | Biotic, predator/herbivore | Increased competitiveness | Uesugi and Kessler (2016) |
| | | Europe | 1770 | 250 | Multiple | Biotic, predator/herbivore | Increased tolerance of herbivory | Liao et al. (2016) |
| *Sonchus oleraceus* | Europe, North Africa | Australia | N/A | N/A | Unknown | Abiotic, temperature | Increased size and germination rates | Ollivier et al. (2020) |
| | | New Zealand | N/A | N/A | Unknown | Abiotic, temperature | Increased size and germination rates | Ollivier et al. (2020) |
| *Sorghum halepense* | Asia and northern Africa | USA | 1800 | 220 | Multiple | Abiotic, temperature and precipitation | Adaptations in germination responses to climate | Fletcher et al. (2020) |
| | | USA | 1800 | 220 | Multiple | Abiotic, temperature | Variable changes in root biomass allocation | Lakoba and Barney (2020) |

| Species | Native range | Introduced range | Intro. date | No. gen. | Intro. history | Selection pressure | Trait | Reference |
|---|---|---|---|---|---|---|---|---|
| *Tamarix ramosissima x chinensis* | Africa, Asia, Southern Europe | USA | 1850 | N/A | Multiple | Biotic, hybridisation | Higher herbivore tolerance | Williams et al. (2014) |
| *Triticum aestivum* | Eurasia | China | N/A | N/A | Unknown | Unknown | More fragile rachis | Ellstrand et al. (2010) |
| *Verbascum thapsus* | Europe | USA | 1750 | 130–260 | Multiple | Abiotic, nutrients | Increased growth and defence | Kumschick et al. (2013) |
| **Microbes** | | | | | | | | |
| *Cryphonectria parasitica* | East Asia | France | 1956 | N/A | Unknown | Abiotic, temperature | Increased cold resistance | Robin et al. (2017) |

Intro. date = date of introduction; No. gen. = number of generations since introduction; Intro. history = introduction history (multiple or single independent introduction(s)).

introduced populations and, as a consequence, many invasive species have evolved clines in climate-related traits (e.g. Mushegian et al., 2021; Table 6.1). Clines allow species to time the progression of key life stages to maximise their fitness under local seasonal conditions. For example, shorter growing seasons at high latitudes frequently lead to the evolution of earlier onset of flowering in plants, and vice versa at low latitudes (Colautti et al., 2010; Colautti and Barrett, 2013). In some instances, clines have evolved in invasive populations despite low environmental variation in their native ranges and limited genetic diversity. For example, burr medic (*Medicago polymorpha*) displays strong clinal variation in flowering time in both its native range in Europe and invasive range in North America (Helliwell et al., 2018). Genomic analyses indicate that North American populations (spanning 80 degrees of latitude) originated from a limited number of native sources (spanning only 11 degrees of latitude) and that they have experienced a strong genetic bottleneck (Helliwell et al., 2018). Despite this, clines in flowering time evolved within 270 generations following the species' introduction into North America.

The immediate availability of suitable habitat, as well as overall habitat heterogeneity, may also trigger rapid evolutionary responses in invasive populations, often observed in traits related to dispersal (e.g. Araspin et al., 2020). The evolution of increased dispersal ability has been identified in numerous invasive animals and plants (e.g. Phillips et al., 2010; Lombaert et al., 2014; Huang et al., 2015; Tabassum and Leishman, 2018; Araspin et al., 2020). Increased dispersal ability is thought to benefit invasive populations by allowing individuals to escape unsuitable habitats (Berthouly-Salazar et al., 2013; Nichols et al., 2020) or, when large and contiguous suitable habitat patches are available, to maximise utilisation of available resources (Williams et al., 2016). Conversely, under heterogenous environmental conditions, decreased dispersal may evolve when recipient habitats are less favourable than the original habitat, leading to geographically structured, and often locally adapted, populations (Lindenmayer and Fischer, 2013). None of the studies included here explicitly tested for the influence of abiotic conditions on the evolution of dispersal-related traits (but see Araspin et al., 2020), but rather, mostly invoked spatial sorting (see Section 6.2.3).

## 6.2.2 Biotic factors

Following their introduction into new environments, exotic species will be free from most biotic interactions from their native range, simultaneously experiencing a whole suite of new ones (Le Roux et al., 2020). These changes may assist (e.g. release from specialist enemies) or hinder (e.g. loss of essential mutualisms) establishment success (Enders et al., 2018). Strong selection on traits involved in interactions between native and invasive species is therefore expected (e.g. Ward et al., 2012; Issaly et al., 2020) and the resulting evolutionary responses in invasive species can be as fast as, or even faster than, adaptations to abiotic conditions (Moran and Alexander, 2014). For example, invasive foxgloves (*Digitalis purpurea*) in South America rapidly evolved adaptations for hummingbird pollination (Mackin et al., 2021). This plant is pollinated by bumblebees in its native European range and, while these bees still pollinate the species in the Americas, strong selection by hummingbirds has led to the rapid evolution of long corolla tubes. This is a typical morphological feature of bird-pollinated flowers for improved pollen transfer. In the case of foxgloves, corolla tubes evolved to be up to 26% longer than in the species' native range (Mackin et al., 2021).

Interspecific competition between resident and invasive species is often thought to result in strong selection (Hodgins et al., 2018). I only identified a few instances where competition was implicated for rapid trait differentiation (Table 6.1). For example, workers of the invasive big-headed ant (*Pheidole megacephala*) show strong divergence in body size that is linked to interspecific competition with resident ants (Wills et al., 2014). In invaded regions with high native ant diversities, *P. megacephala* workers are larger than in invaded areas with depauperate ant faunas. Larger size in areas with high native ant diversities likely evolved to allow big-headed ants to defend their colonies and acquire food under conditions of high interspecific competition (Wills et al., 2014).

The success of many invasive species has been attributed to the release from specialist native enemies (Mitchell and Power, 2003; Catford et al., 2009). For example, invasive plants often suffer less herbivory than both non-invasive exotics (Carpenter and Cappuccino, 2005) and closely related native plants (Agrawal et al., 2005; Hawkes, 2007; but see Chun et al., 2010). As we discussed throughout this book, in the absence of specialist enemies, invasive plants may evolve increased competitive ability Blossey and Nötzold, 1995; also see Chapter 2, Section 2.2.4), whereby they will generally grow larger and produce more offspring but will be less defended against enemies compared with their native range counterparts (Rotter and Holeski, 2018). However, escape from natural enemies in the invasive range may only be temporary. Resident enemies are expected to accumulate on widespread invasive species (e.g. Stricker et al., 2016; Crous et al., 2017;

Rodríguez et al., 2019) and the evolutionary component of such host shifts has been repeatedly demonstrated (e.g. Strauss et al., 2006; Crous et al., 2017; Singer and Parmesan, 2018). The incidence and extent of enemy 'catch-ups' are likely to increase with residence time of the invasive species, its relatedness to native species, and as the abundance of the invader, and thus its potential discovery and exploitation as a novel and uncontested resource, increases. This is eloquently illustrated by invasive Himalayan balsam (*Impatiens glandulifera*). Gruntman et al. (2017) sampled this plant in its invasive range in Central Europe from areas that differed in invasion history, where residence times ranged between 5 and 85 years. They also sampled populations from the native range in India and Pakistan. Using common garden experiments, these authors found that native and 'older' invasive populations have higher herbivore resistance, and to produce more secondary defence compounds, than recently established invasive populations (Gruntman et al., 2017). Field surveys confirmed that old invasive populations experienced more herbivory than recently established ones. Taken together, these findings suggest that, shortly after its arrival in Central Europe, Himalayan balsam underwent EICA, but that enemy defence was regained as the species became more abundant and colonised by resident enemies over time.

Many of the case studies included here identified interspecific hybridisation as underlying rapid evolution (Table 6.1). As we discussed in Chapter 2 (Section 2.2.2), hybridisation often creates novel genotypes which may increase trait variation (Schierenbeck and Ellstrand, 2009; Hodgins et al., 2018) . Moreover, some hybrids display transgressive traits, such as faster growth, larger size (i.e. higher competition), or higher fecundity (i.e. higher reproduction), that have been repeatedly linked to invasion success (Schierenbeck and Ellstrand, 2009; te Beest et al., 2012; Mounger et al., 2021).

Interactions between individuals of the same species may also facilitate evolutionary responses, e.g. when they compete for limited resources or mating opportunities. On the other hand, Allee effects may arise when low population densities negatively impact fitness (Le Roux et al., 2020). For instance, pollen limitation may cause Allee effects in plants that are unable to self-fertilise (Rodger et al., 2013; Lachmuth et al., 2018). It is conceivable that exotic plants may evolve higher levels of self-compatibility to overcome such Allee effects, at least during the initial stages of colonisation (e.g. Rodger et al., 2013; Pannell et al., 2015). On the other hand, and similar to interspecific hybridisation, interbreeding between genetically distinct lineages of the same species following multiple introductions (i.e. genetic admixture) may boost standing genetic variation and, therefore, trait variation for selection to act on (e.g. Irimia et al., 2021; also see Section 6.4 for more details), suggesting a danger to repeated introductions of already-invasive species.

## 6.2.3 Stochastic factors

Invasive species often emerge from bottlenecked populations following founder events (Dlugosch and Parker, 2008). Under these demographic circumstances, trait differentiation between native and invasive populations, or between populations within the invaded range, may result from stochastic processes like genetic drift, rather than from local adaptation (Hodgins et al., 2018). Comparing traits between native and *multiple* and *independent* invasive populations provides one way to more confidently determine the nature of the mechanisms underlying trait differentiation. If trait differentiation is of similar extent and direction in independent invasions, then stochastic processes alone are unlikely to be responsible, as these would have an equal chance of increasing or decreasing trait values in all populations. For example, van Boheemen et al. (2019) found common ragweed, sampled across similar latitudinal gradients in its native North American and invasive ranges in Europe and Australia, to display parallel clines in various traits. Therefore trait differentiation in invasive populations is likely due to selection rather than genetic drift. Similarly, the Indian dwarf mongoose study we discussed at the beginning of this chapter compared traits between native populations and four independent invasive populations to illustrate that sexual selection, rather than genetic drift, likely underlies rapid evolution in invasive populations (Owen and Lahti, 2020). However, as we will discuss later, data from observational studies like these, even when coming from multiple invasive ranges, should be interpreted with caution (see Section 6.5).

Spatial sorting is frequently associated with trait differentiation between invasive populations (Shine et al., 2011; Hodgins et al., 2018). As we discussed in Chapter 2 (Section 2.2.3), the accumulation of dispersive phenotypes at the leading edge of expanding populations may result in strong spatial structure in dispersal-related traits, especially between core and edge populations. This is primarily because of assortative mating between more dispersive individuals at the leading edge (Shine et al., 2011). As with stochastic processes, spatial sorting does not necessarily translate into local adaption. Phillips et al. (2006) famously showed that the evolution of increased dispersal ability in invasive cane toads (*Rhinella marina*) in Australia stemmed from spatial sorting. Within 80 years of the toad's arrival in Australia, daily dispersal rates were between 3 and 10 times higher in edge populations compared with core populations (Alford et al., 2009). These differences have, among others, been attributed to the evolution of longer legs (Phillips et al., 2006), more frequent and further movements in straight paths (Alford et al., 2009), and higher endurance (Llewelyn et al., 2010), in edge toads compared with core toads.

## 6.3 Traits that frequently undergo rapid evolution during invasion

Invasive species with distributions spanning several degrees of latitude often show clinal trait variation (e.g. McGoey et al., 2020; Liu et al., 2020b; Table 6.1). For instance, clines in flowering phenology and/or plant size have been repeatedly found in populations of invasive plants such as Mouse ear cress (*Arabidopsis thaliana*; Samis et al., 2012), California poppy (*Eschscholzia californica*; Leger and Rice, 2007), purple loosestrife (*Lythrum salicaria*; Colautti and Barrett, 2013), and common ragweed (*Ambrosia artemisiifolia*; van Boheemen et al., 2019). Yet, it often remains unknown whether these clines reflect plastic responses, isolation by distance, rapid evolution, the introduced clinal variation from the species' native range, or a combination of these factors. Data from reciprocal transplant experiments (e.g. Colautti and Barrett, 2013) and common garden studies comparing traits between different and independently invaded regions (e.g. Bhattarai et al., 2017; van Boheemen et al., 2019) suggest that they may indeed often be adaptive. As we will discuss in Section 6.5, trait clines also present unique challenges to studies aimed at understanding the roles of adaptive vs stochastic processes in driving rapid evolution in invasive species.

Irrespective of whether traits show clinal variation or not, invasive plants do often grow larger and have higher reproductive output than their native counterparts (Colautti and Lau, 2015). In their seminal paper, Blossey and Nötzold (1995) were among the first to suggest that rapid evolution (via EICA) might underlie these observations. However, traits linked to EICA may also evolve in non-competitive environments, such as highly disturbed habitats (Blumenthal and Hufbauer, 2007). While these habitats are generally invaded by species already at the 'fast end' of the plant economics spectrum (Dawson et al., 2012; Montesinos, 2021), a lack of competition may further select for higher resource capture abilities, faster growth rates, and higher reproduction, to allow invasive species to fully capitalise on (largely uncontested) resources (Dlugosch et al., 2015a).

Animals also often display latitudinal clinal traits. For example, critical photoperiod in insects (the day length required to induce and maintain diapause) usually correlates strongly with latitude (Tyukmaeva et al., 2020), with insects at higher latitudes generally having higher critical photoperiods than those at lower latitudes, an adaptation to the earlier arrival of winter at high latitudes. Urbanski et al. (2012) compared latitudinal variation in photoperiod responses between native populations of the Asian tiger mosquito (*Aedes albopictus*) from Japan and invasive populations in the United States. These authors found that, within just 20 years, invasive populations evolved clines resembling those in Japanese populations. Invasive ectotherms with short generation times, such as mosquitoes, often rapidly evolve clines in climate-related traits.

## 6.4 Introduction history

Interactions between the traits of exotic species, conditions in the new environment, and introduction history determine establishment success and the extent of invasion (Pyšek et al., 2020). Introduction history not only describes the number and genetic composition of sources that founded invasive populations, but also the number of generations that have lapsed since introduction (Estoup and Guillemaud, 2010). Introduction history, via its impacts on genetic and quantitative trait variation, will therefore significantly influence the evolutionary capacity of exotic species.

A large body of population genetics research provides us with an understanding of how introduction histories impact standing genetic diversity in invasive populations (see Chapter 8; e.g. Dlugosch and Parker, 2008; Vicente et al., 2021). For instance, invasions resulting from single introduction events are often characterised by consecutive founder events during range expansion which may result in a serial dilution of genetic diversity (e.g. Michaelides et al., 2018). It is generally assumed that reduced genetic diversity, even when measured at neutral loci, translates into lower adaptive capacity (Reed and Frankham, 2003; Santos et al., 2012; Vilas et al., 2015). Genetic admixture following multiple introductions will therefore benefit invasive populations by increasing genetic diversity (Dlugosch and Parker, 2008; Dlugosch et al., 2015b; Vicente et al., 2021). For the studies included here with known introduction histories, multiple introductions

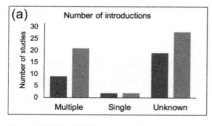
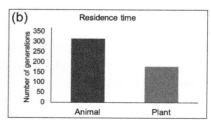
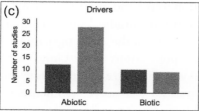
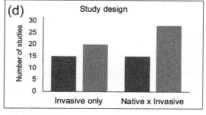

**Fig. 6.1** Summary of 86 selected case studies documenting rapid evolution in invasive populations of 56 plant species (indicated by *red bars* (grey in the print version)) and 30 animal species (indicated by *blue bars* (dark grey in the print version)): (a) Differences in introduction histories are indicated as the number of independent introductions (i.e. multiple, single or unknown); (b) the average number of generations that have lapsed since introduction of species; (c) the frequency by which abiotic and biotic factors were reported as being responsible for rapid evolution; and (d) the type of comparison used to assess rapid evolution, i.e. whether studies compared traits between populations within the invasive range only *or* between populations from the native and invasive range. (For interpretation of the references to colour in this figure legend, the reader is referred to the web version of this article.)

seem to be the rule rather than the exception (Table 6.1; Fig. 6.1). While not explicitly tested in most of these studies, it is reasonable to assume that multiple introductions often involve different native sources (Dlugosch and Parker, 2008; Vicente et al., 2021) and, therefore, that invasive populations consisting of genetic admixtures will harbour substantial genetic diversity. In general, this notion is supported by the number of genetic studies that identify multiple introductions also detecting admixture (e.g. Dlugosch et al., 2015b; Puckett et al., 2016; Hirsch et al., 2021; Vicente et al., 2021; also see Chapter 8, Section 8.2.2). Admixture has important consequences for rapid evolution. First, it will fundamentally reshuffle genetic diversity, from variation among native populations to variation within invasive populations (Colautti and Lau, 2015; Sherpa and Després, 2021). This, in turn, may increase variation in ecologically important traits, and thus the adaptive capacities of populations (Hodgins et al., 2018). Second, compared to parental genotypes, admixed individuals may produce novel or more extreme phenotypes via trait transgression and heterosis (e.g. Lavergne and Molofsky, 2007; Irimia et al., 2021). Third, admixture will lead to an immediate increase in heterozygosity, which may increase short-term fitness via overdominance (i.e. heterozygote advantage) or the masking of deleterious recessive alleles (Hodgins et al., 2018; Irimia et al., 2021; also see Chapter 2, Section 2.2.2). While these benefits of high genetic diversity have not been tested explicitly in any of the studies included here, the data on introduction histories suggest that these advantages are available to invasive species in many instances.

Rapid evolution has also been documented in invasive species that have experienced genetic bottlenecks, in what some have termed a 'genetic paradox' (Allendorf and Lundquist, 2003). Invasive species can only be considered truly paradoxical in this sense if their bottlenecked populations have not succumbed to the negative effects of low genetic variation *and* if they have undergone local adaptation to conditions in the new range (Estoup et al., 2016; Schrieber and Lachmuth, 2017). The burr medic example we discussed earlier represents such a paradox. This species rapidly evolved clinal variation in phenology in its invasive range in North America despite having gone through a population genetic bottleneck (Helliwell et al., 2018). It is important to keep in mind that fitness-related traits are usually quantitative, and therefore their variation may not be as dramatically impacted by demographic processes, such as founder events, as variation at neutral loci that are typically used to test for genetic bottlenecks (Lande, 1988).

It is clear that knowledge of the number of introductions, their putative source(s), and the post-introduction dynamics of populations provide valuable information about the amount of genetic variation present in invasive populations (Dlugosch et al., 2015b; Dlugosch and Parker, 2008; Estoup et al., 2016). This information is also important if we want to distinguish between the roles of preadaptation (also see Chapter 3), selection, or neutral processes in causing trait differentiation. Modern state-of-the-art population genomic approaches, in conjunction with broad geographic sampling in both ranges, now put accurate inferences of introduction histories and post-introduction population dynamics firmly within our reach (Chown et al., 2015; Sherpa and Després, 2021; also see Chapter 8).

## 6.5 Experimental design

Studies searching for evidence of rapid evolution in invasive species often use the analogy of 'ancestor-descendant' comparisons, by studying the genetic basis of phenotypic variation within and among populations from the native and invasive ranges of species. These studies usually involve trait measurements obtained from individual common gardens or reciprocal transplant experiments. However, interpreting trait differences can be challenging at best, and misleading at worst, especially when the introduction histories of species are unknown. This is because studies on trait evolution rarely include the most likely ancestral sources of invasive populations, either because the original founders are extinct or cannot be located, the native range sources cannot be located, or because invasive populations comprise genetic admixtures or are severely bottlenecked (Colautti and Lau, 2015; Hodgins et al., 2018). The ethical issues, logistical complications, and costs involved in conducting experiments between ranges also make reciprocal transplant studies challenging, even when other constraints can be overcome.

Demonstrating that rapid evolution is adaptive requires evidence that 'local' genotypes outperform foreign ones, whether between ranges or between populations within the invasive range. Despite their long history to test for local adaptation, reciprocal transplant experiments between the native and invasive ranges of species have only been occasionally attempted. This is, in part, because these studies require intensive sampling and multiple transplant sites in both ranges in order to account for local adaptation *within* each range, especially when the sources of invasive populations are unknown (Colautti and Lau, 2015). Various meta-analyses illustrate why this matters (e.g. Colautti et al., 2009; Felker-Quinn et al., 2013; Rotter and Holeski, 2018). For example, Colautti et al. (2009) reanalysed data from common garden experiments that compared trait variation between native and invasive populations of 28 plant species. When ignoring latitude, these authors found evidence for larger and more robust plants in invasive ranges compared with native ranges, which could be interpreted as rapid post-introduction evolution. However, when including latitude as covariate in their models, trait differences between ranges not only changed in magnitude but, in some instances, also direction! Thus, in many cases, latitude, and not rapid evolution, was the cause of trait differentiation. This suggests that meaningful estimates of local adaptation based on reciprocal transplant experiments either require many transplant sites across a species' range or transplant sites that fall within the known native range source region(s) of invasive populations. As we will discuss in Chapter 8, genomic analyses provide ways to accurately delimit introduction histories and the putative native sources of invasive populations with remarkable levels of accuracy (Chown et al., 2015; Sherpa and Després, 2021), making transplant experiments between 'ancestral' and invasive populations more achievable.

Many studies use individual common garden experiments to assess the genetic basis of quantitative trait variation in native and invasive populations. Some have compared population differentiation based on quantitative traits to neutral differentiation as a yardstick for (mostly past) selection (hereafter $Q_{ST}$-$F_{ST}$ comparisons; McKay and Latta, 2002; Marchini et al., 2018; Sun and Roderick, 2019). Briefly, quantitative trait variation measured under common garden conditions can be partitioned to estimate population differentiation (i.e. $Q_{ST}$, McKay and Latta, 2002). Such differentiation reflects both neutral processes and past selection. Similarly, standard population genetic analyses of variation at neutral genetic loci, such as microsatellites, can be used to estimate differentiation between populations, i.e. neutral differentiation due to stochastic processes alone ($F_{ST}$). By comparing $Q_{ST}$ and $F_{ST}$ values, one can infer whether stochastic processes alone (i.e. $Q_{ST}=F_{ST}$), disruptive selection (i.e. $Q_{ST}>F_{ST}$), or stabilising selection (i.e. $Q_{ST}<F_{ST}$) underlie trait differentiation (McKay and Latta, 2002). Some researchers have also studied differentiation between populations in the invasive range using $Q_{ST}$-$F_{ST}$ comparisons (Colautti and Lau, 2015). However, these studies are not without limitations. For example, population-level differentiation in the invasive range may stem from independent introductions into different areas (e.g. Keller et al., 2009). Yet, many invasive species spread rapidly from one or a few initial introduction sites, and offer some of the most compelling examples of rapid evolution during invasion as inferred from $Q_{ST}$-$F_{ST}$ comparisons (e.g. Marchini et al., 2018) or reciprocal transplant experiments in the invasive range (e.g. Colautti and Barrett, 2013; Castillo et al., 2021).

## 6.6 Conclusions

Rapid evolution in invasive species is often inferred from observational data and, as a consequence, evidence for local adaptation remains somewhat limited. Studies based on observational data cannot unambiguously determine whether trait differentiation is due to natural selection, even when data from multiple independent invasive ranges are included. Similarly, measuring functional traits of native and invasive populations in individual common garden experiments can only tell us about the genetic basis of differentiation, and not the processes that caused it or whether populations are locally adapted to conditions in their respective ranges. While trait differentiation identified in these experiments may be adaptive, it may also stem from stochastic processes such as founder events and genetic drift.

Reciprocal transplant experiments remain our best approach to test for local adaptation. However, implementing these correctly presents many logistical, ethical, and bureaucratic challenges. These challenges are exacerbated when invasive species have large ranges, unknown introduction histories, and geographically structured populations. There is an urgent need for future studies to incorporate such geographic information when searching for evidence of rapid adaptive evolution during invasion. This can be achieved through population genomic tests of demographic processes (e.g. the occurrence of admixture or genetic bottlenecks) or other aspects of introduction histories (e.g. number and identity of source populations) (Sherpa and Després, 2021). Short-lived annual plants arguably provide the most promising candidates to study local adaptation using reciprocal transplant experiments, as their traits are easy to measure across multiple generations in the field. For some invasive species direct comparisons between founders and their invasive offspring might also be possible (e.g. forestry trial plantations and surrounding invasive populations; e.g. Zenni et al., 2014; Castillo et al., 2021). While opportunities for such founder–invader comparisons are probably rare, they provide unique circumstances to detect rapid evolution, even in species with long generation times (e.g. Castillo et al., 2021).

I only identified one study that investigated rapid evolution in an invasive microbe (Robin et al., 2017). Given their extremely short generation times and sensitivity and responsiveness to (a)biotic conditions, best exemplified by resistance evolution in agricultural and medical environments (Jørgensen et al., 2018), microbes might be particularly good candidates to study rapid evolution during invasion. While there is an ongoing debate as to whether microbes exhibit strong biogeographic structure (Martiny et al., 2006), recent work suggests many do (e.g. Delgado-Baquerizo et al., 2018; Egidi et al., 2019), including microbes known to have been co-introduced into new environments with their invasive hosts (e.g. Ndlovu et al., 2013; Warrington et al., 2019). Under these circumstances, comparing growth rates between invasive microbes and their ancestral strains in a given environment, e.g. on a new host plant, provides exciting and novel opportunities to study rapid evolution, even under controlled laboratory conditions.

In conclusion, evidence of adaptive evolution in invasive species remains modest, not because it rarely occurs, but because of the challenges associated with distinguishing between adaptive and stochastic processes as drivers of evolutionary change. We clearly need more studies to understand how often rapid evolution in invasive species is adaptive.

## References

Agrawal, A.A., Kotanen, P.M., Mitchell, C.E., Power, A.G., Godsoe, W., Klironomos, J., 2005. Enemy release? An experiment with congeneric plant pairs and diverse above-and belowground enemies. Ecology 86, 2979–2989.

Alford, R.A., Brown, G.P., Schwarzkopf, L., Phillips, B.L., Shine, R., 2009. Comparisons through time and space suggest rapid evolution of dispersal behaviour in an invasive species. Wildl. Res. 36, 23–28.

Allendorf, F.W., Lundquist, L.L., 2003. Introduction: population biology, evolution, and control of invasive species. Conserv. Biol. 17, 24–30.

Araspin, L., Serra Martinez, A., Wagener, C., Courant, J., Louppe, V., Padilla, P., Measey, J., Herrel, A., 2020. Rapid shifts in the temperature dependence of locomotor performance in an invasive frog, *Xenopus laevis*, implications for conservation. Integr. Comp. Biol. 60, 456–466.

Arnaud, J.F., Fénart, S., Cordellier, M., Cuguen, J., 2010. Populations of weedy crop–wild hybrid beets show contrasting variation in mating system and population genetic structure. Evol. Appl. 3, 305–318.

Bargielowski, I., Lounibos, L.P., 2014. Rapid evolution of reduced receptivity to interspecific mating in the dengue vector *Aedes aegypti* in response to satyrization by invasive *Aedes albopictus*. Evol. Ecol. 28, 193–203.

Barker, B.S., Andonian, K., Swope, S.M., Luster, D.G., Dlugosch, K.M., 2017. Population genomic analyses reveal a history of range expansion and trait evolution across the native and invaded range of yellow starthistle (*Centaurea solstitialis*). Mol. Ecol. 4, 1131–1147.

Barrett, S.C., Colautti, R.I., Eckert, C.G., 2008. Plant reproductive systems and evolution during biological invasion. Mol. Ecol. 17, 373–383.

Baxter-Gilbert, J., Riley, J.L., Wagener, C., Mohanty, N.P., Measey, J., 2020. Shrinking before our isles: the rapid expression of insular dwarfism in two invasive populations of guttural toad (*Sclerophrys gutturalis*). Biol. Lett. 16, 20200651.

Berthouly-Salazar, C., Hui, C., Blackburn, T.M., Gaboriaud, C., van Rensburg, B.J., van Vuuren, B.J., Le Roux, J.J., 2013. Long-distance dispersal maximizes evolutionary potential during rapid geographic range expansion. Mol. Ecol. 22, 5793–5804.

Bhattarai, G.P., Meyerson, L.A., Anderson, J., Cummings, D., Allen, W.J., Cronin, J.T., 2017. Biogeography of a plant invasion: genetic variation and plasticity in latitudinal clines for traits related to herbivory. Ecol. Monogr. 87, 57–75.

Blossey, B., Nötzold, R., 1995. Evolution of increased competitive ability in plants: a hypothesis. J. Ecol. 83, 887–889.

Blossey, B., Nuzzo, V., Dávalos, A., 2017. Climate and rapid local adaptation as drivers of germination and seed bank dynamics of *Alliaria petiolata* (garlic mustard) in North America. J. Ecol. 105, 1485–1495.

Blumenthal, D.M., Hufbauer, R.A., 2007. Increased plant size in exotic populations: a common-garden test with 14 invasive species. Ecology 88, 2758–2765.

Brennan, A.C., Barker, D., Hiscock, S.J., Abbott, R.J., 2012. Molecular genetic and quantitative trait divergence associated with recent homoploid hybrid speciation: a study of *Senecio squalidus* (Asteraceae). Heredity 108, 87–95.

Card, D.C., Perry, B.W., Adams, R.H., Schield, D.R., Young, A.S., Andrew, A.L., Jezkova, T., Pasquesi, G., Hales, N.R., Walsh, M.R., Rochford, M.R., Mazzotti, F.J., Hart, K.M., Hunter, M.E., Castoe, T.A., 2018. Novel ecological and climatic conditions drive rapid adaptation in invasive Florida Burmese pythons. Mol. Ecol. 27, 4744–4757.

Carpenter, D., Cappuccino, N., 2005. Herbivory, time since introduction and the invasiveness of exotic plants. J. Ecol. 93, 315–321.

Castillo, G., Calahorra-Oliart, A., Núñez-Farfán, J., Valverde, P.L., Arroyo, J., Cruz, L.L., Tapia-López, R., 2019. Selection on tropane alkaloids in native and non–native populations of *Datura stramonium*. Ecol. Evol. 9, 10176–10184.

Castillo, M.L., Schaffner, U., van Wilgen, B.W., Le Roux, J.J., 2021. The contribution of phenotypic traits, their plasticity, and rapid evolution to invasion success: insights from an extraordinary natural experiment. Ecography 44, 1035–1050.

Catford, J.A., Jansson, R., Nilsson, C., 2009. Reducing redundancy in invasion ecology by integrating hypotheses into a single theoretical framework. Divers. Distrib. 15, 22–40.

Chen, Y., Shenkar, N., Ni, P., Lin, Y., Li, S., Zhan, A., 2018. Rapid microevolution during recent range expansion to harsh environments. BMC Evol. Biol. 18, 187.

Chown, S.L., Hodgins, K.A., Griffin, P.C., Oakeshott, J.G., Byrne, M., Hoffmann, A.A., 2015. Biological invasions, climate change and genomics. Evol. Appl. 8, 23–46.

Chun, Y.J., Van Kleunen, M., Dawson, W., 2010. The role of enemy release, tolerance and resistance in plant invasions: linking damage to performance. Ecol. Lett. 13, 937–946.

Chun, Y.J., Le Corre, V., Bretagnolle, F., 2011. Adaptive divergence for a fitness-related trait among invasive *Ambrosia artemisiifolia* populations in France. Mol. Ecol. 20, 1378–1388.

Colautti, R.I., Barrett, S.C.H., 2013. Rapid adaptation to climate facilitates range expansion of an invasive plant. Science 342, 364–366.

Colautti, R.I., Lau, J.A., 2015. Contemporary evolution during invasion: evidence for differentiation, natural selection, and local adaptation. Mol. Ecol. 24, 1999–2017.

Colautti, R.I., Maron, J.L., Barrett, S.C.H., 2009. Common garden comparisons of native and introduced plant populations: latitudinal clines can obscure evolutionary inferences. Evol. Appl. 2, 187–199.

Colautti, R.I., Eckert, C.G., Barrett, S.C., 2010. Evolutionary constraints on adaptive evolution during range expansion in an invasive plant. Proc. R. Soc. B Biol. Sci. 277, 1799–1806.

Colomer-Ventura, F., Martínez-Vilalta, J., Zuccarini, P., Escolà, A., Armengot, L., Castells, E., 2015. Contemporary evolution of an invasive plant is associated with climate but not with herbivory. Funct. Ecol. 29, 1475–1485.

Courant, J., Secondi, J., Guillemet, L., Volette, E., Herrel, A., 2019. Rapid changes in dispersal on a small spatial scale at the range edge of an expanding population. Evol. Ecol. 33, 599–612.

Crous, C.J., Burgess, T.I., Le Roux, J.J., Richardson, D.M., Slippers, B., Wingfield, M.J., 2017. Ecological disequilibrium drives insect pest and pathogen accumulation in non-native trees. AoB Plants 9, plw081.

Darwin, C.R., 1859. The Origin of Species. John Murray, London.

Dawson, W., Rohr, R.P., van Kleunen, M., Fischer, M., 2012. Alien plant species with a wider global distribution are better able to capitalize on increased resource availability. New Phytol. 194, 859–867.

Delgado-Baquerizo, M., Oliverio, A.M., Brewer, T.E., Benavent-González, A., Eldridge, D.J., Bardgett, R.D., Maestre, F.T., Singh, B.K., Fierer, N., 2018. A global atlas of the dominant bacteria found in soil. Science 359, 320–325.

Dlugosch, K.M., Parker, I.M., 2008. Founding events in species invasions: genetic variation, adaptive evolution, and the role of multiple introductions. Mol. Ecol. 17, 431–449.

Dlugosch, K.M., Cang, F.A., Barker, B.S., Andonian, K., Swope, S.M., Rieseberg, L.H., 2015a. Evolution of invasiveness through increased resource use in a vacant niche. Nat. Plants 1, 15066.

Dlugosch, K.M., Anderson, S.R., Braasch, J., Cang, F.A., Gillette, H.D., 2015b. The devil is in the details: genetic variation in introduced populations and its contributions to invasion. Mol. Ecol. 24, 2095–2111.

Egidi, E., Delgado-Baquerizo, M., Plett, J.M., Wang, J., Eldridge, D.J., Bardgett, R.D., Maestre, F.T., Singh, B.K., 2019. A few Ascomycota taxa dominate soil fungal communities worldwide. Nat. Commun. 10, 2369.

Ellstrand, N.C., Heredia, S.M., Leak-Garcia, J.A., Heraty, J.M., Burger, J.C., Yao, L., Nohzadeh-Malakshah, S., Ridley, C.E., 2010. Crops gone wild: evolution of weeds and invasives from domesticated ancestors. Evol. Appl. 3, 494–504.

Enders, M., Hütt, M.-.T., Jeschke, J.M., 2018. Drawing a map of invasion biology based on a network of hypotheses. Ecosphere 9, e02146.

Estoup, A., Guillemaud, T., 2010. Reconstructing routes of invasion using genetic data: why, how and so what? Mol. Ecol. 19, 4113–4130.

Estoup, A., Ravigné, V., Hufbauer, R., Vitalis, R., Gautier, M., Facon, B., 2016. Is there a genetic paradox of biological invasion? Annu. Rev. Ecol. Evol. Syst. 47, 51–72.

Felker-Quinn, E., Schweitzer, J.A., Bailey, J.K., 2013. Meta-analysis reveals evolution in invasive plant species but little support for evolution of increased competitive ability (EICA). Ecol. Evol. 3, 739–751.

Firmat, C., Schliewen, U.K., Losseau, M., Alibert, P., 2012. Body shape differentiation at global and local geographic scales in the invasive cichlid Oreochromis mossambicus. Biol. J. Linn. Soc. 105, 369–381.

Fletcher, R.A., Varnon, K.M., Barney, J.N., 2020. Climate drives differences in the germination niche of a globally distributed invasive grass. J. Plant Ecol. 13, 195–203.

Flory, S.L., Long, F., Clay, K., 2011. Invasive Microstegium populations consistently outperform native range populations across diverse environments. Ecology 92, 2248–2257.

Getman-Pickering, Z.L., terHorst, C.P., Magnoli, S.M., Lau, J.A., 2018. Evolution of increased Medicaco polymorpha size during invasion does not result in increased competitive ability. Oecologia 188, 203–212.

Gruntman, M., Segev, U., Glauser, G., Tielbörger, K., 2017. Evolution of plant defences along an invasion chronosequence: defence is lost due to enemy release—but not forever. J. Ecol. 105, 255–264.

Guo, W.Y., Lambertini, C., Nguyen, L.X., Li, X.Z., Brix, H., 2014. Preadaptation and post–introduction evolution facilitate the invasion of Phragmites australis in North America. Ecol. Evol. 4, 4567–4577.

Hawkes, C.V., 2007. Are invaders moving targets? The generality and persistence of advantages in size, reproduction, and enemy release in invasive plant species with time since introduction. Am. Nat. 170, 832–843.

Hays, W.S.T., Conant, S., 2007. Biology and impacts of Pacific Island invasive species. 1. A worldwide review of effects of the small Indian mongoose, Herpestes javanicus (Carnivora: Herpestidae). Pac. Sci. 61, 3–16.

Helliwell, E.E., Faber-Hammond, J., Lopez, Z.C., Garoutte, A., von Wettberg, E., Friesen, M.L., Porter, S.S., 2018. Rapid establishment of a flowering cline in Medicago polymorpha after invasion of North America. Mol. Ecol. 27, 4758–4774.

Hirsch, H., Richardson, D.M., Pauchard, A., Le Roux, J.J., 2021. Genetic analyses reveal complex introduction histories for the invasive tree Acacia dealbata link around the world. Divers. Distrib. 27, 360–376.

Hodgins, K.A., Rieseberg, L., 2011. Genetic differentiation in life–history traits of introduced and native common ragweed (Ambrosia artemisiifolia) populations. J. Evol. Biol. 24, 2731–2749.

Hodgins, K.A., Bock, D.G., Rieseberg, L.H., 2018. Trait evolution in invasive species. Annu. Plant Rev. Online 1, 1–37.

Huang, F., Peng, S., Chen, B., Liao, H., Huang, Q., Lin, Z., Liu, G., 2015. Rapid evolution of dispersal-related traits during range expansion of an invasive vine *Mikania micrantha*. Oikos 124, 1023–1030.

Hudson, C.M., Vidal-García, M., Murray, T.G., Shine, R., 2020. The accelerating anuran: evolution of locomotor performance in cane toads (*Rhinella marina*, Bufonidae) at an invasion front. Proc. R. Soc. B Biol. Sci. 287, 20201964.

Irimia, R.E., Hierro, J.L., Branco, S., Sotes, G., Cavieres, L.A., Eren, O., Lortie, C.J., French, K., Callaway, R.M., Montesinos, D., 2021. Experimental admixture among geographically disjunct populations of an invasive plant yields a global mosaic of reproductive incompatibility and heterosis. J. Ecol. 00, 1–11.

Issaly, E.A., Sérsic, A.N., Pauw, A., Cocucci, A.A., Traveset, A., Benítez-Vieyra, S.M., Paiaro, V., 2020. Reproductive ecology of the bird-pollinated *Nicotiana glauca* across native and introduced ranges with contrasting pollination environments. Biol. Invasions 22, 485–498.

Jørgensen, P.S., Aktipis, A., Brown, Z., Carrière, Y., Downes, S., Dunn, R.R., Epstein, G., Frisvold, G.B., Hawthorne, D., Gröhn, Y., Tikaramsa Gujar, G., Jasovský, D., Klein, E.-Y., Klein, F., Lhermie, G., Mota-Sanchez, D., Omoto, C., Schlüter, M., Scott, H.M., Wernli, D., Carroll, S.P., 2018. Antibiotic and pesticide susceptibility and the Anthropocene operating space. Nat. Sustain. 1, 632–641.

Keller, S.R., Sowell, D.R., Neiman, M., Wolfe, L.M., Taylor, D.R., 2009. Adaptation and colonization history affect the evolution of clines in two introduced species. New Phytol. 183, 678–690.

Kilkenny, F.F., Galloway, L.F., 2013. Adaptive divergence at the margin of an invaded range. Evolution 67, 722–731.

Kolbe, J.J., Vanmiddlesworth, P.S., Losin, N., Dappen, N., Losos, J.B., 2012. Climatic niche shift predicts thermal trait response in one but not both introductions of the Puerto Rican lizard *Anolis cristatellus* to Miami, Florida, USA. Ecol. Evol. 2, 1503–1516.

Kosmala, G.K., Brown, G.P., Christian, K.A., Hudson, C.M., Shine, R., 2018. The thermal dependency of locomotor performance evolves rapidly within an invasive species. Ecol. Evol. 8, 4403–4408.

Kumschick, S., Hufbauer, R.A., Alba, C., Blumenthal, D.M., 2013. Evolution of fast-growing and more resistant phenotypes in introduced common mullein (*Verbascum thapsus*). J. Ecol. 101, 378–387.

Lachmuth, S., Henrichmann, C., Horn, J., Pagel, J., Schurr, F.M., 2018. Neighbourhood effects on plant reproduction: an experimental–analytical framework and its application to the invasive *Senecio inaequidens*. J. Ecol. 106, 761–773.

Lakoba, V.T., Barney, J.N., 2020. Home climate and habitat drive ecotypic stress response differences in an invasive grass. AoB Plants 12, plaa062.

Lande, R., 1988. Genetics and demography in biological conservation. Science 241, 1455–1460.

Latimer, A.M., Jacobs, B.S., Gianoli, E., Heger, T., Salgado-Luarte, C., 2019. Parallel functional differentiation of an invasive annual plant on two continents. AoB Plants 11, plz010.

Lavergne, S., Molofsky, J., 2007. Increased genetic variation and evolutionary potential drive the success of an invasive grass. Proc. Natl. Acad. Sci. U. S. A. 104, 3883–3888.

Le Gros, A., Clergeau, P., Zuccon, D., Cornette, R., Mathys, B., Samadi, S., 2016. Invasion history and demographic processes associated with rapid morphological changes in the red-whiskered bulbul established on tropical islands. Mol. Ecol. 25, 5359–5376.

Le Roux, J.J., Clusella-Trullas, S., Mokotjomela, T.M., Mairal, M., Richardson, D.M., Skein, L., Wilson, J.R., Weyl, O.L.F., Geerts, S., 2020. Biotic interactions as mediators of biological invasions: insights from South Africa. In: van Wilgen, B.W., Measey, J., Richardson, D.M., Wilson, J.R., Zengeya, T.A. (Eds.), Biological Invasions in South Africa. Springer, Berlin, pp. 387–427.

Lee, C.E., Kiergaard, M., Gelembiuk, G.W., Eads, B.D., Posavi, M., 2011. Pumping ions: rapid parallel evolution of ionic regulation following habitat invasions. Evolution 65, 2229–2244.

Leger, E.A., Rice, K.J., 2007. Assessing the speed and predictability of local adaptation in invasive California poppies (*Eschscholzia californica*). J. Evol. Biol. 20, 1090–1103.

Liao, H., Gurgel, P., Pal, R.W., Hooper, D., Callaway, R.M., 2016. *Solidago gigantea* plants from nonnative ranges compensate more in response to damage than plants from the native range. Ecology 97, 2355–2363.

Lindenmayer, D.B., Fischer, J., 2013. Habitat Fragmentation and Landscape Change: An Ecological and Conservation Synthesis. Island Press, Washington, DC.

Liu, M., Liao, H., Peng, S., 2019. Salt–tolerant native plants have greater responses to other environments when compared to salt–tolerant invasive plants. Ecol. Evol. 9, 7808–7818.

Liu, C., Wolter, C., Xian, W., Jeschke, J.M., 2020a. Most invasive species largely conserve their climatic niche. Proc. Natl. Acad. Sci. U. S. A. 117, 23643–23651.

Liu, W., Zhang, Y., Chen, X., Maung-Douglass, K., Strong, D.R., Pennings, S.C., 2020b. Contrasting plant adaptation strategies to latitude in the native and invasive range of Spartina alterniflora. New Phytol. 226, 623–634.

Llewelyn, J., Phillips, B.L., Alford, R.A., Schwarzkopf, L., Shine, R., 2010. Locomotor performance in an invasive species: cane toads from the invasion front have greater endurance, but not speed, compared to conspecifics from a long-colonised area. Oecologia 162, 343–348.

Lombaert, E., Estoup, A., Facon, B., Joubard, B., Grégoire, J.-.C., Jannin, A., Blin, A., Guillemaud, T., 2014. Rapid increase in dispersal during range expansion in the invasive ladybird *Harmonia axyridis*. J. Evol. Biol. 27, 508–517.

Louppe, V., Lalis, A., Abdelkrim, J., Baron, J., Bed'Hom, B., Becker, A.A.M.J., Catzeflis, F., Lorvelec, O., Zieger, U., Veron, G., 2021. Dispersal history of a globally introduced carnivore, the small Indian mongoose *Urva auropunctata*, with an emphasis on the Caribbean region. Biol. Invasions 23, 2573–2590.

Lowe, S., Browne, M., Boudjelas, S., De Poorter, M., 2004. 100 of the World's worst invasive alien species. A selection from the global invasive species database. In: The Invasive Species Specialist Group (ISSG) a specialist group of the Species Survival Commission (SSC) of the World Conservation Union (IUCN).

Luo, X., Xu, X., Zheng, Y., Guo, H., Hu, S., 2019. The role of phenotypic plasticity and rapid adaptation in determining invasion success of *Plantago virginica*. Biol. Invasions 21, 2679–2692.

Mackin, C.R., Peña, J.F., Blanco, M.A., Balfour, N.J., Castellanos, M.C., 2021. Rapid evolution of a floral trait following acquisition of novel pollinators. J. Ecol. 00, 1–13.

Marchini, G.L., Arredondo, T.M., Cruzan, M.B., 2018. Selective differentiation during the colonization and establishment of a newly invasive species. J. Evol. Biol. 31, 1689–1703.

Martiny, J., Bohannan, B., Brown, J., Colwell, R.K., Fuhrman, J.A., Green, J.L., Horner-Devine, M.N., Kane, M., Adams Krumins, J., Kuske, C.R., Morin, P.J., Naeem, S., Øvreås, L., Reysenbach, A.-L., Smith, V.H., Staley, J.T., 2006. Microbial biogeography: putting microorganisms on the map. Nat. Rev. Microbiol. 4, 102–112.

McGoey, B.V., Hodgins, K.A., Stinchcombe, J.R., 2020. Parallel flowering time clines in native and introduced ragweed populations are likely due to adaptation. Ecol. Evol. 10, 4595–4608.

McKay, J.K., Latta, R.G., 2002. Adaptive population divergence: markers, QTL and traits. Trends Ecol. Evol. 17, 285–291.

Medley, K.A., Westby, K.M., Jenkins, D.G., 2019. Rapid local adaptation to northern winters in the invasive Asian tiger mosquito *Aedes albopictus*: a moving target. J. Appl. Ecol. 56, 2518–2527.

Meimberg, H., Milan, N.F., Karatassiou, M., Espeland, E.K., McKay, J.K., Rice, K.J., 2010. Patterns of introduction and adaptation during the invasion of *Aegilops triuncialis* (Poaceae) into Californian serpentine soils. Mol. Ecol. 19, 5308–5319.

Michaelides, S.N., Goodman, R.M., Crombie, R.I., Kolbe, J.J., 2018. Independent introductions and sequential founder events shape genetic differentiation and diversity of the invasive green anole (*Anolis carolinensis*) on Pacific Islands. Divers. Distrib. 24, 666–679.

Miehls, A.L., Peacor, S.D., McAdam, A.G., 2014. Gape-limited predators as agents of selection on the defensive morphology of an invasive invertebrate. Evolution 68, 2633–2643.

Mitchell, C.E., Power, A.G., 2003. Release of invasive plants from fungal and viral pathogens. Nature 421, 625–627.

Mittan, C.S., Zamudio, K.R., 2019. Rapid adaptation to cold in the invasive cane toad *Rhinella marina*. Conserv. Physiol. 7, coy075.

Montesinos, D., 2021. Fast invasives fastly become faster: invasive plants align largely with the fast side of the plant economics spectrum. J. Ecol. 00, 1–5. https://doi.org/10.1111/1365-2745.13616.

Montesinos, D., Santiago, G., Callaway, R.M., 2012. Neo–allopatry and rapid reproductive isolation. Am. Nat. 180, 529–533.

Moran, E.V., Alexander, J.M., 2014. Evolutionary responses to global change: lessons from invasive species. Ecol. Lett. 17, 637–649.

Mounger, J., Ainouche, M.L., Bossdorf, O., Cavé-Radet, A., Li, B., Parepa, M., Salmon, A., Yang, J., Richards, C.L., 2021. Epigenetics and the success of invasive plants. Philos. Trans. R. Soc. B Biol. Sci. 376, 20200117.

Mushegian, A.A., Neupane, N., Batz, Z., Mogi, M., Tuno, N., Toma, T., Miyagi, I., Ries, L., Armbruster, P.A., 2021. Ecological mechanism of climate-mediated selection in a rapidly evolving invasive species. Ecol. Lett. 24, 698–707.

Ndlovu, J., Richardson, D.M., Wilson, J.R.U., Le Roux, J.J., 2013. Co-invasion of south African ecosystems by an Australian legume and its rhizobial symbionts. J. Biogeogr. 40, 1240–1251.

Nichols, B.S., Leubner-Metzger, G., Jansen, V.A.A., 2020. Between a rock and a hard place: adaptive sensing and site-specific dispersal. Ecol. Lett. 23, 1370–1379.

Novy, A., Flory, S.L., Hartman, J.M., 2013. Evidence for rapid evolution of phenology in an invasive grass. J. Evol. Biol. 26, 443–450.

Oduor, A.M., Leimu, R., van Kleunen, M., 2016. Invasive plant species are locally adapted just as frequently and at least as strongly as native plant species. J. Ecol. 104, 957–968.

Ollivier, M., Kazakou, E., Corbin, M., Sartori, K., Gooden, B., Lesieur, V., Thomann, T., Martin, J.-F., Tixier, M.S., 2020. Trait differentiation between native and introduced populations of the invasive plant *Sonchus oleraceus* L. (Asteraceae). NeoBiota 55, 85–115.

O'Neill, E.M., Hearn, E.J., Cogbill, J.M., Kajita, Y., 2017. Rapid evolution of a divergent ecogeographic cline in introduced lady beetles. Evol. Ecol. 31, 695–705.

Owen, M.A., Lahti, D.C., 2015. Sexual dimorphism and condition dependence in the anal pad of the small Indian mongoose (*Herpestes auropunctatus*). Can. J. Zool. 93, 397–402.

Owen, M.A., Lahti, D.C., 2020. Rapid evolution by sexual selection in a wild, invasive mammal. Evolution 74, 740–748.

Pannell, J.R., Auld, J.R., Brandvain, Y., Burd, M., Busch, J.W., Cheptou, P.-.O., Conner, J.K., Goldberg, E.E., Grant, A.-.G., Grossenbacher, D.L., Hovick, S.M., Igic, B., Kalisz, S., Petanidou, T., Randle, A.M., de Casas, R.R., Pauw, A., Vamosi, J.C., Winn, A.A., 2015. The scope of Baker's law. New Phytol. 208, 656–667.

Parepa, M., Fischer, M., Krebs, C., Bossdorf, O., 2014. Hybridization increases invasive knotweed success. Evol. Appl. 7, 413–420.

Parker, G.A., 2020. Conceptual developments in sperm competition: a very brief synopsis. Philos. Trans. R. Soc. B Biol. Sci. 375, 20200061.

Phillips, B.L., Brown, G.P., Webb, J.K., Shine, R., 2006. Invasion and the evolution of speed in toads. Nature 439, 803.

Phillips, B.L., Brown, G.P., Shine, R., 2010. Evolutionarily accelerated invasions: the rate of dispersal evolves upwards during the range advance of cane toads. J. Evol. Biol. 23, 2595–2601.

Prentis, P.J., Wilson, J.R., Dormontt, E.E., Richardson, D.M., Lowe, A.J., 2008. Adaptive evolution in invasive species. Trends Plant Sci. 13, 288–294.

Puckett, E.E., Park, J., Combs, M., Blum, M.J., Bryant, J.E., Caccone, A., Costa, F., Deinum, E.E., Esther, A., Himsworth, C.G., Keightley, P.D., Ko, A., Lundkvist, A., McElhinney, L.M., Morand, S., Robins, J., Russell, J., Strand, T.M., Suarez, O., Yon, L., Munshi-Southet, J., 2016. Global population divergence and admixture of the brown rat (*Rattus norvegicus*). Proc. R. Soc. B Biol. Sci. 283, 1–9.

Pyšek, P., Bacher, S., Kühn, I., Novoa, A., Catford, J.A., Hulme, P.E., Pergl, J., Richardson, D.M., Wilson, J.R.U., Blackburn, T.M., 2020. MAcroecological framework for invasive aliens (MAFIA): disentangling large-scale context dependence in biological invasions. NeoBiota 62, 407–461.

Ramula, S., Kalske, A., 2020. Introduced plants of *Lupinus polyphyllus* are larger but flower less frequently than conspecifics from the native range: results of the first year. Ecol. Evol. 10, 13742–13751.

Reed, D.H., Frankham, R., 2003. Correlation between fitness and genetic diversity. Conserv. Biol. 17, 230–237.

Renaud, S., Gomes Rodrigues, H., Ledevin, R., Pisanu, B., Chapuis, J.-.L., Hardouin, E.A., 2015. Fast response to anthropogenic disturbances. Biol. J. Linn. Soc. 114, 513–526.

Reznick, D.N., Ghalambor, C.K., 2001. The population ecology of contemporary adaptations: what empirical studies reveal about the conditions that promote adaptive evolution. Genetica 112, 183–198.

Reznick, D.N., Losos, J., Travis, J., 2019. From low to high gear: there has been a paradigm shift in our understanding of evolution. Ecol. Lett. 22, 233–244.

Ridley, C.E., Ellstrand, N.C., 2010. Rapid evolution of morphology and adaptive life history in the invasive California wild radish (*Raphanus sativus*) and the implications for management. Evol. Appl. 3, 64–76.

Robin, C., Andanson, A., Saint-Jean, G., Fabreguettes, O., Dutech, C., 2017. What was old is new again: thermal adaptation within clonal lineages during range expansion in a fungal pathogen. Mol. Ecol. 26, 1952–1963.

Rodger, J.G., van Kleunen, M., Johnson, S.D., 2013. Pollinators, mates and Allee effects: the importance of self-pollination for fecundity in an invasive lily. Funct. Ecol. 27, 1023–1033.

Rodríguez, J., Thompson, V., Rubido-Bará, M., Cordero-Rivera, A., González, L., 2019. Herbivore accumulation on invasive alien plants increases the distribution range of generalist herbivorous insects and supports proliferation of non-native insect pests. Biol. Invasions 21, 1511–1527.

Rotter, M.C., Holeski, L.M., 2018. A meta-analysis of the evolution of increased competitive ability hypothesis: genetic-based trait variation and herbivory resistance trade-offs. Biol. Invasions 20, 2647–2660.

Samis, K.E., Murren, C.J., Bossdorf, O., Donohue, K., Fenster, C.B., Malmberg, R.L., Purugganan, M.D., Stinchcombe, J.R., 2012. Longitudinal trends in climate drive flowering time clines in North American *Arabidopsis thaliana*. Ecol. Evol. 2, 1162–1180.

Santos, J., Pascual, M., Simões, P., Fragata, I., Lima, M., Kellen, B., Santos, M., Marques, A., Rose, M.R., Matos, M., 2012. From nature to the laboratory: the impact of founder effects on adaptation. J. Evol. Biol. 25, 2607–2622.

Schierenbeck, K.A., Ellstrand, N.C., 2009. Hybridization and the evolution of invasiveness in plants and other organisms. Biol. Invasions 11, 1093–1105.

Schrieber, K., Lachmuth, S., 2017. The genetic paradox of invasions revisited: the potential role of inbreeding × environment interactions in invasion success. Biol. Rev. 92, 939–952.

Schrieber, K., Wolf, S., Wypior, C., Höhlig, D., Hensen, I., Lachmuth, S., 2017. Adaptive and non-adaptive evolution of trait means and genetic trait correlations for herbivory resistance and performance in an invasive plant. Oikos 126, 572–582.

Seiter, S., Ohsaki, N., Kingsolver, J., 2013. Parallel invasions produce heterogenous patterns of life history adaptation: rapid divergence in an invasive insect. J. Evol. Biol. 26, 2721–2728.

Sherpa, S., Després, L., 2021. The evolutionary dynamics of biological invasions: a multi-approach perspective. Evol. Appl. 14, 1463–1484.

Sherpa, S., Guéguen, M., Renaud, J., Blum, M., Gaude, T., Laporte, F., Akiner, M., Alten, B., Aranda, C., Barre-Cardi, H., Bellini, R., Bengoa Paulis, M., Chen, X.G., Eritja, R., Flacio, E., Foxi, C., Ishak, I.H., Kalan, K., Kasai, S., Montarsi, F., Després, L., 2019. Predicting the success of an invader: niche shift versus niche conservatism. Ecol. Evol. 9, 12658–12675. https://doi.org/10.1002/ece3.5734.

Sherrard, M.E., Maherali, H., 2012. Local adaptation across a fertility gradient is influenced by soil biota in the invasive grass, Bromus inermis. Evol. Ecol. 26, 529–544.

Shine, R., Brown, G.P., Phillips, B.L., 2011. An evolutionary process that assembles phenotypes through space rather than through time. Proc. Natl. Acad. Sci. U. S. A. 108, 5708–5711.

Singer, M.C., Parmesan, C., 2018. Lethal trap created by adaptive evolutionary response to an exotic resource. Nature 557, 238–241.

Smith, S.E., Palkovacs, E.P., Weidel, B.C., Bunnell, D.B., Jones, A.W., Bloom, D.D., 2020. A century of intermittent eco–evolutionary feedbacks resulted in novel trait combinations in invasive Great Lakes alewives (Alosa pseudoharengus). Evol. Appl. 13, 2630–2645.

Strauss, S.Y., Lau, J.A., Carroll, S.P., 2006. Evolutionary responses of natives to introduced species: what do introductions tell us about natural communities? Ecol. Lett. 9, 357–374.

Stricker, K.B., Harmon, P.F., Goss, E.M., Clay, K., Flory, L.S., 2016. Emergence and accumulation of novel pathogens suppress an invasive species. Ecol. Lett. 19, 469–477.

Sultan, S.E., Matesanz, S., 2015. An ideal weed: plasticity and invasiveness in Polygonum cespitosum. Ann. N. Y. Acad. Sci. 1360, 101–119.

Sultan, S.E., Horgan-Kobelski, T., Nichols, L.M., Riggs, C.E., Waples, R.K., 2013. A resurrection study reveals rapid adaptive evolution within populations of an invasive plant. Evol. Appl. 6, 266–278.

Sun, Y., Roderick, G.K., 2019. Rapid evolution of invasive traits facilitates the invasion of common ragweed, Ambrosia artemisifolia. J. Ecol. 107, 2673–2687.

Tabassum, S., Leishman, M.R., 2018. Have your cake and eat it too: greater dispersal ability and faster germination towards range edges of an invasive plant species in eastern Australia. Biol. Invasions 20, 1199–1210.

te Beest, M., Le Roux, J.J., Richardson, D.M., Brysting, A.K., Suda, J., Kubesová, M., Pysek, P., 2012. The more the better? The role of polyploidy in facilitating plant invasions. Ann. Bot. 109, 19–45.

Turner, K.G., Fréville, H., Rieseberg, L.H., 2015. Adaptive plasticity and niche expansion in an invasive thistle. Ecol. Evol. 5, 3183–3197.

Tyukmaeva, V., Lankinen, P., Kinnunen, J., Kauranen, H., Hoikkala, A., 2020. Latitudinal clines in the timing and temperature-sensitivity of photoperiodic reproductive diapause in Drosophila montana. Ecography 43, 759–768.

Uesugi, A., Kessler, A., 2016. Herbivore release drives parallel patterns of evolutionary divergence in invasive plant phenotypes. J. Ecol. 104, 876–886.

Urbanski, J., Mogi, M., O'Donnell, D., DeCotiis, M., Toma, T., Armbruster, P., 2012. Rapid adaptive evolution of photoperiodic response during invasion and range expansion across a climatic gradient. Am. Nat. 179, 490–500.

van Boheemen, L.A., Bou-Assi, S., Uesugi, A., Hodgins, K.A., 2019. Rapid growth and defence evolution following multiple introductions. Ecol. Evol. 9, 7942–7956.

Vicente, S., Máguas, C., Richardson, D.M., Trindade, H., Wilson, J.R.U., Le Roux, J.J., 2021. Highly diverse and highly successful: invasive Australian acacias have not experienced genetic bottlenecks globally. Ann. Bot. 128, 149–157.

Vilas, A., Pérez-Figueroa, A., Quesada, H., Caballero, A., 2015. Allelic diversity for neutral markers retains a higher adaptive potential for quantitative traits than expected heterozygosity. Mol. Ecol. 24, 4419–4432.

Ward, M., Johnson, S.D., Zalucki, M.P., 2012. Modes of reproduction in three invasive milkweeds are consistent with Baker's rule. Biol. Invasions 14, 1237–1250.

Warrington, S., Ellis, A., Novoa, A., Wandrag, E.M., Hulme, P.E., Duncan, R.P., Valentine, A., Le Roux, J.J., 2019. Cointroductions of Australian acacias and their rhizobial mutualists in the Southern Hemisphere. J. Biogeogr. 46, 1519–1531.

Wendling, C.C., Wegner, K.M., 2015. Adaptation to enemy shifts: rapid resistance evolution to local *Vibrio* spp. in invasive Pacific oysters. Proc. R. Soc. B Biol. Sci. 282, 20142244.

Westley, P.A., Ward, E.J., Fleming, I.A., 2013. Fine–scale local adaptation in an invasive freshwater fish has evolved in contemporary time. Proc. R. Soc. B Biol. Sci. 280, 20122327.

Williams, W.I., Friedman, J.M., Gaskin, J.F., Norton, A.P., 2014. Hybridization of an invasive shrub affects tolerance and resistance to defoliation by a biological control agent. Evol. Appl. 7, 381–393.

Williams, J.L., Snyder, R.E., Levine, J.M., 2016. The influence of evolution on population spread through patchy landscapes. Am. Nat. 188, 15–26.

Wills, B.D., Moreau, C.S., Wray, B.D., Hoffmann, B.D., Suarez, A.V., 2014. Body size variation and caste ratios in geographically distinct populations of the invasive big-headed ant, *Pheidole megacephala* (Hymenoptera: Formicidae). Biol. J. Linn. Soc. 113, 423–438.

Xu, X., Wolfe, L., Diez, J., Zheng, Y., Guo, H., Hu, S., 2019. Differential germination strategies of native and introduced populations of the invasive species *Plantago virginica*. NeoBiota 43, 101–118.

Zenni, R.D., Bailey, J.K., Simberloff, D., 2014. Rapid evolution and range expansion of an invasive plant are driven by provenance-environment interactions. Ecol. Lett. 17, 727–735.

# CHAPTER 7

# Evolutionary impacts of invasive species on native species

## Chapter Outline

**7.1** Introduction     136
**7.2** Direct evolutionary impacts     140
   **7.2.1** Native plants adapting to compete with invasive plants     140
   **7.2.2** Native animals adapting to compete with invasive animals     141
   **7.2.3** Native animals adapting to invasive predators     142
   **7.2.4** Native predators adapting to invasive prey     143
   **7.2.5** Native species adapting to invasive mutualists     145
**7.3** Genetic impacts via hybridisation     147
**7.4** Indirect impacts     148
**7.5** Invasive species as drivers of extinction and speciation     149
**7.6** Conclusion     152
References     153

## Abstract

This chapter describes the evolutionary consequences of biological invasions for native species. As abundant resources (e.g. hosts or prey) or threats (e.g. predators), invasive species impose strong selection pressures on native species that may lead to rapid evolutionary responses. This chapter considers the direct impacts of invasive species on natives through various interaction types (e.g. competition, predation, mutualisms) and how natives respond evolutionarily to these. Indirect effects by invaders and hybridisation between native and invasive are also briefly discussed as drivers of evolutionary change in native species.

**Keywords:** Allelopathy, Competition, Extinction, Hybridisation, Mutualism, Predator–prey interactions, Rapid evolution

*Contemporary evolution in response to anthropogenic change appears to be increasingly common, and biological invasions will be a chief theater in which such evolution plays out.*

***Carroll (2007)***

*The Evolutionary Ecology of Invasive Species*
https://doi.org/10.1016/B978-0-12-818378-6.00002-4

## 7.1 Introduction

The distributions of some exotic species are truly remarkable. I have not been to many parts of the world without seeing weedy ribwort plantain (*Plantago lanceolata*). From the sidewalks in Cape Town, along hiking trails on Réunion Island, to my backyard in Sydney, this European species can be found almost anywhere. Most people would not consider ribwort plantain to be a particularly noteworthy invasive plant, often only found as scattered individuals in highly disturbed anthropic habitats. Yet, this species provides us with one of the most fascinating examples of how an invasive species can impact the evolutionary dynamics of native species.

Ribwort plantain was accidentally introduced into North America, probably shortly after the arrival of the first European settlers (Mack, 2003). In 1993 Michael Singer and colleagues reported that the native butterfly, the Edith's checkerspot (*Euphydryas editha*), seemingly developed a preference for weedy ribwort plantain as its new primary host at Schneider's Meadow in Nevada. While these butterflies can utilise numerous native plant species, they are usually monophagous at a particular location. Before the arrival of ribwort plantain, the butterfly almost exclusively fed on the native maiden blue-eyed Mary (*Collinsia parviflora*) at Schneider's Meadow. During spring, butterfly larvae feed intensely for 2 weeks or so, before *Collinsia* plants start to die off, forcing them into diapause. In order to enhance offspring survival, female butterflies often seek out plants in cool moist areas for oviposition, where the onset of plant senescence is likely to be delayed. The following spring, exposure to sunlight breaks diapause and larvae resume feeding and develop by basking in sunlight on the bare ground around *Collinsia* plants. Larval body temperature is typically 10–12 °C above ambient temperatures, and their development fastest at 30–35 °C (Singer and Parmesan, 2018). This represents a paradox, as post-diapause larvae no longer prefer the cool moist areas chosen by their mothers, where thermal conditions are now unfavourable. For female butterflies this represents a trade-off between offspring survival and maternal fecundity. That is, while laying eggs in cooler areas may increase their fecundity, extended larval development on plants with delayed senescence also causes delays in the emergence of adults, possibly until such time when *Collinsia* plants are no longer available (Singer and Parmesan, 2010). The arrival of ribwort plantain in Nevada provided Edith's checkerspots with an opportunity to partly overcome this dilemma.

Ribwort plantain survives throughout most of the summer in the Western United States. Therefore oviposition on it provided Edith's check-erspot larvae with a valuable food source throughout their pre-diapause development, leading to higher survival into adulthood. While larval growth is around 18% slower on ribwort plantain compared to on *Collinsia* plants (Thomas et al., 1987), the constraints associated with feeding on the short-lived native plant made ribwort plantain an attractive alternative host to the butterfly. Selection for oviposition on ribwort plantain was strong. For example, within less than a decade, preference for ribwort plantain by the butterfly increased from less than 10% to over 50% (Singer et al., 1993). This evolution occurred during a time when Schneider's Meadow was used as a cattle farm and when the vegetation was regularly grazed (Singer and Parmesan, 2018). In 2005 the meadow changed ownership and cattle were removed, causing grasses to quickly grow lush. By 2007 most ribwort plantain plants, on which butterflies were by now 100% reliant, became embedded in dense grass cover (Singer and Parmesan, 2018). Before the extraction of cattle, 'open' ribwort plantain plants provided Edith check-erspot eggs with thermal conditions that were on average 13.4 °C above ambient temperatures, which are optimal for the butterfly's development. Around grass-embedded ribwort plantain plants, however, temperatures dropped by around 7.4 °C, which was too cold for development (Singer and Parmesan, 2018). Moreover, post-diapause larvae were unable to bask in sunlight as no bare ground was available around grass-embedded ribwort plantain. The rapidly evolved preference for ribwort plantain meant that Edith checkerspots could not switch back to their original *Collinsia* host and searches for eggs, larvae, and adults at Schneider's Meadow between 2008 and 2012 yielded nothing—the population was extinct!

As dominant members of communities, invasive species often cause direct evolutionary impacts on native species (Cox, 2004; Strauss et al., 2006; Carroll, 2007; Berthon, 2015; Reznick et al., 2019). Invasive species also often cause dramatic, and often rapid, changes in abiotic and biotic con-ditions, resulting in indirect evolutionary impacts on native species (e.g. Lankau, 2012; Berthon, 2015). For instance, in south-eastern Australia the invasive seaweed, *Caulerpa taxifolia*, lowers water flow rates and dissolved oxygen levels, creating anoxic sediments that have higher silt content and abundance of large phytoplankton (Gribben et al., 2009; McKinnon et al., 2009). These changes have led to the rapid evolution of longer and broader shells in the native bivalve, *Anadara trapezia*, presumably for it to cope with the altered sediment conditions and food resources created by *Caulerpa* invasion (Wright et al., 2012).

Invasive species are frequently encountered by native species as novel resources (e.g. as prey items) or threats (e.g. as predators) that may lead to strong directional selection (Cox, 2004; Carroll, 2007). The strength of such selection will depend on the encounter frequency between, and the levels of eco-evolutionary experience (EEE) shared by, interacting invasive and native species (Saul et al., 2013; Saul and Jeschke, 2015; Fig. 7.1). Eco-evolutionary experience (EEE) is expected to influence selection strength, and therefore partly, the extent and rate by which rapid evolution in native species occurs (Fig. 7.1). As we discussed in Chapter 3, in the context of invasion ecology, EEE describes the exposure of both native and exotic species to biotic interactions over evolutionary timescales and emphasises the role of traits that evolved in previous environments (i.e. preadaptations) in driving the establishment success and adaptability of introduced species (Saul et al., 2013; Saul and Jeschke, 2015; Heger et al., 2019). The EEE of exotic species will therefore determine how quickly their populations become widespread, and the form and strength of their interactions with native species (Carroll, 2007). Selection is expected to be particularly strong on native species that share moderate to high EEE with invasive species, e.g. when native herbivores colonise invasive plants that are closely related to their native host(s). Similarly, native species that share high EEE with the ecological conditions created by invasive species may more easily adapt mechanisms to avoid them (Carroll, 2011). Host shifts by soapberry bugs onto invasive sapindaceous host plants we discussed in Chapter 3 (Section 3.3) exemplify how EEE may influence rapid evolutionary responses in natives to invasive species. As their name suggests, these bugs share high EEE with soapberry plants (family Sapindaceae) (Carroll and Loye, 2012). For example, in the United States, the native soapberry bug, *Jadera haematoloma*, have adopted exotic goldenrain trees (*Koelreuteria elegans*). Compared to the bug's native *Cardiospermum corindum* balloon vine host, goldenrain trees have much smaller fruit capsules and, in response, the bug rapidly evolved shorter proboscides (beaks) to more efficiently feed on the seeds of the new host (Carroll and Boyd, 1992). The close phylogenetic relationship between goldenrain trees and *J. haematoloma*'s native balloon vine host, likely provided this bug with adequate EEE to readily adopt the new host plant. Similar host shifts, and subsequent evolutionary responses in soapberry bugs, have been reported from elsewhere in the world (Carroll et al., 2005; Andres et al., 2013; Foster et al., 2019).

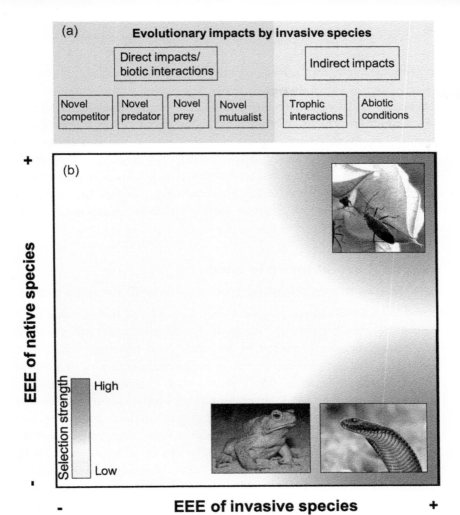

**(a)** **Evolutionary impacts by invasive species**

| Direct impacts/ biotic interactions | | | | Indirect impacts | |
|---|---|---|---|---|---|
| Novel competitor | Novel predator | Novel prey | Novel mutualist | Trophic interactions | Abiotic conditions |

**(b)**

**EEE of native species** (+ / –)

Selection strength: High — Low

**EEE of invasive species** (– / +)

**Fig. 7.1** (a) Selection on native species by invasive species can stem from various direct and indirect invader-mediated impacts. (b) For direct impacts, selection strength will partly reflect the levels of eco-evolutionary experience (EEE) shared between invasive and native species (also see Chapter 3). When they share low EEE, invasive species may exert minimal selection on native species (and vice versa), primarily because interaction frequency is expected to be low. Similarly, native species that share EEE with inexperienced invasive species may experience nominal selection. Native species lacking EEE with experienced invasive species, on the other hand, are expected to experience strong selection. For instance, the native red-bellied black snake (*Pseudechis porphyriacus*) in Australia shares little EEE to experienced invasive cane toads (*Rhinella marina*; bottom right). Cane toads are highly poisonous and exert strong selection on predators such as red-bellied black snakes. Similarly, strong selection on natives may result when both interacting partners share high EEE with each other, as was the case for native *Leptocoris* soapberry bugs (in the subfamily Serinethinae) and their newly adopted invasive *Cardiospermum* balloon vine host plants (*top right*). Credit (panel b): Image of soapberry bug on balloon vine (top right): Scott Carroll. Image of cane toad (bottom right): Matthew Greenlees. Image of red-bellied black snake (bottom right): Matt Clancy.

Notwithstanding the roles of selection strength and EEE, the evolutionary responses of native species will only be adaptive if trait variation has a genetic basis and if it translates into fitness differences between historical ecological conditions and those created by invasive species. From a genetic perspective, the evolutionary impacts of hybridisation between native and invasive species have been frequently documented (Rhymer and Simberloff, 1996; Todesco et al., 2016). We have touched on examples of invasive species causing evolution in native species throughout this book. In this chapter we will discuss different aspects of such evolutionary responses in more detail.

## 7.2 Direct evolutionary impacts

### 7.2.1 Native plants adapting to compete with invasive plants

In invaded habitats, native plants not only have to deal with familiar constraints like specialist enemies, but possibly also with novel ones, such as toxic phytochemicals produced by invasive plants (Zhang et al., 2021). To coexist with invasive species, native plants could evolve to suppress or tolerate them (Stotz et al., 2016). Different traits modulate plant competitiveness, and how they affect competition may be highly context dependent (Wang et al., 2010; Baron et al., 2015). Despite this, traits such as faster early growth, greater height, larger seed size, higher investment in root growth, and increased resistance to allelochemicals have been repeatedly linked with the ability of native plants to tolerate, or even suppress, invasive species (Callaway et al., 2005; Leger, 2008; Goergen et al., 2011; Lankau, 2012; Conti et al., 2018; Semchenko et al., 2018). Despite their usually complex genetic backgrounds, many of these traits have been shown to undergo rapid evolution in native species in response to competition with invasives (e.g. see Callaway et al., 2005; Berthon, 2015; Oduor, 2021).

Spotted knapweed (*Centaurea stoebe*) was accidently introduced into North America as a contaminant of imported seeds and solid ship ballast in the late 1800s. The species is now considered one of the most damaging invasive plants on the continent, where it outcompetes native species and often forms near monocultures. Two studies have tested whether the strong competitive ability of this species has triggered evolutionary responses in the native bluebunch wheatgrass (*Pseudoroegneria spicata*), a species that often co-occurs with invasive spotted knapweed (Fletcher et al., 2016; Gibson et al., 2018). Both studies collected individual grasses from knapweed-invaded and uninvaded habitats in Montana and grew these under competition with the invader. While the ability to suppress knapweed was similar between experienced and inexperienced bluebunch wheatgrass individuals, experienced ones were more tolerant of the invader (Fletcher et al., 2016; Gibson et al., 2018), a likely result of rapid evolution of higher leaf and root biomass production in invaded habitats (Gibson et al., 2018).

As we have discussed, one way by which invasive species can outcompete natives is through the production of phytochemicals that inhibit or suppress native species. The high competitiveness of spotted knapweed is partly due to such allelopathy (Callaway and Aschehoug, 2000; Callaway and Ridenour, 2004). Callaway et al. (2005) were first to describe evolution in native species to tolerate knapweed allelopathy. These authors collected ramets of five native grass species from areas with a 20- to 30-year history of knapweed invasion as well as from nearby areas with no history of invasion, in Western Montana. Grasses from these two areas were grown in a glasshouse experiment under competition with spotted knapweed and soil treatments that included the addition of activated carbon. Activated carbon binds organic molecules and therefore removes knapweed allelochemicals from soils. The authors also exposed seeds collected in the field from all five grass species to (±)-catechin, a major allelochemical found in high concentrations in spotted knapweed-invaded soils that is absent in uninvaded soils (Bais et al., 2003). Their results showed that, while growth responses of all five grass species varied in their susceptibility to spotted knapweed allelochemicals, growth of experienced individuals from invaded areas was always less affected than that of inexperienced individuals. The latter was mirrored by germination success of grass seeds exposed to (±)-catechin (Callaway et al., 2005). Coupled together, these results support the rapid evolution of these native grasses in the response to the allelopathic impacts of spotted knapweed. Since Callaway et al.'s (2005) work, many authors have found evidence for rapid evolutionary responses by native plants to the allelochemicals of invasive plants (e.g. Mealor and Hild, 2007; Lankau, 2012; Huang et al., 2018).

## 7.2.2 Native animals adapting to compete with invasive animals

Unlike plants, there have only been few studies testing for rapid evolutionary responses in native animals due to competition with invasive species (e.g. Bourke et al., 1999; Sidorovich et al., 1999). The brown anole lizard (*Anolis sagrei*) provides one example. This lizard is native to Cuba and the Bahamas but has been widely introduced to other parts of the world. For example, it reached southern Florida in the USA in the 1970s, rapidly spreading as far as Southern Georgia, Texas, Louisiana, Mississippi, and Alabama, where it outcompetes many native lizard species. Stuart et al. (2014) wanted to know whether such competition has caused rapid evolution in the native green anole (*A. carolinensis*). Brown and green anoles compete strongly as they use the same habitats (trees, from the ground to the crown) and share highly similar ecologies (Losos, 2009). Observational data suggest that, where they co-occur, green anoles always perch higher up in trees than brown anoles (Schettino et al., 2010; Edwards and Lailvaux, 2012). In 1995 Stuart et al. (2014) identified six small islands in Florida where, at the time, only green anoles were present. They then intentionally introduced brown anoles to three 'treatment islands' while leaving three islands as 'controls'. Perch heights of both lizard species were measured during repeat visits to all islands over the next 39 months. After only 3 months of releasing brown anoles, green anoles on treatment islands had significantly higher perch heights compared to lizards on control islands, a pattern that held throughout the study's duration. Moreover, this shift to a more arboreal lifestyle in green anoles coincided with the rapid evolution of larger toepads with more lamellae, traits that enable lizards to grip on slender branches higher up on trees (Stuart et al., 2014). Comparisons between green anoles on islands that have been naturally invaded by brown anoles and green anoles on nearby uninvaded islands mimicked results of the 1995 introduction experiment (Stuart et al., 2014). Population genetic analyses confirmed that these behavioural and morphological changes in green anoles were indeed the result of multiple independent evolutionary events rather than being demographic anomalies such as non-random migration among invaded islands (Stuart et al., 2014).

## 7.2.3 Native animals adapting to invasive predators

The American red swamp crayfish (*Procambarus clarkii*) is one of the most widely introduced freshwater species in the world (CABI—Invasive Species Compendium, 2021). In Europe, this species has not only impacted aquatic biodiversity through predation, but also as a disease vector and ecosystem engineer (Souty-Grosset et al., 2016). More recently, evolutionary impacts on native species have also been documented. In Lombardy, Italy, this crayfish was first detected in the early 2000s, thereafter quickly becoming established in waterbodies throughout the region (Ficetola et al., 2011; Lo Parrino et al., 2020). Melotto et al. (2020) investigated evolutionary responses of the native Italian agile frog (*Rana latastei*) to predation by the crayfish. These authors compared tadpole development before (2003) and after (2017) American red swamp crayfish invasion across a wide geographic area in Lombardy. Prior to invasion, tadpoles showed pronounced between-population variation in development time, a consequence of historical adaptation to local climatic conditions. For example, data from 2003 showed that tadpoles in cooler foothill sites in Northern Lombardy had faster development times compared with those from warmer lowland sites further south. This is a common observation in ectotherms, as their metabolism generally slows down at low temperatures, leading to reduced developmental rates. As a consequence, populations in cold environments often evolve faster intrinsic development times to overcome the developmental constraints associated with low temperatures. This geographic signal in developmental rates of Italian agile frogs disappeared by 2017, 14 years after the arrival of the American red-clawed crayfish (Melotto et al., 2020). Instead, tadpole developed significantly faster in crayfish-invaded ponds compared with uninvaded ponds, irrespective of whether these ponds were located in foothill or lowland areas. These changes likely evolved to allow earlier metamorphosis, thereby reducing the frog's exposure to crayfish predation. Remarkably, this evolution occurred over just 3–6 frog generations, suggesting very strong predator-mediated selection, that overtook climate as the most important driver of geographic variation in the frog's development.

## 7.2.4 Native predators adapting to invasive prey

In their role as potential prey for native predators, invasive species can be either harmful or palatable. Native predators may therefore naturally avoid or consume these invasive species, or acquire the ability to do so via rapid evolution. As we discussed previously, Australian predators and invasive cane toads (*Rhinella marina*) illustrate just how quickly such evolutionary responses may occur. Cane toads produce a potent cocktail of toxins that they excrete through glands on their shoulders. Importantly, the chemical constituents of these toxins differ from those produced by native Australian anurans (Daly et al., 1987), and therefore most Australian predators lack appreciable levels of EEE to cane toad toxins. Despite this, cane toads are frequently attacked and consumed by native predators, presumably because of their superficial resemblance to Australian frogs. As we discussed in Chapter 3, the amount of toxin cane toads produce varies throughout their life cycle and displays strong positive allometry (Hayes et al., 2009). It is therefore reasonable to expect larger mature (and therefore more poisonous) toads to exert the strongest selection pressure on native predators. This is well illustrated by the toad's evolutionary impacts on native Australian snakes. Snakes are gape-limited predators, so the size of their heads determines the size of prey they can consume. Snake head size, in turn, has strong negative allometry and, hence, the chance of ingesting a cane toad large enough to cause fatal poisoning will be lower in larger-bodied than smaller-bodied snakes. The evolution of smaller head size (i.e. gape size) is therefore likely to occur in response to toad poisoning. To test this hypothesis, Phillips and Shine (2004) collected morphological data for four native Australian snake species, spanning a period of 80 years since the arrival of cane toads. Two of the snake species, which are known to be vulnerable to cane toad poisoning, the red-bellied black snake (*Pseudechis porphyriacus*) and the common tree snake (*Dendrelaphis punctulatus*), have indeed evolved smaller heads and larger body sizes in relation to the time since introduction of cane toads. In contrast, two other snake species that are less vulnerable to toad poisoning, the swamp snake (*Hemiaspis signata*) and common keelback snake (*Tropidonophis mairii)*, did not show morphological changes in response to cane toad exposure. Swamp snakes already have unusually small heads that make them incapable of ingesting toads that are large enough to kill them (Phillips et al., 2003). Keelbacks, on the other hand, are large-bodied snakes and have normal-sized heads but have a high tolerance to cane toad toxins. Keelbacks also display a preference for native frogs over cane toads, and when feeding on the latter, prefer smaller toads over larger ones. These observations are even true when keelbacks have never encountered cane toads before (Llewelyn et al., 2010, 2012). It is thought that this snake's Asian ancestry, and thus sympatric distribution with toads, provide it with sufficient EEE with cane toads. The same is probably true for common tree snakes, as they are able to avoid cane toads when native prey items are available (Llewelyn et al., 2018). So why have tree snakes evolved smaller head sizes in response to cane toads given their EEE with toads? The fact that tree snakes do occasionally attack toads and consume them may still allow selection to occur, albeit 'softer' than for predators like red-bellied black snake that lack any EEE to toads (Llewelyn et al., 2018). These examples illustrate an important point; that the responses of native species to invasion may be affected by both contemporary adaptation and historical preadaptations.

Mammals have not escaped the impacts of cane toads in Australia. For instance, numbers of the endangered carnivorous marsupial, the northern quoll (*Dasyurus hallucatus*), have rapidly declined following cane toad invasion (Shine, 2010). Despite being susceptible to toad toxins, northern quolls still attack and consume them. Kelly and Phillips (2019) found evidence to suggest that northern quoll populations have experienced strong selection for cane toad avoidance as a consequence, a behaviour that is strongly heritable in this species.

## 7.2.5 Native species adapting to invasive mutualists

Mutualist interactions often play key roles in determining the establishment success of introduced exotic species (Richardson et al., 2000), as well as their ecological impacts once they become widespread (Traveset and Richardson, 2006, 2011; Le Roux, 2020). Invasive species frequently form mutualisms with native species, sometimes with evolutionary consequences for one of the, or both, interacting partners. Here, we discuss such evolutionary impacts on native species by examining mutualistic relationships plants have with nitrogen-fixing bacteria and seed dispersers.

The success of legumes as invasive species has been partly attributed to their ability to form symbioses with nitrogen-fixing rhizobia (Parker, 2001; Le Roux et al., 2017). Rhizobia are soil bacteria capable of forming specialised structures, called nodules, on the roots of most legumes. Inside root nodules rhizobia fix atmospheric nitrogen into ammonium that legumes can utilise, in return receiving carbon-rich photo-assimilates from the host plant. Co-introductions of exotic legumes and their rhizobia into new environments are surprisingly common (e.g. Ndlovu et al., 2013; Horn et al., 2014; Le Roux et al., 2017; Warrington et al., 2019). These 'familiar' associations may benefit introduced legumes, partly because the specificity and efficiency of the legume–rhizobium symbiosis is modulated by complex molecular communication (Perret et al., 2000), and thus the genotypes of interacting plant and bacterial partners (Barrett et al., 2015). In general, nodulation is initiated when legume roots deposit (iso)flavonoids into the rhizosphere which stimulate the expression of specific rhizobial symbiotic genes, known as nodulation genes (Le Roux et al., 2017). These symbiotic genes are located on mobile genetic elements, such as plasmids or genetic islands (Rogel et al., 2011), that may be transferred between different rhizobial species, and even genera, through conjugation and horizontal gene transfer (HGT; Lemaire et al., 2015). Therefore rhizobia with different identities (i.e. based on their core genomes) may harbour identical nodulation genes (Le Roux et al., 2017).

When exotic legumes and rhizobia are co-introduced, HGT between native and exotic rhizobia may have important consequences for the evolutionary dynamics of native rhizobia. Symbiotic genes carried by rhizobia are the products of historical selection and co-evolution with their hosts. However, when co-introduced with legumes, there is no guarantee that exotic rhizobia will be well adapted to local environmental conditions. Resident native rhizobia, on the other hand, will not only be adapted to local conditions but are also likely to be more abundant than co-introduced rhizobia, at least initially. Native rhizobia may therefore be highly competitive for nodulation of introduced exotic legumes *if* they can acquire symbiotic genes from co-introduced rhizobia via HGT. Many researchers have found this to be the case for agricultural and invasive legumes. For example, birdsfoot trefoil (*Lotus corniculatus*) was co-introduced with a single rhizobium strain into New Zealand as a pasture plant. A few years later, the legume was found to nodulate with a wide variety of rhizobia, most of which carried symbiotic genes, but not core genomes, of the strain that was originally co-introduced with the plant (Sullivan et al., 1995). Similarly, in China, rhizobia of invasive black locust (*Robinia pseudoacacia*) have core genomes corresponding to those of Chinese rhizobia, but symbiotic genes acquired via HGT from rhizobia in the native North American range of the legume (Wei et al., 2009). More recently, Horn et al. (2014) found a diverse group of native rhizobia to have evolved the ability to nodulate invasive Scotch broom (*Cytisus scoparius*) in the United States after acquiring symbiotic genes from European Scotch broom rhizobia via HGT. To date, no studies have tested how these changes in the genetic make-up and host preference of native rhizobia impact native legumes; however, it is likely that symbiotic nitrogen fixation will be negatively affected (Le Roux et al., 2016, 2017).

Numerous examples of invasive animals acting as seed dispersers of native plants exist (e.g. Heleno et al., 2013; Vizentin-Bugoni et al., 2019; Le Roux et al., 2020). Spurge olive (*Cneorum tricoccon*), a plant whose native range in Europe has dramatically contracted over the last few decades, is currently restricted to the Balearic Islands (Riera et al., 2002). Seeds of the plant are primarily dispersed by two native lizards: the Balearic lizard (*Podarcis lilfordi*) in the eastern Balearic Islands and the Ibiza wall lizard (*P. pityusensis*) in the South-western Balearic Islands (Traveset, 1995; Riera et al., 2002). *Podarcis lifordi* has recently gone extinct on some Balearic Islands, with remnant populations only found on surrounding islets, while *P. pityusensis* is currently restricted to only a few islands (Cooper Jr. and Peréz-Mellado, 2012). Numerous invasive carnivores, such as the pine marten (*Martes martes*), are found on the Balearic Islands. Pine martens also feed on olive spurge fruit and acts as an effective seed disperser for the species. Traveset et al. (2019) compared the size of olive spurge seeds in 15 populations on different Balearic Islands. Within each population, seed dispersal was *only* by one of the native lizards or the invasive pine marten. Comparisons between seeds collected directly from plants in the field and from faeces each population's dispersal agent showed that pine martens substantially modify the selection regime exerted by the two native lizards on seed size. That is, pine martens exerted strong positive directional selection by consuming larger fruit containing larger seeds more frequently than smaller fruits with small seeds, while the two lizards imposed negative directional selection. Even after controlling for maternal effects and local climatic conditions, Traveset et al. (2019) found that seed size variation between olive spurge populations is best explained by whether they were dispersed by native or invasive animals.

## 7.3 Genetic impacts via hybridisation

Because exotic species often share little, or no, co-evolutionary history with native congeners, reproductive barriers may be weaker between them than among native congeners. In some cases, this has led to hybridisation, causing significant evolutionary impacts (Rhymer and Simberloff, 1996; Todesco et al., 2016) and rapid declines of native populations (Rhymer and Simberloff, 1996; Wolf et al., 2001; Todesco et al., 2016). One way through which such impacts may occur is introgression (i.e. when hybrid offspring backcross to one or both parental species), which dilutes gene pools and purge them of locally adapted genotypes (Allendorf et al., 2001; Olden et al., 2004; Schulte et al., 2012). In the long term, this may lead to the extinction of native populations (Rhymer and Simberloff, 1996; Wolf et al., 2001; Todesco et al., 2016).

A textbook example of the evolutionary impacts of invasive on native species via hybridisation is the Mallard duck (*Anas platyrhynchos*). This duck has been widely introduced around the world for hunting and ornamental purposes (Long, 1981) and has since become invasive in Australia, the Hawaiian Islands, New Zealand, South Africa, and the mainland USA (Rhymer et al., 1994; Fowler et al., 2009; Guay and Tracey, 2009; Stephens et al., 2020). Invasive Mallards hybridise with several native duck species, e.g. the Florida mottled duck (*A. fulvigula* subspecies *fulvigula*), the American black duck (*A. rubripes*), the New Zealand grey duck (*A. superciliosa* subspecies *superciliosa*), the endangered endemic Hawaiian duck (*A. wyvilliana*), and the African yellow-billed duck (*A. undulata*) (Rhymer et al., 1994; Kulikova et al., 2004; Mank et al., 2004; Fowler et al., 2009; Stephens et al., 2020). Many of these hybrids are fertile and genetic introgression has been reported in some instances (e.g. in the American black duck, Mank et al., 2004; New Zealand grey duck, Rhymer et al., 1994; yellow-billed duck, Stephens et al., 2020). In New Zealand, extensive introgression has led to the virtual elimination of the New Zealand grey duck (Rhymer et al., 1994).

Intriguing examples where hybridisation between invasive and native species acts as 'bridges' for gene flow between native congeners also exist. For example, the white sucker fish (*Catostomus commersonii*) is highly invasive in the Colorado River Basin in the United States, where it competes strongly with two native congeners; the flannelmouth sucker (*C. latipinnis*) and bluehead sucker (*C. discobolus*). Hybrids between white and flannelmouth suckers have been identified and genetic data suggest that shortly after the appearance of these hybrids, introgression between the two native suckers occurred (McDonald et al., 2008). Hybridisation between the two native suckers was not known prior to the introduction of white suckers or, if present, must have resulted in maladapted or infertile F1 hybrids (McDonald et al., 2008). Mandeville et al. (2015) found putative hybrids between the two native fishes to have around 50% bluehead sucker genetic ancestry and varying proportions of flannelmouth and white sucker genes, suggesting that gene flow between the two native species involves interbreeding between flannelmouth x white sucker F1 hybrids and pure bluehead suckers. This genetic pattern is intriguing and suggests that invasive white suckers facilitate gene flow between the two native fishes (McDonald et al., 2008).

A recent review on the ecological, evolutionary, and genetic factors affecting the extinction risk reported that 27 out of 69 studies identified hybridisation with invasive species as posing an immediate extinction threat to native species (Todesco et al., 2016). The ever-increasing number of exotic species being introduced globally (Seebens et al., 2020) suggests that hybridisation between invasive and native species will continue to be a significant evolutionary threat to native biodiversity.

## 7.4 Indirect impacts

Indirect impacts of invasive species on native species are hard to predict and difficult to quantify, with only a few known cases of natives responding evolutionarily to such impacts (Berthon, 2015). On example is the allelopathic chemicals produced by some invasive plants that may indirectly impact plant–plant interactions. For example, compounds originating from soil microbial degradation or transformation of chemicals produced by invasive plants may negatively affect the growth of native plants (Inderjit and Weiner, 2001; Inderjit and van der Putten, 2010). Allelochemicals may also directly impact soil biota which, in turn, may change the way it interacts with native plants. For example, invasive garlic mustard (*Alliaria petiolata*) inhibit native plants by interfering with their mycorrhizal fungal mutualists via allelopathy (Lankau, 2012). Like all crucifers (family Brassicaceae), garlic mustard does not form mycorrhizal associations. Lankau et al. (2009) found that the production of sinigrin, a major allelochemical of garlic mustard, evolved rapidly in the species' invasive range in the United States, where individuals in leading-edge populations produce higher quantities of this chemical than those in core populations. Lankau (2012) also found that garlic mustard produces more sinigrin in the presence of dense populations of native competitors, suggesting that selection from interspecific competition drove rapid evolution of enhanced allelopathy. Using reciprocal transplant experiments, Lankau (2012) tested whether different populations of native clearweed (*Pilea pumila*) differed in their ability to compete against garlic mustard and if so, whether mycorrhization played a role. They sampled and germinated clearweed seeds from high and low sinigrin-producing garlic mustard populations and grew these under competition with the invader. Data from these experiments suggest that clearweed from invaded areas with high sinigrin production was more tolerant of the garlic mustard, and maintained higher levels of mycorrhization, than clearweed from invaded areas where sinigrin production was low. These results not only illustrate how indirect impacts can cause rapid evolution in native species, but also that co-evolutionary dynamics between invasive and native species can establish (Lankau, 2012).

## 7.5 Invasive species as drivers of extinction and speciation

The severity of the evolutionary impacts of invasive species on native species may fall anywhere along a continuum between extinction and speciation. Invasive species may drive native species to extinction, especially when they cannot adapt fast enough to the adverse conditions created by invaders (Blackburn et al., 2019; Le Roux et al., 2019). This typically involves selection pressures by the invader with which native species share little, or no, EEE, such as competitive effects (Gilbert and Levine, 2013), novel diseases (Lips, 2016), and predation (Doherty et al., 2016). For example, the rosy wolf snail (*Euglandina rosea*), a voracious predator of land snails and slugs, was presumed to be an excellent biological control agent for invasive snails. For this reason, it was introduced to numerous South Pacific islands to control invasive giant African land snails (*Achatina fulica*), but instead, it has decimated endemic snail populations. These endemic snails lack EEE to predatory snails and at least 134 species went extinct following the introduction of the rosy wolf snail (Regnier et al., 2009).

As we discussed before, hybridisation between native and invasive species represents a significant evolutionary impact. In some instances, this can lead to the formation of new species (Thomas, 2015). For example, invasive American smooth cordgrass (*Spartina alterniflora*) and native small cordgrass (*S. maritima*) hybridised in England and gave rise to sterile hybrids (Gray et al., 1991, also see Chapter 4, Section 4.7). Fertility in this hybrid lineage was restored through genome doubling and led to the formation of a new allododecaploid species, *S. anglica*, now itself considered a major invasive species around the world.

Another example of invasive species causing speciation in native species involves numerous East Asian *Lonicera* honeysuckles that have been introduced into North America over the last 250 years, which have hybridised, giving rise to highly invasive populations (Barnes and Cottam, 1974). Hybrid honeysuckles have been readily adopted by two native tephritid fruit flies, *Rhagoletis mendax* and *R. zephyria*, the specialist parasites of native blueberries (*Vaccinium corymbosum*) and snowberries (*Symphoricarpos albus* var. *laevigatus*), respectively (Schwarz et al., 2007). Blueberries, snowberries, and invasive honeysuckles often occur in sympatry, providing opportunities for *R. mendax* and *R. zephyria* to come in contact and mate. Schwarz et al. (2005) found flies on honeysuckles to be hybrids between *R. mendax* and *R. zephyria* that are reproductively isolated, and genetically distinct, from both parental fly species. *Lonicera* flies also prefer hybrid honeysuckles as host plants over blueberries and snowberries (Fig. 7.2). Honeysuckles therefore facilitated hybrid speciation, first by breaking down reproductive barriers between *R. mendax* and *R. zephyria*, and second, by facilitating ecological isolation between the three fly taxa (Schwarz et al., 2007).

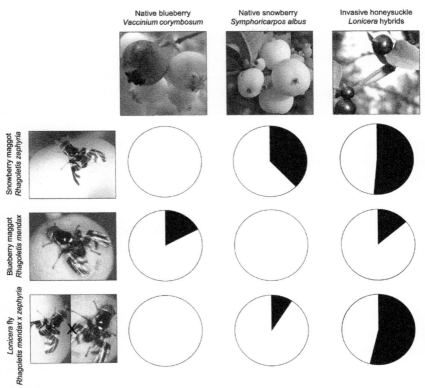

**Fig. 7.2** Hybrid speciation in *Rhagoletis* fruit flies following the introduction of exotic honeysuckles (*Lonicera* species) into North America. In the United States, invasive hybrid honeysuckles act as host plants for the two native fly species, *R. mendax* and *R. zephyria*, and facilitated hybridisation between them. Schwarz et al. (2007) exposed these two fly species and their hybrid offspring to fruit of only the original host plant of *R. mendax* (i.e. blueberry), *R. zephyria* (i.e. snowberry), or hybrid *Lonicera* honeysuckles. The proportion of females that used each host fruit at least once is shown in black for each host-fly combination (redrawn from Schwarz et al., 2007). These results indicate that parental fly species cannot utilise each other's native host plants, but that both can utilise invasive honeysuckles. The new invasive host plant, therefore, provided these two fly species with mating opportunities. Hybrid *Lonicera* flies also show higher preference for their new host than for blueberries and snowberries, resulting in ecological isolation between different fly taxa on different host plant species (Schwarz et al., 2005, 2007). *Credit: The following photos were sourced from Wikimedia Commons under the Creative Commons Share-Alike Licence (CC-BY-SA 3.0): Rhagoletis mendax (Jerry A. Payne), Lonicera fruit (Qwert1234), snowberry fruit (H. Zell), blueberry fruit (Green Yoshi). Image of R. zephyria: Ken Sproule.*

## 7.6 Conclusion

A growing body of research on the evolutionary consequences of invasive species suggests that natives often undergo rapid evolution in response to the opportunities or adverse conditions created by invasion. Knowing which invasive species are more likely to exert selection on which native species, or which native species are more likely to respond evolutionarily to invasion, is not straightforward. However, our fundamental understanding of evolutionary processes can go some way in predicting whether native species are likely to experience strong selection from invasive species (Fig. 7.1). Historical selection that shaped existing trait variation in both native and invasive species, i.e. their EEE, ultimately determines the form and frequency of invader–native species interactions (Saul et al., 2013; Saul and Jeschke, 2015). Selection is therefore expected to be particularly strong when native species have preadaptations for interacting with invasive species. Similarly, invasive species with high EEE to evolutionary naïve native species may exert strong selection on natives.

Native species that are unable to adapt to invasive species (whether due to a lack of EEE or genetic diversity) may suffer severe population declines, or ultimately, extinction. Both of these outcomes will have flow-on consequences for the evolution of native communities. Yet, our understanding of the evolutionary responses of native species to invasion generally comes case studies on individual species, and rarely from studies on multi-trophic interactions. Most habitats are also invaded by multiple species, which may lead to complex selection pressures on native species. Considering this, it is important that future research focuses on multiple interacting species to better understand the evolutionary impacts of invasive species.

Reports on the evolutionary responses of native species to invasive species have mostly examined the direct effects of invaders on natives. In contrast, evolutionary response of native species to the cascading indirect effects of invasive species has received almost no attention, yet, these impacts must be commonplace. For instance, if a native parasitoid population rapidly increases in response to a new compatible exotic insect host, another native insect in the web may suffer subsequently as a result of increased parasitoid attack. While indirect impacts by invasive species are difficult to detect and measure, we need more studies at the community level to understand how multiple and indirect interactions with invasive species impact the evolution of native species.

# References

Allendorf, F.W., Leary, R.F., Spruell, P., Wenburg, J.K., 2001. The problems with hybrids: setting conservation guidelines. Trends Ecol. Evol. 16, 613–622.

Andres, J.A., Thampy, P.R., Mathieson, M.T., Loye, J., Zalucki, M.P., Dingle, H., Carroll, S.P., 2013. Hybridization and adaptation to introduced balloon vines in an Australian soapberry bug. Mol. Ecol. 22, 6116–6130.

Bais, H.P., Vepachedu, R., Gilroy, S., Callaway, R.M., Vivanco, J.M., 2003. Allelopathy and exotic plants: from genes to invasion. Science 301, 1377–1380.

Barnes, W.J., Cottam, G., 1974. Some autecological studies of the *Lonicera* × *bella* complex. Ecology 55, 40–50.

Baron, E., Richirt, J., Villoutreix, R., Amsellem, L., Roux, F., 2015. The genetics of intra- and interspecific competitive response and effect in a local population of an annual plant species. Funct. Ecol. 29, 1361–1370.

Barrett, L.G., Bever, J.D., Bissett, A., Thrall, P.H., 2015. Partner diversity and identity impacts on plant productivity in *Acacia*–rhizobial interactions. J. Ecol. 103, 130–142.

Berthon, K., 2015. How do native species respond to invaders? Mechanistic and trait-based perspectives. Biol. Invasions 17, 2199–2211.

Blackburn, T.M., Bellard, C., Ricciardi, A., 2019. Alien versus native species as drivers of recent extinctions. Front. Ecol. Environ. 17, 203–207.

Bourke, P., Magnan, P., Rodríguez, M.A., 1999. Phenotypic responses of lacustrine brook charr in relation to the intensity of interspecific competition. Evol. Ecol. 13, 19–31.

CABI—Invasive Species Compendium, 2021. Datasheet: *Procambarus clarkii*. Downloaded from https://www.cabi.org/isc/datasheet/67878.

Callaway, R.M., Aschehoug, E.T., 2000. Invasive plants versus their new and old neighbors: a mechanism for exotic invasion. Science 290, 521–523.

Callaway, R.M., Ridenour, W.M., 2004. Novel weapons: a biochemically based hypothesis for invasive success and the evolution of increased competitive ability. Front. Ecol. Environ. 2, 436–443.

Callaway, R.M., Ridenour, W.M., Laboski, T., Weir, T., Vivanco, J.M., 2005. Natural selection for resistance to the allelopathic effects of invasive plants. J. Ecol. 93, 576–583.

Carroll, S.P., 2007. Natives adapting to invasive species: ecology, genes, and the sustainability of conservation. Ecol. Res. 22, 892–901.

Carroll, S.P., 2011. Conciliation biology: the eco-evolutionary management of permanently invaded biotic systems. Evol. Appl. 4, 184–199.

Carroll, S.P., Boyd, C., 1992. Host race radiation in the soapberry bug—natural history with the history. Evolution 46, 1052–1069.

Carroll, S.P., Loye, J.E., 2012. Soapberry bug (Hemiptera: Rhopalidae: Serinethinae) native and introduced host plants: biogeographic background of anthropogenic evolution. Ann. Entomol. Soc. Am. 105, 671–684.

Carroll, S.P., Loye, J.E., Dingle, H., Mathieson, M., Famula, T.R., Zalucki, M.P., 2005. And the beak shall inherit—evolution in response to invasion. Ecol. Lett. 8, 944–951.

Conti, L., Block, S., Parepa, M., Münkemüller, T., Thuiller, W., Acosta, A.T.R., van Kleunen, M., Dullinger, S., Essl, F., Dullinger, I., Moser, D., Klonner, G., Bossdorf, O., Carboni, M., 2018. Functional trait differences and trait plasticity mediate biotic resistance to potential plant invaders. J. Ecol. 106, 1607–1620.

Cooper Jr., W.E., Peréz-Mellado, V., 2012. Historical influence of predation pressure on escape by *Podarcis* lizards in the Balearic Islands. Biol. J. Linn. Soc. 107, 254–268.

Cox, G.W., 2004. Alien Species and Evolution. Island Press, Washington, DC.

Daly, J.W., Myers, C.W., Whittaker, N., 1987. Further classification of skin alkaloids from Neotropical poison frogs (Dendrobatidae) with a general survey of toxic/noxious substances in the amphibia. Toxicon 25, 1023–1095.

Doherty, T.S., Glen, A.S., Nimmo, D.G., Ritchie, E.G., Dickman, C.R., 2016. Invasive predators and global biodiversity loss. Proc. Natl. Acad. Sci. U. S. A. 113, 11261–11265.

Edwards, J.R., Lailvaux, S.P., 2012. Display behavior and habitat use in single and mixed populations of *Anolis carolinensis* and *Anolis sagrei* lizards. Ethology 118, 494–502.

Ficetola, G.F., Siesa, M.E., Manenti, R., Bottoni, L., De Bernardi, F., Padoa-Schioppa, E., 2011. Early assessment of the impact of alien species: differential consequences of an invasive crayfish on adult and larval amphibians. Divers. Distrib. 17, 1141–1151.

Fletcher, R.A., Callaway, R.M., Atwater, D.Z., 2016. An exotic invasive plant selects for increased tolerance, but not competitive suppression, in a native grass. Oecologia 181, 499–505.

Foster, J.D., Ellis, A.G., Foxcroft, L.C., Carroll, S.P., Le Roux, J.J., 2019. The potential evolutionary impact of invasive balloon vines on native soapberry bugs in South Africa. NeoBiota 49, 19–35.

Fowler, A.C., Eadie, J.M., Engilis, A., 2009. Identification of endangered Hawaiian ducks (*Anas wyvilliana*), introduced north American mallards (*A. platyrhynchos*) and their hybrids using multilocus genotypes. Conserv. Genet. 10, 1747–1758.

Gibson, A., Nelson, C.R., Atwater, D.Z., 2018. Response of bluebunch wheatgrass to invasion: differences in competitive ability among invader-experienced and invader-naïve populations. Funct. Ecol. 32, 1857–1866.

Gilbert, B., Levine, J.M., 2013. Plant invasions and extinction debts. Proc. Natl. Acad. Sci. U. S. A. 110, 1744–1749.

Goergen, E.M., Leger, E.A., Espeland, E.K., 2011. Native perennial grasses show evolutionary response to *Bromus tectorum* (Cheatgrass) invasion. PLoS One 6, e18145.

Gray, A.J., Marshall, D.F., Raybould, A.F., 1991. A century of evolution in *Spartina anglica*. Adv. Ecol. Res. 21, 1–62.

Gribben, P.E., Wright, J.T., O'Connor, W.A., Doblin, M.A., Eyre, B., Steinberg, P.D., 2009. Reduced performance of native infauna following recruitment to a habitat-forming invasive marine alga. Oecologia 158, 733–745.

Guay, P.J., Tracey, J.P., 2009. Feral Mallards: a risk for hybridisation with wild Pacific black ducks in Australia? Vic. Nat. 126, 87–91.

Hayes, R.A., Crossland, M.R., Hagman, M., Capon, R.J., Shine, R., 2009. Ontogenetic variation in the chemical defenses of cane toads (*Bufo marinus*): toxin profiles and effects on predators. J. Chem. Ecol. 35, 391–399.

Heger, T., Bernard-Verdier, M., Gessler, A., Greenwood, A.D., Grossart, H.P., Hilker, M., Keinath, S., Kowarik, I., Kueffer, C., Marquard, E., Müller, J., Niemeier, S., Onandia, G., Petermann, J.S., Rillig, M.C., Rödel, M.-O., Saul, W.-C., Schittko, C., Tockner, K., Joshi, J., Jeschke, J.M., 2019. Towards an integrative, eco-evolutionary understanding of ecological novelty: studying and communicating inter-linked effects of global change. Bioscience 69, 888–899.

Heleno, R.H., Olesen, J.M., Nogales, M., Vargas, P., Traveset, A., 2013. Seed dispersal networks in the Galápagos and the consequences of alien plant invasions. Proc. R. Soc. B Biol. Sci. 280, 1750.

Horn, K., Parker, I.M., Malek, W., Rodríguez-Echeverría, S., Parker, M.A., 2014. Disparate origins of *Bradyrhizobium* symbionts for invasive populations of *Cytisus scoparius* (Leguminosae) in North America. FEMS Microbiol. Ecol. 89, 89–98.

Huang, F., Lankau, R., Peng, S., 2018. Coexistence via coevolution driven by reduced allelochemical effects and increased tolerance to competition between invasive and native plants. New Phytol. 218, 357–369.

Inderjit, Weiner, J., 2001. Plant allelochemical interference or soil chemical ecology? Perspectives in Plant Ecology. Evol. Syst. 4, 3–12.

Inderjit, van der Putten, W.H., 2010. Impacts of soil microbial communities on exotic plant invasions. Trends Ecol. Evol. 25, 512–519.

Kelly, E., Phillips, B.L., 2019. Targeted gene flow and rapid adaptation in an endangered marsupial. Conserv. Biol. 33, 112–121.

Kulikova, I.V., Zhuravlev, Y.N., McCracken, K.G., 2004. Asymmetric hybridisation and sex-biased gene flow between eastern spot-billed ducks (*Anas zonorhyncha*) and mallards (*A. platyrhynchos*) in the Russian Far East. Auk 121, 930–949.

Lankau, R.A., 2012. Coevolution between invasive and native plants driven by chemical composition and soil biota. Proc. Natl. Acad. Sci. U. S. A. 109, 11240–11245.

Lankau, R.A., Nuzzo, V., Spyreas, G., Davis, A.S., 2009. Evolutionary limits ameliorate the negative impact of an invasive plant. Proc. Natl. Acad. Sci. U. S. A. 106, 15362–15367.

Le Roux, J.J., 2020. Molecular ecology of plant–microbial interactions during invasions: progress and challenges. In: Traveset, A., Richardson, D.M. (Eds.), Plant Invasions: The Role of Biotic Interactions. Invasive Species Series, CABI, UK, pp. 340–362.

Le Roux, J.J., Mavengere, N., Ellis, A.G., 2016. The structure of legume–rhizobium interaction networks and their response to tree invasions. AoB Plants 8, plw038.

Le Roux, J.J., Hui, C., Keet, J.-.H., Ellis, A.G., 2017. Co-introduction vs ecological fitting as pathways to the establishment of effective mutualisms during biological invasions. New Phytol. 215, 1354–1360.

Le Roux, J.J., Hui, C., Castillo, M.L., Iriondo, J.M., Keet, J.-.H., Khapugin, A., Médail, A.F., Rejmánek, M., Theron, G., Yannelli, F.A., Hirsch, H., 2019. Recent anthropogenic plant extinctions differ in biodiversity hotspots and coldspots. Curr. Biol. 29, 2912–2918.

Le Roux, J.J., Clusella-Trullas, S., Mokotjomela, T.M., Mairal, M., Richardson, D.M., Skein, L., Wilson, J.R., Weyl, O.L.F., Geerts, S., 2020. Biotic interactions as mediators of biological invasions: insights from South Africa. In: van Wilgen, B.W., Measey, J., Richardson, D.M., Wilson, J.R., Zengeya, T.A. (Eds.), Biological Invasions in South Africa. Springer, Berlin, pp. 387–427.

Leger, E.A., 2008. The adaptive value of remnant native plants in invaded communities: an example from the great basin. Ecol. Appl. 18, 1226–1235.

Lemaire, B., Van Cauwenberghe, J., Chimphango, S., Stirton, C., Honnay, O., Smets, E., Muasya, A.M., 2015. Recombination and horizontal transfer of nodulation and ACC deaminase (acdS) genes within alpha and Betaproteobacteria nodulating legumes of the cape fynbos biome. FEMS Microbiol. Ecol. 91, fiv118.

Lips, K.R., 2016. Overview of chytrid emergence and impacts on amphibians. Philos. Trans. R. Soc. B 371, 20150465.

Llewelyn, J., Schwarzkopf, L., Alford, R., Shine, R., 2010. Something different for dinner? Responses of a native Australian predator (the keelback snake) to an invasive prey species (the cane toad). Biol. Invasions 12, 1045–1051.

Llewelyn, J., Bell, K., Schwarzkopf, L., Alford, R.A., Shine, R., 2012. Ontogenetic shifts in a prey's chemical defence influence feeding responses of a snake predator. Oecologia 169, 965–973.

Llewelyn, J., Choyce, N.C., Phillips, B.L., Webb, J.K., Pearson, D.J., Schwarzkopf, L., Shine, R., 2018. Behavioural responses of an Australian colubrid snake (*Dendrelaphis punctulatus*) to a novel toxic prey item (the cane toad *Rhinella marina*). Biol. Invasions 20, 2507–2516.

Lo Parrino, E., Ficetola, G.F., Manenti, R., Falaschi, M., 2020. Thirty years of invasion: the distribution of the invasive crayfish *Procambarus clarkii* in Italy. Biogeographia 35, 43–50.

Long, J.L., 1981. Introduced Birds of the World: The Worldwide History, Distribution and Influence of Birds Introduced to New Environments. David & Charles, London.

Losos, J.B., 2009. Lizards in an Evolutionary Tree: Ecology and Adaptive Radiation of Anoles. University of California Press, Berkeley, US.

Mack, R.N., 2003. Plant naturalizations and invasions in the eastern United States: 1634–1860. Ann. Mo. Bot. Gard. 90, 77–90.

Mandeville, E.G., Parchman, T.L., McDonald, D.B., Buerkle, C.A., 2015. Highly variable reproductive isolation among pairs of *Catostomus* species. Mol. Ecol. 24, 1856–1872.

Mank, J.E., Carlson, J.E., Brittingham, M.C., 2004. A century of hybridisation: decreasing genetic distance between American black ducks and mallards. Conserv. Genet. 5, 395–403.

McDonald, D.B., Parchman, T.L., Bower, M.R., Hubert, W.A., Rahel, F.J., 2008. An introduced and a native vertebrate hybridize to form a genetic bridge to a second native species. Proc. Natl. Acad. Sci. U. S. A. 105, 10837–10842.

McKinnon, J.G., Gribben, P.E., Davis, A.R., Jolley, D.F., Wright, J.T., 2009. Differences in soft-sediment macro- benthic assemblages invaded by *Caulerpa taxifolia* compared to uninvaded habitats. Mar. Ecol. Prog. Ser. 380, 59–71.

Mealor, B.A., Hild, A.L., 2007. Post-invasion evolution of native plant populations: a test of biological resilience. Oikos 116, 1493–1500.

Melotto, A., Manenti, R., Ficetola, G.F., 2020. Rapid adaptation to invasive predators overwhelms natural gradients of intraspecific variation. Nat. Commun. 11, 3608.

Ndlovu, J., Richardson, D.M., Wilson, J.R., Le Roux, J.J., 2013. Co-invasion of South African ecosystems by an Australian legume and its rhizobial symbionts. J. Biogeogr. 40, 1240–1251.

Oduor, A.M.O., 2021. Native plant species show evolutionary responses to invasion by *Parthenium hysterophorus* in an African savanna. New Phytol. https://doi.org/10.1111/nph.17574. In press.

Olden, J.D., Poff, N.L.R., Douglas, M.R., Douglas, M.E., Fausch, K.D., 2004. Ecological and evolutionary consequences of biotic homogenization. Trends Ecol. Evol. 19, 18–24.

Parker, M.A., 2001. Mutualism as a constraint on invasion success for legumes and rhizobia. Divers. Distrib. 7, 125–136.

Perret, X., Staehelin, C., Broughton, W.J., 2000. Molecular basis of symbiotic promiscuity. Microbiol. Mol. Biol. Rev. 64, 180–201.

Phillips, B.L., Shine, R., 2004. Adapting to an invasive species: toxic cane toads induce morphological change in Australian snakes. Proc. Natl. Acad. Sci. U. S. A. 101, 17150–17155.

Phillips, B.L., Brown, G.P., Shine, R., 2003. Assessing the potential impact of cane toads on Australian snakes. Conserv. Biol. 17, 1738–1747.

Regnier, C., Fontaine, B., Bouchet, P., 2009. Not knowing, not recording, not listing: numerous unnoticed mollusk extinctions. Conserv. Biol. 23, 1214–1221.

Reznick, D.N., Losos, J., Travis, J., 2019. From low to high gear: there has been a paradigm shift in our understanding of evolution. Ecol. Lett. 22, 233–244.

Rhymer, J.M., Simberloff, D., 1996. Extinction by hybridization and introgression. Annu. Rev. Ecol. Syst. 27, 83–109.

Rhymer, J.M., Williams, M.J., Braun, M.J., 1994. Mitochondrial analysis of gene flow between New Zealand mallards (*Anas platyrhynchos*) and Grey ducks (*A. superciliosa*). Auk 111, 970–978.

Richardson, D.M., Allsopp, N., D'Antonio, C.M., Milton, S.J., Rejmánek, M., 2000. Plant invasions—the role of mutualisms. Biol. Rev. 75, 65–93.

Riera, N., Traveset, A., García, O., 2002. Breakage of mutualisms by alien species: the case of *Cneorum tricoccon* L. in the Balearic Islands (western Mediterranean Sea). J. Biogeogr. 29, 713–719.

Rogel, M.A., Ormeno-Orrillo, E., Martinez-Romero, E.M., 2011. Symbiovars in rhizobia reflect bacterial adaptation to legumes. Syst. Appl. Microbiol. 34, 96–104.

Saul, W.-C., Jeschke, J.M., 2015. Eco-evolutionary experience in novel species interactions. Ecol. Lett. 18, 236–245.

Saul, W.-C., Jeschke, J.M., Heger, T., 2013. The role of eco-evolutionary experience in invasion success. NeoBiota 17, 57–74.

Schettino, L.R., Losos, J.B., Hertz, P.E., de Queiroz, K., Chamizo, A.R., Leal, M., González, V.R., 2010. The anoles of Soroa: aspects of their ecological relationships. Breviora 520, 1–22.

Schulte, U., Veith, M., Hochkirch, A., 2012. Rapid genetic assimilation of native wall lizard populations (*Podarcis muralis*) through extensive hybridisation with introduced lineages. Mol. Ecol. 21, 4313–4326.

Schwarz, D., Matta, B.M., Shakir-Botteri, N.L., McPheron, B.A., 2005. Host shift to an invasive plant triggers rapid animal hybrid speciation. Nature 436, 546–549.

Schwarz, D., Shoemaker, K.D., Botteri, N.L., McPheron, B.A., 2007. A novel preference for an invasive plant as a mechanism for animal hybrid speciation. Evolution 61, 245–256.

Seebens, H., Bacher, S., Blackburn, T.M., Capinha, C., Dawson, W., Dullinger, S., Genovesi, P., Hulme, P.E., van Kleunen, M., Kühn, I., Jeschke, J.M., Lenzner, B., Liebhold, A.M., Pattison, Z., Pergl, J., Pyšek, P., Winter, M., Essl, F., 2020. Projecting the continental accumulation of alien species through to 2050. Glob. Chang. Biol. 27, 970–982.

Semchenko, M., Lepik, A., Abakumova, M., Zobel, K., 2018. Different sets of belowground traits predict the ability of plant species to suppress and tolerate their competitors. Plant Soil 424, 157–169.

Shine, R., 2010. The ecological impact of invasive cane toads (*Bufo marinus*) in Australia. Q. Rev. Biol. 85, 253–291.

Sidorovich, V., Kruuk, H., Macdonald, D.W., 1999. Body size and interactions between European and American mink (*Mustela lutreola* and *M. vison*) in Eastern Europe. J. Zool. 248, 521–527.

Singer, M.C., Parmesan, C., 2010. Phenological asynchrony between herbivorous insects and their hosts: signal of climate change or pre-existing adaptive strategy? Philos. Trans. R. Soc. B 365, 3161–3176.

Singer, M.C., Parmesan, C., 2018. Lethal trap created by adaptive evolutionary response to an exotic resource. Nature 557, 238–241.

Singer, M.C., Thomas, C.D., Parmesan, C., 1993. Rapid human-induced evolution of insect—host associations. Nature 366, 681–683.

Souty-Grosset, C., Anastacio, P., Aquiloni, L., Banha, F., Choquer, J., Chucholl, C., Tricarico, E., 2016. The red swamp crayfish *Procambarus clarkii* in Europe: impacts on aquatic ecosystems and human well-being. Limnologica 58, 78–96.

Stephens, K., Measey, J., Reynolds, C., Le Roux, J.J., 2020. Occurrence and extent of hybridisation between the invasive mallard duck and native yellow-billed duck in South Africa. Biol. Invasions 22, 693–707.

Stotz, G.C., Gianoli, E., Cahill Jr., J.F., 2016. Spatial pattern of invasion and the evolutionary responses of native plant species. Evol. Appl. 9, 939–951.

Strauss, S.Y., Lau, J.A., Carroll, S.P., 2006. Evolutionary responses of natives to introduced species: what do introductions tell us about natural communities? Ecol. Lett. 9, 357–374.

Stuart, Y.E., Campbell, T.S., Hohenlohe, P.A., Reynolds, R.G., Revell, L.J., Losos, J.B., 2014. Rapid evolution of a native species following invasion by a congener. Science 346, 463–466.

Sullivan, J.T., Patrick, H.N., Lowther, W.L., Scott, D.B., Ronson, C.W., 1995. Nodulating strains of *rhizobium loti* arise through chromosomal symbiotic gene transfer in the environment. Proc. Natl. Acad. Sci. U. S. A. 92, 8985–8989.

Thomas, C.D., 2015. Rapid acceleration of plant speciation during the anthropocene. Trends Ecol. Evol. 30, 448–455.

Thomas, C.D., Ng, D., Singer, M.C., Mallet, J.L.B., Parmesan, C., Billington, H.L., 1987. Incorporation of a European weed into the diet of a North American herbivore. Evolution 41, 892–901.

Todesco, M., Pascual, M.A., Owens, G.L., Ostevik, K.L., Moyers, B.T., Hübner, S., Heredia, S.M., Hahn, M.A., Caseys, C., Bock, D.G., Rieseberg, L.H., 2016. Hybridisation and extinction. Evol. Appl. 9, 892–908.

Traveset, A., 1995. Seed dispersal of *Cneorum tricoccon* by lizards and mammals in the Balearic archipelago. Acta Oecol. 16, 171–178.

Traveset, A., Richardson, D.M., 2006. Biological invasions as disruptors of plant reproductive mutualisms. Trends Ecol. Evol. 21, 208–216.

Traveset, A., Richardson, D.M., 2011. Mutualisms—key drivers of invasions… key casualties of invasions. In: Richardson, D.M. (Ed.), Fifty Years of Invasion Ecology. The Legacy of Charles Elton. Wiley-Blackwell, Oxford, pp. 143–160.

Traveset, A., Escribano-Avila, G., Gómez, J.M., Valido, A., 2019. Conflicting selection on *Cneorum tricoccon* (Rutaceae) seed size caused by native and alien seed dispersers. Evolution 73, 2204–2215.

Vizentin-Bugoni, J., Tarwater, C.E., Foster, J.T., Drake, D.R., Gleditsch, J.M., Hruska, A.-M., Kelley, J.P., Sperry, J.H., 2019. Structure, spatial dynamics, and stability of novel mutualistic networks. Science 364, 78–82.

Wang, P., Stieglitz, T., Zhou, D.W., Cahill Jr., J.F., 2010. Are competitive effect and response two sides of the same coin, or fundamentally different? Funct. Ecol. 24, 196–207.

Warrington, S., Ellis, A., Novoa, A., Wandrag, E.M., Hulme, P.E., Duncan, R.P., Valentine, A., Le Roux, J.J., 2019. Cointroductions of Australian acacias and their rhizobial mutualists in the southern hemisphere. J. Biogeogr. 46, 1519–1531.

Wei, G., Chen, W., Zhu, W., Chen, C., Young, J.P.W., Bontemps, C., 2009. Invasive *Robinia pseudoacacia* in China is nodulated by *Mesorhizobium* and *Sinorhizobium* species that share similar nodulation genes with native American symbionts. FEMS Microbiol. Ecol. 68, 320–328.

Wolf, D.E., Takebayashi, N., Rieseberg, L.H., 2001. Predicting the risk of extinction through hybridisation. Conserv. Biol. 15, 1039–1053.

Wright, J.T., Gribben, P.E., Byers, J.E., Monro, K., 2012. Invasive ecosystem engineer selects for different phenotypes of an associated native species. Ecology 93, 1262–1268.

Zhang, Z., Liu, Y., Yuan, L., Weber, E., van Kleunen, M., 2021. Effect of allelopathy on plant performance: a meta-analysis. Ecol. Lett. 24, 348–362.

# CHAPTER 8

# Invasion genetics: Molecular genetic insights into the spatial and temporal dynamics of biological invasions

## Contents

| | | |
|---|---|---|
| **8.1** | Introduction | 160 |
| **8.2** | The genetics of invasive species | 165 |
| | **8.2.1** Resolving uncertain and problematic taxonomies | 165 |
| | **8.2.2** Inferring invasion histories | 166 |
| | **8.2.3** Comparing apples with apples | 168 |
| | **8.2.4** The genetics of colonisation | 169 |
| | **8.2.5** The genetics of dispersal | 171 |
| **8.3** | Genetic signatures of adaptation | 174 |
| | **8.3.1** Insights from trait data and neutral genetic makers | 174 |
| | **8.3.2** Insights from genomics and transcriptomics | 175 |
| | **8.3.3** Whole-genome sequencing analyses of invasive species | 177 |
| **8.4** | Concluding remarks and future directions | 178 |
| | References | 179 |

## Abstract

This chapter provides an overview of the field of invasion genetics: the use of population genetics and genomics, phylogeography, and phylogenetics to better understand the ecology and evolutionary biology of invasive species. Continuous advances in molecular biology have made genetic tools central in addressing applied (e.g. invasive species management) as well fundamental (e.g. the evolution of invasiveness) questions in invasion biology. This chapter summarises how genetic studies have helped us to solve taxonomic uncertainties, introduction histories, and to grasp the evolutionary dynamics of invasive species. Case studies are provided to illustrate the contribution of invasion genetics to our understanding of how demographic processes and natural selection influence exotic species during different stages of invasion. The chapter also discusses the emerging role of whole-genome sequencing and transcriptomics in understanding invasive species biology. Food for thought for future research is provided.

*The Evolutionary Ecology of Invasive Species*
https://doi.org/10.1016/B978-0-12-818378-6.00007-3

**Keywords:** Gene flow, Genomics, Introduction history, Invasion genetics, Molecular systematics, Next-generation sequencing, Phylogeography, Propagule pressure, $Q_{ST}$-$F_{ST}$ comparisons, Transcriptomics

*Successful establishment of populations in new locations thus likely always results in novel selection pressures on the population, hence genetic change and adaptation. Genetic correlations, new genetic variation introduced by hybridization, founder effects ... must all contribute to the speed and direction of evolution. The challenge for students of the genetics of colonizing species will be to work out how they are integrated and their relative importance.*

**Price and Sol (2008)**

## 8.1 Introduction

It is hard to believe that *The Genetics of Colonizing Species* was published only 56 years ago (Baker and Stebbins, 1965). Some have argued that this seminal publication marked the dawn of 'invasion genetics' as an independent research field (see Box 8.1; Barrett, 2015). *The Genetics of Colonizing Species*

---

**BOX 8.1 A brief overview of key molecular innovations that contributed to the development of *invasion genetics* as an independent research discipline.**

Many have argued that the publication of Baker and Stebbins (1965) *The Genetics of Colonizing Species* marked the birth of 'invasion genetics' (e.g. Barrett, 2015). Since then, the field has exploded, especially over the last two and a half decades. This, in part, reflects the leaps and bounds made in molecular technologies and analytical approaches (Fig. 8.1). Numerous excellent reviews on the molecular approaches routinely used in invasion genetics are available (e.g. Hufbauer, 2004; Le Roux and Wieczorek, 2009; Lawson Handley et al., 2011; Chown et al., 2015; Rius et al., 2015; Sherpa and Després, 2021) and only key developments are outlined here.

Shortly after the publication of *The Genetics of Colonizing Species*, the discovery of protein electrophoresis in 1966 marked the start of the allozyme 'revolution'. Allozymes are variants of the same enzyme that differ structurally and are encoded by different alleles of the same gene. The ease of separating these variants using electrophoresis led to a surge in population genetic studies (e.g. see Hamrick et al., 1979), but these rarely involved invasive species. For instance, it was only about a decade after their discovery that the first allozyme studies on invasive species appeared (e.g. Wu and Jain, 1978).

---

*Continued*

**BOX 8.1  A brief overview of key molecular innovations that contributed to the development of *invasion genetics* as an independent research discipline—cont'd**

A major innovation in population genetics was the first description of genetic variation in the mitochondrial genome (mtDNA) (Avise et al., 1979a, 1979b). Because of its maternal inheritance and general lack of recombination, mtDNA provide unique perspectives on population genetic variation and structure (Allendorf, 2017). Unlike nuclear genomic regions, genealogies based on mtDNA are not 'mixed up' or 'diluted', meaning that their spatial structure is explicit and easy to quantify, in what has become known as the field of phylogeography (Avise, 2000). In conjunction with nuclear genetic information, phylogeographic studies based on mtDNA sequencing data have made important contributions to invasion biology, especially in disentangling the introduction histories of invasive species and identifying hybridisation (see Section 8.2.2; Le Roux and Wieczorek, 2009). For instance, Darling et al. (2008) investigated the global introduction history of the European shore crab (*Carcinus maenas*) based on mitochondrial and nuclear genetic data. These authors found different invasive populations to vary greatly in genetic diversity and that no correlation exists between genetic diversity and invasiveness. For example, bottlenecked populations in the United States have expanded their range over 2000 km, while genetically highly diverse populations in Cape Town, South Africa, remain restricted to two harbours (Darling et al., 2008; Mabin et al., 2017).

Various 'fragment-based' approaches for estimating genetic diversity (e.g. amplified fragment length polymorphisms, restriction fragment length polymorphisms, etc.; see Le Roux and Wieczorek, 2009 for more details) have emerged since the 1980s, but it was the discovery of microsatellites in the early 1990s that made them markers of choice in population genetic analyses, including in invasive species (Goldstein and Pollock, 1997; Fig. 8.1). Microsatellite markers provided three major advantages over their predecessors. First, large numbers of loci could be screened in virtually any organism. Second, the use of PCR amplification meant that microsatellites could be genotyped without destructive sampling and from a variety of biological materials like faeces, gut contents, pollen grains, etc. Lastly, unlike previous fragment markers, microsatellites are co-dominant and, therefore, can be included in analyses based on allele frequencies.

Analyses based on microsatellite loci provided biologists with opportunities to infer aspects of introduction histories, such as the occurrence of genetic bottlenecks (e.g. Tsutsui et al., 2000; Hardesty et al., 2012), multiple introductions (e.g. Genton et al., 2005; Vicente et al., 2021), and genetic admixture (e.g. Thompson et al., 2015). These markers also

*Continued*

---

**BOX 8.1 A brief overview of key molecular innovations that contributed to the development of *invasion genetics* as an independent research discipline—cont'd**

provided new ways to understand landscape-scale dispersal (e.g. Berthouly-Salazar et al., 2013; Medley et al., 2015) and to determine the sources of invasive populations (e.g. Valade et al., 2009; Ascunce et al., 2011). The increased ease of developing species-specific microsatellite markers, together with advances in analytical and statistical methods, likely played an important role in the surge in population genetic studies on invasive species since the late 1990s (Fig. 8.1).

The advent of high-throughput next-generation sequencing (NGS) technologies led to the genomics era, providing tools to estimate genetic diversity and structure at unprecedented levels of detail, including in non-model organisms. For example, NGS genomic studies typically sample tens of thousands of variants across a genome (Morin et al., 2004). Population genomics is conceptually different from population genetics in that it involves the sampling of a mapped genome at high enough resolution to detect the effects of microevolutionary forces (i.e. selection, genetic drift, mutation, migration, gene flow, inbreeding) on particular genomic regions (Allendorf, 2017).

---

provided us with the first synthesis of the "evolutionary game" that exotic species have to play in order to deal "…not with some simple defined change in selective conditions, but with a whole new ecological system in which the species has to find a place for itself" (Baker and Stebbins, 1965, p. 1). Given their often-short generation times and rapid spread under novel environmental conditions, it is surprising that evolutionary biologists only much later caught up on the idea that invasions are "more informative than most laboratory experimental work" to study evolution (Baker and Stebbins, 1965, p. 1; also see Chapters 2 and 5; Mooney and Cleland, 2001; Reznick and Ghalambor, 2001; Lee, 2002; Moran and Alexander, 2014; Reznick et al., 2019). This has now changed. The genetics of invasive species, in particular, have received much research attention in recent years (Fig. 8.1).

It is evident that fundamental insights into biological invasions can be gained from genetic studies (Le Roux and Wieczorek, 2009; Bock et al., 2015; Chown et al., 2015; Chan and Briski, 2017; Sherpa and Després, 2021). Not only do we now appreciate the fact that many invaders evolve in their new ranges, but also that such evolution occurs in populations that are genetically dynamic over space and time. For example, genetic diversity

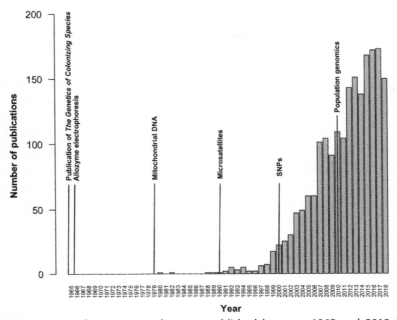

**Fig. 8.1** Number of peer-reviewed papers published between 1965 and 2018 that include the exact phrase 'invasion genetics', retrieved from a Google Scholar search on 30 September 2019. Major molecular genetic developments following the publication of *The Genetics of Colonizing Species* (Baker and Stebbins, 1965) are indicated on the timeline.

is low in some introduced populations (see Chapters 2 and 5; e.g. Frankham, 2005; Facon et al., 2011; Bodt et al., 2020; Havenga et al., 2020) while other introduced populations show little or no reduction in standing genetic diversity compared to their native sources (e.g. van Boheemen et al., 2017; Vicente et al., 2021). The latter often stem from multiple introductions that bring together divergent gene pools from different sources (Dlugosch and Parker, 2008; Cristescu, 2015; Vicente et al., 2021). Sometimes invasive populations themselves may act as 'bridgeheads', i.e. sources for additional introductions (e.g. Lombaert et al., 2010; Bertelsmeier et al., 2018; Blumenfeld et al., 2021).

How demographic processes influence genetic variation, and therefore the evolutionary capacity of introduced populations, was intensely debated in *The Genetics of Colonizing Species*. Standing genetic variation in introduced populations will influence how natural selection and non–adaptive processes affect microevolution (see Chapter 2, Sections 2.2.1–2.2.3 for a detailed discussion). On the one hand, the effects of genetic drift will be strong in small bottlenecked populations while, on the other, large and genetically diverse populations will be more buffered against drift and will have high adaptive capacities (Estoup and Guillemaud, 2010). It is not always clear whether the link between invasiveness and multiple introductions reflects the effects of density dependence (e.g. Allee effects) or high adaptability stemming from high genetic diversity. The benefits of high genetic diversity are, however, intuitive. In addition to multiple introductions, interspecific hybridisation may 'rescue' inbred populations of exotic species (Ellstrand and Schieren-beck, 2000; Schierenbeck and Ellstrand, 2009; Mesgaran et al., 2016). The phenotypic effects of hybridisation are often dramatic and have been repeatedly linked to invasion success (also see Chapter 2, Section 2.2.2; Chapter 4, Section 4.7; te Beest et al., 2012; Hovick and Whitney, 2014; Glisson and Larkin, 2021).

Many of the early genetic studies on invasive species focussed on the impacts of introduction histories on neutral genetic diversity (Tsutsui et al., 2001; Genton et al., 2005; Le Roux and Wieczorek, 2009; Hardesty et al., 2012). Later studies used genetic data to model more complex demo-graphic scenarios (e.g. Boissin et al., 2012; Benazzo et al., 2015; Cristescu, 2015, also see Sections 8.2.2–8.2.4). In conjunction with trait data, neutral genetic markers may also provide us with insights into rapid evolution and local adaptation during invasion via comparisons of population genetic structure based on trait (i.e. $Q_{ST}$) and genetic marker data (i.e. $F_{ST}$) (see Section 8.3.1 for more details; Merila and Crnokrak, 2001; McKay and Latta, 2002; Marchini and Arredondo, 2018; Sun and Roderick, 2019). More recently, genomic analyses of next-generation sequencing data not only provide detailed information on introduction histories of invasive spe-cies (Chown et al., 2015; Sherpa and Després, 2021), but also tools to iden-tify genomic regions that are under selection (e.g. Andrew et al., 2018; Stuart et al., 2021), ways to differentiate between the effects of genetic drift and selection on genomic variation (Oh et al., 2021), and means to assess the genetic basis of phenotypic plasticity (Rius et al., 2015). A concurrent explo-sion in transcriptome analyses have provided us with detailed information on gene expression in invasive populations (Chown et al., 2015; Rius et al., 2015; Guo et al., 2018). From a theoretical viewpoint, invasion genetics studies have offered many valuable insights into the nuts and bolts of con-temporary evolution, such as the speed at which it occurs and the effects of demographic processes on evolutionary change invasive species (Bock et al., 2015; Wu et al., 2019). From an applied perspective, the field has advanced our understanding of the attributes that makes some species invasive and others not (Lee, 2002; Lawson Handley et al., 2011; Bock et al., 2015; Uesugi et al., 2020). This chapter synthesises the contribution of invasion genetics studies to key areas in the broader field of invasion biology.

## 8.2 The genetics of invasive species

### 8.2.1 Resolving uncertain and problematic taxonomies

Sound taxonomy is not only vital to understand the ecological and evolutionary characteristics of successful invasive species but also for the effective management of invasive populations (Pyšek et al., 2013). However, identifying exotic species is not always straight forward. Poor or outdated taxonomy in the native range (e.g. Le Roux et al., 2006; Castillo et al., 2021), the introduction of multiple cryptic taxa (e.g. Stepien and Tumeo, 2006; Jarić et al., 2018; Castillo et al., 2021), post-introduction hybridisation (e.g. Stephens et al., 2020; Castillo et al., 2021), among others, can complicate accurate identification of invasive species (Le Roux and Wieczorek, 2009). From an applied perspective, taxonomic misidentification could, in the worst-case scenario, hamper management strategies, such as biological control (see Section 8.2.2; Smith et al., 2018; Jourdan et al., 2019; Castillo et al., 2021). Accurate identification of invasive species is also critical if we want to understand the historical (i.e. preadaptations) and contemporary (e.g. admixture, hybridisation, rapid adaptation) evolutionary processes that underlie their success.

What exactly defines a species has been the subject of intense debate (Wilkins, 2011) and a discussion on the merits of various species concepts falls outside the scope of this book. Irrespective of how we define species, the processes that generate or eliminate them are dynamic and ongoing. For example, the Port Jackson willow (*Acacia saligna*), a native from Western Australia, is highly invasive in many parts of the world (Thompson et al., 2015). In Australia, this tree is in the incipient stages of speciation and populations represent several subspecies (Maslin and McDonald, 2004; Thompson et al., 2012). Genetic studies have identified structured lineages that correspond to different subspecies in Australia (Millar et al., 2008, 2011; Thompson et al., 2012). Moreover, different subspecies appear to be adapted to different bioclimatic conditions in Australia, suggesting that it is possible to predict differences in their invasive distributions based on climate suitability in nonnative areas and the genetic identity of the tree (Thompson et al., 2011). Despite this, climatic niche models were unable to predict the known distribution of Port Jackson willow in South Africa (Thompson et al., 2011).

Genetic analyses found South African populations to be more distinct from all *A. saligna* subspecies than would predicted by demographic processes alone, such as strong founder events or multiple independent introductions followed by admixture (Thompson et al., 2012, 2015). In fact, South African populations have a close genetic relationship with planted acacias in Australia (Thompson et al., 2012). Thompson et al. (2012) also found deep phylogenetic divergence between South African populations and Port Jackson willow subspecies, leading these authors to conclude that invasive populations likely arose from introgressive hybridisation. In the absence of this crucial piece of information, the apparent niche differentiation between South African and Australian populations could have been interpreted as post-introduction evolution and a niche shift.

As we discussed throughout this book, hybridisation often precedes successful invasion (see Chapter 2, Section 2.2.2; Schierenbeck and Ellstrand, 2009; Ellstrand and Schierenbeck, 2000; te Beest et al., 2012; Hovick and Whitney, 2014). Hybridisation between invasive and native species represents a significant threat to native biodiversity through genetic swamping and the purging of locally adapted native genotypes (Allendorf et al., 2001; Olden et al., 2004). Molecular genetic analyses are particularly useful for detecting hybridisation and introgression, especially as it is often difficult to identify hybrids and their backcrosses based on morphology alone (Allendorf et al., 2001; Devillard et al., 2014).

As we discussed in Chapter 7 (Section 7.3), the Mallard Duck (*Anas platyrhynchos*) is the poster child for the evolutionary impacts on natives via introgressive hybridisation. Invasive Mallards readily hybridise with several closely related *Anas* species (Mank et al., 2004; Rhymer et al., 1994; Fowler et al., 2009; Stephens et al., 2020). For example, in South Africa, hybridisation between Mallards and the native yellow-billed duck (*A. undulata*) has led to genetic introgression (Stephens et al., 2020). Stephens et al. (2020) also found geographically isolated yellow-billed duck populations (as far as 1000 km apart) to have frequent migration between them, which may result in an intriguing situation whereby hybrid genotypes can disperse over vast distances, causing the introgression of Mallard genes into far-off yellow-billed duck populations, even when invasive Mallards are absent in these populations (Stephens et al., 2020).

## 8.2.2 Inferring invasion histories

The introduction histories of invasive species have enjoyed much research attention (Estoup and Guillemaud, 2010; Cristescu, 2015; Sherpa and

Després, 2021). Historic observational data (e.g. presence/absence records) have been used to estimate invasion routes (e.g. Suarez et al., 2001; Hui et al., 2012) and, in some instances, to trace contemporary range expansions (e.g. see Tatem et al., 2006; Hui et al., 2012; Ceschin et al., 2018). However, such data are often scant or incomplete (Estoup and Guillemaud, 2010) and cannot provide information on the native sources of invasive populations. Many invasive species also have large native range distributions and may stem from complex introduction scenarios (e.g. Hirsch et al., 2021), which limits the use of observational data to understand introduction histories. For example, when using such data it would be impossible to distinguish between a single introduction event of high propagule numbers and multiple introduction events of moderate propagule size (Le Roux et al., 2011; Lombaert et al., 2010).

Genetic studies can provide detailed information on the sources (e.g. Ascunce et al., 2011; Javal et al., 2019), propagule pressure (e.g. Genton et al., 2005; Janáč et al., 2017; Brandes et al., 2019), and invasion routes (e.g. Cristescu et al., 2001; Javal et al., 2019; Sherpa et al., 2019) of invasive species. While a positive link between genetic diversity and invasion success is not always apparent, we do know that many widespread invaders stem from multiple independent introductions (Dlugosch and Parker, 2008; Brandes et al., 2019; Vicente et al., 2021). Multiple introductions not only increase genetic diversity but also boost population numbers and sizes, and therefore buffer them against the effects of demographic processes and environmental stochasticity (Vicente et al., 2021). For example, genetic analysis of St. John's wort (*Hypericum perforatum*) found multiple independent introductions from across the species' native range to have founded invasive populations in the United States (Maron et al., 2004). Subsequent admixture resulted in substantial genetic variation within invasive populations. Between 12 and 15 generations later, invasive populations evolved increased growth performance and reproduction (Maron et al., 2004). Detailed knowledge on introduction histories, and the demographic processes that underlie them, can also aid management, e.g. by identifying the most likely routes and origins of propagules or their transport vectors (Hampton et al., 2004; Le Roux and Wieczorek, 2009; Chown et al., 2015).

Biological control reunites invasive species with their specialist natural enemies (also see Chapter 9, Section 9.2) and is currently considered the most cost-effective and sustainable management option against widespread invaders (Messing and Wright, 2006; van Wilgen et al., 2012). Knowledge of the geographic source(s) and genetic diversity of invasive populations may be crucial for successful biocontrol (Le Roux and Wieczorek, 2009; Smith et al., 2018). Identifying host-specific biocontrol agents is technically challenging, time consuming, and expensive. Species with large native ranges are also expected to show strong population genetic structure due to local adaptation, including to their co-evolved specialist enemies. Several authors have therefore argued that failure to correctly identify the native source(s) of invasive populations may hamper biocontrol efforts (Rosen, 1986; Le Roux and Wieczorek, 2009; Smith et al., 2018). For example, the native range distribution of Old-World climbing fern (*Lygodium microphyllum*) includes parts of tropical Africa, Asia, Australia, and some western Pacific islands (CABI, 2019a). This fern is an aggressive invasive species in southern Florida in the United States. Phylogeographic analysis showed that native populations are highly structured (Goolsby et al., 2006), information that was critical for selection of the best-fit genotype of the eriophyid mite biocontrol agent, *Floracarus perrepae*. A native population of the fern in Queensland, Australia, was found to be the exact genetic match of invasive populations in Florida and, unsurprisingly, mites from Queensland performed considerably better on Florida ferns than mites collected from other native range regions (Goolsby et al., 2006). This serves as a powerful example of the importance of knowing the sources of invasive populations when exploring and testing biological control agents (also see Jourdan et al., 2019; Paterson et al., 2019).

### 8.2.3 Comparing apples with apples

Some of the most compelling evidence for rapid evolution in invasive species comes from reciprocal transplant experiments in the invasive range (e.g. Colautti and Barrett, 2013; Medley et al., 2019; Castillo et al., 2021; Simón-Porcar et al., 2021). These studies aim to test whether local genotypes have higher fitness than 'foreign' genotypes at particular locations, and therefore whether populations are locally adapted. On the other hand, many evolutionary studies test hypotheses framed around predictions of trait differentiation between invasive and native populations (Colautti and Lau, 2015; Hodgins et al., 2018).

Knowing the native sources of invasive species allows comparisons between 'ancestral lineages' and their invasive 'descendants'. This information is critical when studying rapid evolution since shifts in trait values in invasive populations can often result from stochastic processes rather than selection. For instance, the subsampling of genetic diversity present in a source population, followed by serial founder events and genetic bottlenecks in the invaded range, can lead to dramatic changes in trait variation in invasive populations (Keller and Taylor, 2008). Knowledge on the source(s) of invasive populations is also important to account for strong spatial or clinal structure in the native range. For example, Barker et al. (2017) inferred the invasion history of the yellow starthistle (*Centaurea solstitialis*) in the Americas. This species has a large native range that spans large parts of Eurasia. Barker et al.'s (2017) genomic analyses indicated that the species first reached South America and that this region acted as a bridgehead for North American invasions. Trait comparisons between invasive populations and their western European sources also suggest that yellow starthistle in the Americas evolved larger size on at least two independent occasions (Dlugosch et al., 2015). However, plants from the Americas, eastern Europe, and Asia are of similar size (Barker et al., 2017). This study illustrates why identifying the sources of invasive species is crucial when studying rapid evolution. For example, if Barker et al. (2017) for some reason only included yellow starthistle populations from the Americas and Asia in their experiments, they would have reached a completely different conclusion.

## 8.2.4 The genetics of colonisation

Many introduced populations experience some reduction in genetic diversity due to founder events and genetic bottlenecks (Dlugosch and Parker, 2008). This may decrease their evolutionary capacities and cause inbreeding depression (see Chapter 2, Section 2.2.1). Low genetic diversity will also exacerbate the effects of stochastic processes like genetic drift on trait variation, which may negatively affect the performance of introduced populations (i.e. drift load, Lynch et al., 1995a, 1995b). Many authors have argued that these genetic constraints may explain why some exotic species experience long lag phases before becoming invasive or fail to establish (Sakai et al., 2001). Unfortunately, we do not have genetic data for most failed introductions, making it hard to make robust inferences about the presumed link between genetic diversity and establishment success. We do, however, know of instances where inbred invasive populations survive and undergo adaptation to novel conditions, in what has been referred to as a 'genetic paradox' (Allendorf and Lundquist, 2003; Frankham, 2005; Hufbauer, 2008).

Invasion genetics enable researchers to estimate the amount of standing genetic diversity that is present in introduced populations. Many studies have compared genetic diversity between invasive populations and their sources as proxies for propagule pressure (Dlugosch and Parker, 2008; Estoup and Guillemaud, 2010; Cristescu, 2015). While the general assumption that invasive populations will have lower genetic variation than their sources seems to hold up, a survey of 80 species across various taxonomic groups found that, on average, invasive populations only have about 15% less allelic variation and 18.7% lower expected heterozygosity than native range populations (Dlugosch and Parker, 2008). Some of the world's most widespread invasive species have no genetic diversity. For instance, a single clone of fountain grass (*Pennisetum setaceum*) has successfully invaded vast areas in Australia, Namibia, South Africa, and the United States (Poulin et al., 2005; Le Roux et al., 2007). Conversely, some invasive populations have higher levels of genetic variation than native range populations. For example, a recent analysis found Australian *Acacia* species to often harbour higher genetic diversity in their invasive than native populations (Vicente et al., 2021; Fig. 8.2). Multiple introductions from different sources, followed by extensive admixture, probably led to these elevated levels of genetic diversity, a phenomenon that appears to be commonplace (Novak and Mack, 1993; Kolbe et al., 2004, 2007; Genton et al., 2005; Dlugosch and Parker, 2008; Le Roux et al., 2011; Vicente et al., 2021).

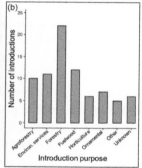
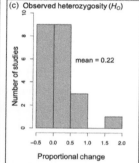

**Fig. 8.2** (a) Australian acacias, such as *Acacia longifolia*, are considered some of the world's worst invasive species. (b) *Acacia* species have been introduced around the world for a variety of reasons, often under circumstances that called for high propagule pressure. (c) A recent review of 20 case studies (of 11 *Acacia* species) found genetic diversity in invasive populations to often be as high, or higher, than in native range populations (see Vicente et al., 2021 for more details). For instance, proportional change in observed heterozygosity ($H_O$), estimated as ($H_O$ in invaded range $- H_O$ in native source)/$H_O$ in native source, suggests that invasive populations have, on average, 22% higher heterozygosity than native range populations. *(Drawn from data reported in Vicente, S., Máguas, C., Richardson, D.M., Trindade, H., Wilson, J.R.U., Le Roux, J.J., 2021. Highly diverse and highly successful: invasive Australian acacias have not experienced genetic bottlenecks globally. Ann. Bot. 23, 149-157; Photo credit (panel a): Johannes Le Roux.)*

It remains difficult to separate the contributions of the number of individuals versus the number of genetic variants to the positive link often observed between propagule pressure and invasion success (Simberloff, 2009). While high propagule pressure may assist the establishment of exotic species by buffering their populations against environmental and demographic stochasticity, it also often underlies high genetic diversity (Bock et al., 2015; Vicente et al., 2021). It is conceivable that propagule pressure also increases the chances of introducing preadapted genotypes. For example, as we discussed earlier, subspecies of the Port Jackson willow (*Acacia saligna*) occupy distinct climate zones in native Australia (see Section 8.2.1). Introductions of this species around the world have been characterised by high propagule pressure, often involving numerous subspecies and extensive admixture (Thompson et al., 2012, 2015). This may have enhanced the introduction of diverse genotypes that are preadapted to different climate conditions. Others have also suggested that high propagule pressure may enhance complementarity, whereby the facilitation between different genotypes, or trait differences between them, may enhance the performance of genetically diverse populations compared with populations that lack genetic diversity (Crawford and Whitney, 2010; Forsman, 2014; Le Roux et al., 2014; Bock et al., 2015).

## 8.2.5 The genetics of dispersal

Attributes of dispersal, measured by traditional methods such as mark-recapture studies, have been valuable in understanding biological invasions, but only tell half the story. Dispersal dynamics, e.g. the frequency of long-distance versus short-distance dispersal, may not be informative about absolute range expansions, especially in heterogeneous environments. This is because successful range expansion ultimately depends on the probability of propagules reaching suitable habitats to establish viable populations. Similarly, habitat suitability alone may not tell us much about range expansions if dispersal prohibits propagules from reaching suitable habitats (Le Roux et al., 2010). Therefore measures of both dispersal abilities and habitat suitability are needed to obtain realistic predictions of range expansion (Kearney and Porter, 2009; Cruzan and Hendrickson, 2020). Measuring dispersal directly in the field, especially rare long-distance events, remains challenging. Tracking the movement and spatial distribution of genes (i.e. gene flow) has provided us with sophisticated ways to indirectly infer dispersal (Ouborg et al., 1999; Le Roux and Wieczorek, 2009; Cruzan and

Hendrickson, 2020; Mairal et al., 2021). For example, if neutral population genetic structure of a species is known, then statistical (demographic) models can be applied to determine the amount of dispersal, strength of genetic drift, etc., needed to explain the observed genetic structure. Applying these approaches, along with some basic knowledge on organismal biology, can provide remarkably accurate insights into dispersal. For example, Bélouard et al. (2019) studied landscape-scale dispersal of the invasive American red swamp crayfish (*Procambarus clarkii*) among waterbodies in northwest France. Despite their study area only covering 15 km$^2$ these authors found crayfish populations to be highly structured. They also identified a clear pattern of colonisation, linked to restricted habitat connectivity, whereby isolated waterbodies harboured crayfish populations with low effective population sizes and low dispersal rates (Bélouard et al., 2019).

Genetic studies on invasive species have also provided answers to important theoretical questions in dispersal ecology. For example, Reid's paradox states that many species expand their ranges faster than would be predicted by dispersal rates alone (Clark et al., 1998). This paradox was partly solved by theory showing that dispersal rates can increase towards the leading edge of expanding populations (Le Corre et al., 1997), a notion supported by studies on invasive species such as cane toads (see Chapter 2, Section 2.1; Travis and Dytham, 2002; Phillips et al., 2010; Huang et al., 2015). Another theoretical puzzle is how genetic diversity is maintained during rapid range expansions (Bialozyt et al., 2006). Standard models of diffusive dispersal (i.e. short-distance dispersal) predict an erosion of genetic diversity at the leading edge due to successive founder events (Klopfstein et al., 2006; Travis et al., 2007) whereas models of stratified dispersal (i.e. short-distance dispersal with occasional long-distance dispersal) show a radically different result: an increase in genetic diversity at the leading edge (Bialozyt et al., 2006). This certainly seems to be the case for some invasive species. For example, Berthouly-Salazar et al. (2013) found an increase in both dispersal rates and genetic diversity towards the leading edge of invasive European starlings (*Sturnus vulgaris*) in South Africa (Fig. 8.3). These authors inferred dispersal dynamics by estimating the rate of change (i.e. slope) in the relationship between population genetic differentiation and geographic distance (i.e. isolation by distance, IBD) along the species' route of invasion. A low rate of change, or no change, in IBD indicates that dispersal is constant, whereas a high rate of change would indicate that dispersal rates are changing. These authors found a decrease in IBD slopes of populations to correlate strongly with

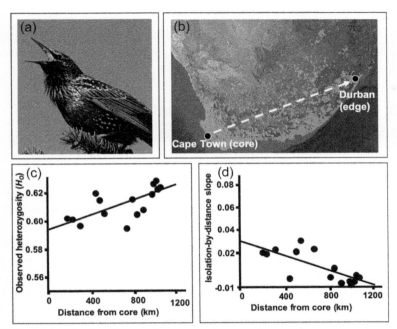

**Fig. 8.3** (a) European starlings (*Sturnus vulgaris*) were introduced into Cape Town, South Africa, in the late 19th century. Since then the species has spread towards the north-eastern parts of the country as indicated by the *white arrow* in panel b. Berthouly-Salazar et al. (2013) found genetic diversity to increase towards the leading edge of the species (panel c) as a consequence of accelerated dispersal along the route of expansion (shown by decrease in IBD slope along invasion route; panel d). *(Photograph (panel a): sourced from Wikimedia Commons under the Creative Commons Share-Alike License (CC-BY-SA 3.0). Original author: Michael Graf.)*

their distance from the core, suggesting that higher dispersal evolved along the invasion route of this species (Fig. 8.3).

Accelerated dispersal in European starlings also appears to be linked to unfavourable environmental conditions, such as changes in rainfall (Berthouly-Salazar et al., 2013). These inferences fit historical occurrence records of the species in South Africa well. For example, based on observational data, Hui et al. (2012) estimated that starlings dispersed between 4 and 8 km/year in high winter rainfall areas close to Cape Town, the initial introduction site of starlings in South Africa. They also found dispersal to increase more than two-fold, to between 8 and 32 km/year, 150 km east of Cape Town, an area where winter precipitation dramatically changes (Hui et al., 2012). More recently, Stuart et al. (2021) found evidence for genomic regions that are under strong selection and that are related to responses in precipitation in invasive European starlings populations in Australia. Taken together, these results suggest that increased dispersal in European starlings in South Africa may have evolved to allow escape from suboptimal habitat conditions (Hui et al., 2012; Berthouly-Salazar et al., 2013).

## 8.3 Genetic signatures of adaptation

### 8.3.1 Insights from trait data and neutral genetic makers

In conjunction with trait data, population genetic structure based on neutral markers have provided us with important insights into rapid evolution during invasion. Neutral genetic structure (i.e. $F_{ST}$; Wright, 1965) reflects population differentiation stemming from stochastic and demographic processes. Conversely, population differentiation in quantitative traits (i.e. $Q_{ST}$; Merila and Crnokrak, 2001) reflects structure due to both neutral processes and past selection. Significant differences between $Q_{ST}$ and $F_{ST}$ estimates for the same populations therefore provide insights into the processes underlying trait differentiation (Leinonen et al., 2013). When $Q_{ST} = F_{ST}$, then traits variation is neutral and a product of genetic drift alone. When $Q_{ST} > F_{ST}$, then trait differentiation is higher than expected by chance alone. Under these circumstances, the magnitude of the difference between $Q_{ST}$ and $F_{ST}$ reflects the degree of local adaptation, i.e. divergent selection (Leinonen et al., 2013). Lastly, when $Q_{ST} < F_{ST}$ then stabilising selection is presumed, i.e. the same phenotype is favoured in different environments (Leinonen et al., 2013).

One of the criticisms of $Q_{ST}$–$F_{ST}$ comparisons is that they cannot tell us whether populations are locally adapted to current environmental conditions as they provide snapshots of *past* evolution (Keller and Taylor, 2008; Whitlock, 2008). Colautti and Lau (2015) provide an in-depth discussion on various other shortcomings of $Q_{ST}$–$F_{ST}$ comparisons in the context of studying rapid evolution in invasive species. These authors rightly argue that interactions between introduction history (i.e. the number sources and independent introductions) and population genetic structure in the native range may lead to $Q_{ST}$–$F_{ST}$ inferences that are difficult to interpret (Colautti and Lau, 2015). Despite these drawbacks, $Q_{ST}$–$F_{ST}$ comparisons can, with careful experimental design, provide evidence to support rapid evolution (Leinonen et al., 2013).

Marchini and Arredondo (2018) used $Q_{ST}$–$F_{ST}$ comparisons to study rapid evolution in invasive populations of slender false brome (*Brachypodium sylvaticum*). Since summers in the grass' invasive range in

Oregon in the United States are more arid than in the native range in Europe, North Africa, and the Middle East, these authors tested for rapid evolution in traits related to drought tolerance. They also used population genetic analyses to infer the introduction history of the species into the United States. This allowed them to weigh individual $Q_{ST}$ estimates relative to their source population's contribution to invasive populations (Marchini and Arredondo, 2018). Therefore false inferences of selection from trait values of native range populations that were not sources of invasive populations were minimised (Colautti and Lau, 2015). Marchini and Arredondo (2018) found divergence between invasive and native slender false brome populations in some of the traits, but not across the entire invasive range of the species, suggesting that these traits underwent evolution shortly after the species' introduction into the United States and prior to it becoming widespread.

## 8.3.2 Insights from genomics and transcriptomics

The advent of NGS technologies opened the door for studies to investigate genetic diversity and gene expression at the genome level with astonishing levels of detail. For example, by comparing genetic diversity across the genome (or transcriptome) of a species it is also possible to identify loci that display atypical patterns of variation (Andrews and Luikart, 2014). Loci with significantly higher or lower genetic differentiation than that expected under neutrality are considered 'outliers' and, thus, to be under selection. Even though tens of thousands of polymorphic loci are typically screened when using NGS approaches, less than 5% are usually outliers (Hoffmann and Willi, 2008). A major drawback of outlier analyses is that they seldom tell us about the selection pressures that cause these genetic patterns. However, as more annotated whole-genome sequences become available (see Section 8.3.3), so do reliable gene annotations which, in turn, can provide important insights into the nature of selection experienced by invasive species.

The large geographic ranges of many invasive species mean that different populations are likely exposed to different environmental conditions. This makes invasions ideal systems to study evolution at the genome or transcriptome level. For example, Yadav et al. (2019) investigated how environmental factors shaped gene flow and adaptation among widespread populations of the pest grasshopper, *Phaulacridium vittatum*, in Australia. These authors sampled grasshoppers across a 900-km latitudinal gradient in eastern

Australia. While their genomic analyses identified high levels of connectivity, these authors also identified annual temperature as an important barrier to gene flow in this species (Yadav et al., 2019). This, they argued, is consistent with the idea that dispersal in ectotherms is generally influenced by temperature (Yadav et al., 2019). Out of the more than 11,000 single nucleotide polymorphisms (SNPs) they screened, Yadav et al. (2019) identified only 242 outlier loci. They were able to annotate four of these loci, and even these limited data confirmed temperature as an important selection pressure. For example, one locus was annotated as Uridine diphosphate-glucuronosyl transferase, an enzyme known for its role in UV shielding and pigmentation in insects (Yadav et al., 2019).

As we discussed throughout this book, invasive cane toads (*Rhinella marina*) in Australia evolved higher dispersal abilities in leading-edge populations via spatial sorting (Phillips et al., 2006, 2010). Others have since documented heritable differences in other traits between core and leading-edge populations, including in behaviour (Gruber et al., 2017), thermal dependency of dispersal (Kosmala et al., 2018), and immunity (Brown et al., 2015). However, the contribution of spatial sorting, genetic drift, or selection to these differences remains largely unknown. Numerous studies have reported low levels of standing genetic diversity in Australian cane toad populations (e.g. Slade and Moritz, 1998; Leblois et al., 2000). To better understand the processes driving trait diversification in cane toads, Selechnik et al. (2019) compared transcriptome variation in spleen and brain tissues of toads sampled from across Australia, French Guiana (the native range), and Hawaii (the bridgehead that founded invasive Australian populations). Variation in spleen transcriptomes identified three major population genetic clusters: (1) a native French Guianan cluster, (2) a cluster including Hawaiian populations and 'old' Australian populations close to the original introduction site, and (3) a cluster consisting of Australian toads in leading-edge populations and from areas between core and leading-edge populations (i.e. 'intermediate toads'). Transcriptome data from brain tissue, excluding French Guianan populations, largely resembled the structure between Hawaiian and Australian populations based on spleen data. Genetic diversity estimates from brain transcriptomes also confirmed a reduction in diversity between the Hawaiian bridgehead/old Australian and the leading-edge/ 'intermediate' populations, especially at outlier loci. Hawaiian and core populations represent the source and initial introduction stages of the Australian invasion, respectively, and are found in areas where environmental conditions are similar to those in the toad's native range (Selechnik et al., 2019). Conversely, intermediate and leading-edge populations in Australia are found in areas that are markedly warmer and more arid than areas in the toad's native range. Selechnik et al. (2019) argued that reduced outlier locus variation in these invasive populations is a likely consequence of directional selection from harsh climate conditions, supported by the fact that these loci were frequently annotated as heat stress and dehydration resistance genes. Therefore, despite the known roles of genetic drift and spatial sorting in rapid evolution in cane toads (Shine et al., 2011), this study also highlights the possible role of selection.

### 8.3.3 Whole-genome sequencing analyses of invasive species

High-throughput NGS technologies have revolutionised the biological sciences. As NGS costs have gone down, publicly available whole-genome data have gone up (Ellegren, 2014). For example, the National Centre for Biotechnology Information (NCBI; http://www.ncbi.nlm.nih.gov) currently lists 60,399 complete or draft genome sequence assemblies for 16,271 eukaryotes, 326,685 prokaryotes, 42,580 viruses, 28,948 plasmids, and 19,051 organelles (accessed 26 April 2021). Most of these whole-genome sequences are from non-model organisms, including invasive species like cane toads (Edwards et al., 2018), Argentine ants (*Linepithema humile*; Smith et al., 2011), Fire ants (*Solenopsis invicta*; Wurm et al., 2011), castor bean (*Ricinus communis* Chan et al., 2010), and Mallard ducks (Huang et al., 2013).

Whole-genome sequences promise a more in-depth understanding of the genetic attributes of successful invasive species. They also provide a means to infer the processes underlying molecular evolution (neutral vs deterministic), their intensity (e.g. strength of bottlenecks or selection), the nature of selection (i.e. stabilising, directional or disruptive), and candidate markers to study population genomics and adaptation. The invasive Argentine ant provides a good example. The phenomenal invasion success of this ant (CABI, 2019b) has enjoyed a lot of ecological and evolutionary research attention across all levels of biological organisation, from genes to populations, and across various temporal and spatial scales (e.g. Brandt et al., 2009; Sagata and Lester, 2009; Hu et al., 2017; Cabrera et al., 2021). The ant's tolerance to a wide range of environmental conditions has contributed to its invasion success. In invasive ranges, Argentine ant workers of neighbouring nests also lack aggression towards each, leading to the formation of supercolonies that can span thousands of kilometres (Giraud et al., 2002). Supercolonies are absent in the native range and their formation in invaded regions has been linked to strong genetic bottlenecks (Tsutsui et al., 2000).

Given the low genetic diversity in supercolonies, biologists have struggled to understand how invasive Argentine ants deal with pathogens and parasites (Tsutsui et al., 2000). Smith et al. (2011) published the first draft genome for the Argentine ant, which provided some answers to this puzzle. These authors were able to annotate 202 genes involved in immune responses, a number much lower than previously reported from model insects like *Drosophila melanogaster*. These authors therefore hypothesised that hygienic behaviour and chemical defences likely play important roles in supressing pathogens and parasites in supercolonies. The generalist diet of Argentine ants may also expose them to a wide range of toxic compounds. Smith et al. (2011) identified 111 cytochrome P450 genes in the Argentine ant genome. Of these, 69 were from the so-called CYP3 clan P450s, at the time the highest number discovered in any insect genome. CYP3 P450s are involved in the oxidative detoxification of xenobiotics and their high abundance in the Argentine ant genome suggests that the ant is preadapted to many of the toxins it may encounter as a result of its broad diet or chemical control (Smith et al., 2011). This example illustrates how whole-genome sequencing can be informative about a species' traits and the processes that shaped them, and ultimately, how these influence invasion success.

## 8.4 Concluding remarks and future directions

The field of invasion genetics has made significant contributions over the last two decades to our understanding of the mechanisms underlying successful invasion and the dynamics of invasive species. The field has been instrumental in solving taxonomic uncertainties in invasive taxa, reconstructing invasion histories, and inferring how demographic processes and selection shape genetic and trait variation during invasion. There are many unanswered questions in invasion biology that genetic studies can help us answer (Bock et al., 2015). For instance, whether the link between propagule pressure and invasion success reflect genetic- or density-dependent constraints on population growth remains largely unknown. While some researchers have manipulated levels of the genetic diversity (e.g. Crawford and Whitney, 2010) and others levels of propagule pressure (e.g. Von Holle and Simberloff, 2005), we need experiments that combine these approaches to disentangle their relative contributions to establishment success and invasion (e.g. see Hovick and Whitney, 2019). Similarly, we lack convincing evidence for whether lag phases are the result of low population densities, the time needed to reshuffle standing genetic variation, or both. Here, genomic analyses of historical collections, such as museum or herbarium samples, in conjunction with estimates of contemporary genetic diversity may be particularly useful (e.g. see Martin et al., 2014). Understanding the genetic architecture of invasiveness remains a contentious topic. Bock et al. (2015) suggested two approaches that may help us understand such complex genetic attributes. First, 'top-down' approaches could entail identifying heritable trait variation that differs between invasive and non-invasive genotypes and the loci and alleles that characterise them. This can be done using approaches routinely employed in crop genetics, e.g. candidate gene analyses and genome-wide association studies or quantitative trait mapping. Alternatively, a 'bottom-up' approach could entail the use of transcriptomic data to identify changes in gene expression or allele frequencies between invasive and non-invasive genotypes. Annotation of these loci can be used to make inferences about the traits likely involved and the selection pressures that influence them.

In conclusion, the field of invasion genetics is blossoming and, together with rapid advances in affordable technical and analytical approaches, is likely to continue doing so. Many unknowns remain in invasion biology and genomic studies hold great promise to answer some of these questions.

## References

Allendorf, F.W., 2017. Genetics and the conservation of natural populations: allozymes to genomes. Mol. Ecol. 26, 420–430.

Allendorf, F.W., Lundquist, L.L., 2003. Introduction: population biology, evolution, and control of invasive species. Conserv. Biol. 17, 24–30.

Allendorf, F.W., Leary, R.F., Spruell, P., Wenburg, J.K., 2001. The problems with hybrids: Setting conservation guidelines. Trends Ecol. Evol. 16, 613–622.

Andrew, S.C., Jensen, H., Hagen, I.J., Lundregan, S., Griffith, S.C., 2018. Signatures of genetic adaptation to extremely varied Australian environments in introduced European house sparrows. Mol. Ecol. 27, 4542–4555.

Andrews, K.R., Luikart, G., 2014. Recent novel approaches for population genomics data analysis. Mol. Ecol. 23, 1661–1667.

Ascunce, M.S., Yang, C.-.C., Oakey, J., Calcaterra, L., Wu, W.J., Shih, C.J., Goudet, J., Ross, K.G., Shoemaker, D., 2011. Global invasion history of the fire ant Solenopsis invicta. Science 331, 1066–1068.

Avise, J.C., 2000. Phylogeography: The History and Formation of Species. Harvard University Press, Cambridge, MA.

Avise, J.C., Giblin-Davidson, C., Laerm, J., Patton, J.C., Lansman, R.A., 1979a. Mitochondrial DNA clones and matriarchal phylogeny within and among geographic populations of the pocket gopher, Geomys pinetis. Proc. Natl. Acad. Sci. U. S. A. 76, 6694–6698.

Avise, J.C., Lansman, R.A., Shade, R.O., 1979b. The use of restriction endonucleases to measure mitochondrial DNA sequence relatedness in natural populations. I. Population structure and evolution in the genus *Peromyscus*. Genetics 92, 279–295.

Baker, H.G., Stebbins, G.L. (Eds.), 1965. The Genetics of Colonizing Species. Academic Press, New York.

Barker, B.S., Andonian, K., Swope, S.M., Dlugosch, K.M., 2017. Population genomic analyses reveal a history of range expansion and trait evolution across the native and invaded range of yellow starthistle (*Centaurea solstitialis*). Mol. Ecol. 26, 1131–1147.

Barrett, S.C.H., 2015. Foundations of invasion genetics: the Baker and Stebbins legacy. Mol. Ecol. 24, 1927–1941.

Bélouard, N., Paillisson, J.-.M., Oger, A., Besnard, A.-.L., Petit, E.J., 2019. Genetic drift during the spread phase of a biological invasion. Mol. Ecol. 28, 4375–4387.

Benazzo, A., Ghirotto, S., Vilaça, S., Hoban, S., 2015. Using ABC and microsatellite data to detect multiple introductions of invasive species from a single source. Heredity 115, 262–272.

Bertelsmeier, C., Ollier, S., Liebhold, A.M., Brockerhoff, E.G., Ward, D., Keller, L., 2018. Recurrent bridgehead effects accelerate global alien ant spread. Proc. Natl. Acad. Sci. U. S. A. 115, 5486–5491.

Berthouly-Salazar, C., Hui, C., Blackburn, T.M., Gaboriaud, C., Van Rensburg, B.J., Jansen van Vuuren, B., Le Roux, J.J., 2013. Long-distance dispersal maximizes evolutionary potential during rapid geographic range expansion. Mol. Ecol. 22, 5793–5804.

Bialozyt, R., Ziegenhagen, B., Petit, R.J., 2006. Contrasting effects of long distance seed dispersal on genetic diversity during range expansion. J. Evol. Biol. 19, 12–20.

Blumenfeld, A.J., Eyer, P.A., Husseneder, C., Mo, J., Johnson, L.N.L., Wang, C., Kenneth Grace, J., Chouvenc, T., Wang, S., Vargo, E.L., 2021. Bridgehead effect and multiple introductions shape the global invasion history of a termite. Commun. Biol. 4, 196.

Bock, D.G., Caseys, C., Cousens, R.D., Hahn, M.A., Heredia Sariel Hübner, S.M., Turner, K.A., Whitney, K.D., Rieseberg, L.H., 2015. What we still don't know about invasion genetics. Mol. Ecol. 24, 2277–2297.

Bodt, L.H., Rollins, L.A., Zichello, J.M., 2020. Contrasting mitochondrial diversity of European starlings (*Sturnus vulgaris*) across three invasive continental distributions. Ecol. Evol. 10, 10186–10195.

Boissin, E., Hurley, B., Wingfield, M.J., Vasaitis, R., Stenlid, J., Davis, C., de Groot, P., Ahumada, R., Carnegie, A., Goldarazena, A., Klasmer, P., Wermelinger, B., Slippers, B., 2012. Retracing the routes of introduction of invasive species: the case of the *Sirex noctilio* woodwasp. Mol. Ecol. 21, 5728–5744.

Brandes, U., Furevik, B.B., Nielsen, L.R., Kjær, E.D., Rosef, L., Fjellheim, S., 2019. Introduction history and population genetics of intracontinental scotch broom (*Cytisus scoparius*) invasion. Divers. Distrib. 25, 1773–1786.

Brandt, M., van Wilgenburg, E., Tsutsui, N.D., 2009. Global-scale analyses of chemical ecology and population genetics in the Argentine ant. Mol. Ecol. 18, 997–1005.

Brown, G.P., Phillips, B.L., Dubey, S., Shine, R., 2015. Invader immunology: invasion history alters immune system function in cane toads (*Rhinella marina*) in tropical Australia. Ecol. Lett. 18, 57–65.

CABI, 2019a. *Lygodium microphyllum*. Invasive Species Compendium. CABI, Wallingford, UK. https://www.cabi.org/isc/datasheet/110270. (Accessed August 2019).

CABI, 2019b. *Linepithema humile*. Invasive Species Compendium. CABI, Wallingford, UK. https://www.cabi.org/isc/datasheet/30839. (Accessed August 2019).

Cabrera, E., Rivas Fontan, I., Hoffmann, B.D., Josens, R., 2021. Laboratory and field insights into the dynamics and behavior of Argentine ants, *Linepithema humile*, feeding from hydrogels. Pest Manag. Sci. https://doi.org/10.1002/ps.6368.

Castillo, M.L., Schaffner, U., van Wilgen, B.W., Montaño, N.M., Bustamante, R.O., Cosacov, A., Mathese, M.J., Le Roux, J.J., 2021. Genetic insights into the globally invasive and taxonomically problematic tree genus *Prosopis*. AoB PLANTS 13, plaa069.

Ceschin, S., Abati, S., Ellwood, N.T.W., Zuccarello, V., 2018. Riding invasion waves: Spatial and temporal patterns of the invasive *Lemna minuta* from its arrival to its spread across Europe. Aquat. Bot. 150, 1–8.

Chan, F.T., Briski, E., 2017. An overview of recent research in marine biological invasions. Mar. Biol. 164, 121.

Chan, A.P., Crabtree, J., Zhao, Q., Lorenzi, H., Orvis, J., Puiu, D., Melake-Berhan, A., Jones, K.M., Redman, J., Chen, G., Cahoon, E.B., Gedil, M., Stanke, M., Haas, B.J., Wortman, J.R., Fraser-Liggett, C.M., Ravel, J., Rabinowicz, P.D., 2010. Draft genome sequence of the oilseed species *Ricinus communis*. Nat. Biotechnol. 28, 951–956.

Chown, S.L., Hodgins, K.A., Griffin, P.C., Oakeshott, J.G., Byrne, M., Hoffmann, A.A., 2015. Biological invasions, climate change and genomics. Evol. Appl. 8, 23–46.

Clark, J.S., Fastie, C., Hurtt, G., Jackson, S.T., Johnson, C., King, G.A., Lewis, M., Lynch, J., Pacala, S., Prentice, I.C., Schupp, E.W., Webb III, T., Wyckoff, P., 1998. Reid's paradox of rapid plant migration: dispersal theory and interpretation of paleoecological records. Bioscience 48, 13–24.

Colautti, R.I., Barrett, S.C.H., 2013. Rapid adaptation to climate facilitates range expansion of an invasive plant. Science 342, 364–366.

Colautti, R., Lau, J.A., 2015. Contemporary evolution during invasion: evidence for differentiation, natural selection, and local adaptation. Mol. Ecol. 24, 1999–2017.

Crawford, K.M., Whitney, K.D., 2010. Population genetic diversity influences colonization success. Mol. Ecol. 19, 1253–1263.

Cristescu, M., 2015. Genetic reconstructions of invasion history. Mol. Ecol. 24, 2212–2225.

Cristescu, M.E., Hebert, P.D.N., Witt, J.D.S., MacIsaac, H.J., Grigorovich, I.A., 2001. An invasion history for *Cercopagis pengoi* based on mitochondrial gene sequences. Limnol. Oceanogr. 46, 224–229.

Cruzan, M.B., Hendrickson, E.C., 2020. Landscape genetics of plants: challenges and opportunities. Plant Commun. 1, 100100.

Darling, J.A., Bagley, M.J., Roman, J., Tepolt, C.K., Geller, J.B., 2008. Genetic patterns across multiple introductions of the globally invasive crab genus *Carcinus*. Mol. Ecol. 17, 4992–5007.

Devillard, S., Jombart, T., Léger, F., Pontier, D., Say, L., Ruette, S., 2014. How reliable are morphological and anatomical characters to distinguish European wildcats, domestic cats and their hybrids in France? J. Zool. Syst. Evol. Res. 52, 154–162.

Dlugosch, K.M., Parker, I.M., 2008. Founding events in species invasion: genetic variation, adaptive evolution, and the role of multiple introductions. Mol. Ecol. 17, 431–449.

Dlugosch, K.M., Cang, F.A., Barker, B.S., Andonian, K., Swope, S.M., Rieseberg, L.H., 2015. Evolution of invasiveness through increased resource use in a vacant niche. Nat. Plants 1, 15066.

Edwards, R.J., Tuipulotu, D.E., Amos, T.G., O'Meally, D., Richardson, M.F., Russell, T.-L., Vallinoto, M., Carneiro, M., Ferrand, N., Wilkins, M.R., Sequeira, F., Rollins, L.A., Holmes, E.C., Shine, R., White, P.A., 2018. Draft genome assembly of the invasive cane toad, *Rhinella marina*. GigaScience 7, giy095.

Ellegren, H., 2014. Genome sequencing and population genomics in non-model organisms. Trends Ecol. Evol. 29, 51–63.

Ellstrand, N.C., Schierenbeck, K.A., 2000. Hybridization as a stimulus for the evolution of invasiveness in plants? Proc. Natl. Acad. Sci. U. S. A. 97, 7043–7050.

Estoup, A., Guillemaud, T., 2010. Reconstructing routes of invasion using genetic data: why, how and so what? Mol. Ecol. 19, 4113–4130.

Facon, B., Hufbauer, R.A., Tayeh, A., Loiseau, A., Lombaert, E., Vitalis, R., Guillemaud, T., Lundgren, J.G., Estoup, A., 2011. Inbreeding depression is purged in the invasive insect *Harmonia axyridis*. Curr. Biol. 21, 424–427.

Forsman, A., 2014. Effects of genotypic and phenotypic variation on establishment are important for conservation, invasion, and infection biology. Proc. Natl. Acad. Sci. U. S. A. 111, 302–307.

Fowler, A.C., Eadie, J.M., Engilis, A., 2009. Identification of endangered Hawaiian ducks (*Anas wyvilliana*), introduced North American mallards (*A. platyrhynchos*) and their hybrids using multilocus genotypes. Conserv. Genet. 10, 1747–1758.

Frankham, R., 2005. Invasion biology: resolving the genetic paradox in invasive species. Heredity 94, 385.

Genton, B., Shykoff, J.A., Giraud, T., 2005. High genetic diversity in French invasive populations of common ragweed *Ambrosia artemisiifolia* as a consequence of multiple sources of introduction. Mol. Ecol. 14, 4275–4285.

Giraud, T., Pedersen, J.S., Keller, L., 2002. Evolution of supercolonies: the Argentine ants of southern Europe. Proc. Natl. Acad. Sci. U. S. A. 99, 6075–6079.

Glisson, W.J., Larkin, D.J., 2021. Hybrid watermilfoil (*Myriophyllum spicatum* × *Myriophyllum sibiricum*) exhibits traits associated with greater invasiveness than its introduced and native parental taxa. Biol. Invasions. https://doi.org/10.1007/s10530-021-02514-7 (In Press).

Goldstein, D.B., Pollock, D.D., 1997. Launching microsatellites: a review of mutation processes and methods of phylogenetic inference. J. Hered. 88, 335–342.

Goolsby, J.A., De Barro, P.J., Makinson, J.R., Pemberton, R.W., Hartley, D.M., Frohlicj, D.R., 2006. Matching the origin of an invasive weed for the selection of a herbivore haplotype for a biological control programme. Mol. Ecol. 16, 287–297.

Gruber, J., Brown, G.P., Whiting, M.J., Shine, R., 2017. Is the behavioural divergence between range-core and range-edge populations of cane toads (*Rhinella marina*) due to evolutionary change or developmental plasticity? R. Soc. Open Sci. 4, 766.

Guo, W., Liu, Y., Ng, W.L., Liao, P.-C., Huang, B.-H., Li, W., Li, C., Shi, X., Huang, Y., 2018. Comparative transcriptome analysis of the invasive weed *Mikania micrantha* with its native congeners provides insights into genetic basis underlying successful invasion. BMC Genomics 19, 392.

Hampton, J.O., Spencer, P.B.S., Alpers, D.L., Twigg, L.E., Woolnough, A.P., Doust, J., Higgs, T., Pluske, J., 2004. Molecular techniques, wildlife management and the importance of genetic population structure and dispersal: a case study with feral pigs. J. Appl. Ecol. 41, 735–743.

Hamrick, J.L., Linhart, Y.B., Mitton, J.B., 1979. Relationships between life history characteristics and electrophoretically detectable genetic variation in plants. Annu. Rev. Ecol. Syst. 10, 173–200.

Hardesty, B.D., Le Roux, J.J., Rocha, O.J., Meyer, J.-Y., Westcott, D., Wieczorek, A.M., 2012. Getting here from there: testing the genetic paradigm underpinning introduction histories and invasion success. Divers. Distrib. 18, 147–157.

Havenga, M., Wingfield, B.D., Wingfield, M.J., Dreyer, L.L., Roets, F., Chen, S., Aylward, J., 2020. Low genetic diversity and strong geographic structure in introduced populations of the Eucalyptus foliar pathogen *Teratosphaeria destructans*. Plant Pathol. 69, 1540–1550.

Hirsch, H., Richardson, D.M., Pauchard, A., Le Roux, J.J., 2021. Genetic analyses reveal complex introduction histories for the invasive tree *Acacia dealbata* Link around the world. Divers. Distrib. 27, 360–376.

Hodgins, K.A., Bock, D.G., Rieseberg, L., 2018. Trait evolution in invasive species. Annu. Plant Rev. Online 1, 1–37.

Hoffmann, A.A., Willi, Y., 2008. Detecting genetic responses to environmental change. Nat. Rev. Genet. 9, 421–432.

Hovick, S.M., Whitney, K.D., 2014. Hybridisation is associated with increased fecundity and size in invasive taxa: meta-analytic support for the hybridisation-invasion hypothesis. Ecol. Lett. 17, 1464–1477.

Hovick, S.M., Whitney, K.D., 2019. Propagule pressure and genetic diversity enhance colonization by a ruderal species: a multi-generation field experiment. Ecol. Monogr. 89, e01368.

Hu, Y., Holway, D.A., Łukasik, P., Chau, L., Kay, A.D., LeBrun, E.G., Miller, K.A., Sanders, J.G., Suarez, A.V., Russell, J.A., 2017. By their own devices: invasive Argentine ants have shifted diet without clear aid from symbiotic microbes. Mol. Ecol. 26, 1608–1630.

Huang, Y., Li, Y., Burt, D.W., Chen, H., Zhang, Y., Qian, W., Kim, H., Gan, S., Zhao, Y., Li, J., Yi, K., Feng, H., Zhu, P., Li, B., Liu, Q., Fairley, S., Magor, K.E., Du, Z., Hu, X., Goodman, L., Li, N., 2013. The duck genome and transcriptome provide insight into an avian influenza virus reservoir species. Nat. Genet. 45, 776–783.

Huang, F., Peng, S., Chen, B., Liao, H., Huang, Q., Lin, Z., Liu, G., 2015. Rapid evolution of dispersal-related traits during range expansion of an invasive vine *Mikania micrantha*. Oikos 124, 1023–1030.

Hufbauer, R.A., 2004. Population genetics of invasions: can we link neutral markers to management? Weed Technol. 18, 1522–1527.

Hufbauer, R.A., 2008. Biological invasions: paradox lost and paradise gained. Curr. Biol. 18, R246–R247.

Hui, C., Roura-Pascual, N., Brotons, L., Robinson, R.A., Evans, K.L., 2012. Flexible dispersal strategies in native and non-native ranges: environmental quality and the "good–stay, bad–disperse" rule. Ecography 35, 1024–1032.

Janáč, M., Bryja, J., Ondračková, M., Mendel, J., Jurajda, P., 2017. Genetic structure of three invasive gobiid species along the Danube-Rhine invasion corridor: similar distributions, different histories. Aquat. Invasions 12, 551–564.

Jarić, I., Heger, T., Monzon, F.C., Jeschke, J.M., Kowarik, I., McConkey, K.R., Pyšek, P., Sagouis, A., Essl, F., 2018. Crypticity in biological invasions. Trends Ecol. Evol. 34, 291–302.

Javal, M., Lombaert, E., Tsykun, T., Courtin, C., Kerdelhué, C., Prospero, S., Roques, A., Roux, G., 2019. Deciphering the worldwide invasion of the Asian long-horned beetle: a recurrent invasion process from the native area together with a bridgehead effect. Mol. Ecol. 28, 951–967.

Jourdan, M., Thomann, T., Kriticos, D.J., Bon, M.-C., Sheppard, A., Baker, G.H., 2019. Sourcing effective biological control agents of conical snails, *Cochlicella acuta*, in Europe and north Africa for release in southern Australia. Biol. Control 134, 1–14.

Kearney, M.R., Porter, W.P., 2009. Mechanistic niche modelling: combining physiological and spatial data to predict species' ranges. Ecol. Lett. 12, 334–350.

Keller, S.R., Taylor, D.R., 2008. History, chance and adaptation during biological invasion: separating stochastic phenotypic evolution from response to selection. Ecol. Lett. 11, 852–866.

Klopfstein, S., Currat, M., Excoffier, L., 2006. The fate of mutations surfing on the wave of a range expansion. Mol. Biol. Evol. 23, 482–490.

Kolbe, J.J., Glor, R.E., Schettino, L.R., Lara, A.C., Larson, A., Losos, J.B., 2004. Genetic variation increases during biological invastion by a Cuban lizard. Nature 431, 177–181.

Kolbe, J.J., Glor, R.E., Schettino, L.R., Lara, A.C., Larson, A., Losos, J.B., 2007. Multiple sources, admixture, and genetic variation in introduced *Anolis* lizard populations. Conserv. Biol. 21, 1612–1625.

Kosmala, G.K., Brown, G.P., Christian, K.A., Hudson, C.M., Shine, R., 2018. The thermal dependency of locomotor performance evolves rapidly within an invasive species. Ecol. Evol. 8, 4403–4408.

Lawson Handley, L.J., Estoup, A., Evans, D.M., Thomas, C.E., Lombaert, E., Facon, B., Aebi, A., Roy, H.E., 2011. Ecological genetics of invasive alien species. BioControl 56, 409–428.

Le Corre, V., Machon, N., Petit, R.J., Kremer, A., 1997. Colonization with long-distance seed dispersal and genetic structure of maternally inherited genes in forest trees: a simulation study. Genet. Res. 69, 117–125.

Le Roux, J., Wieczorek, A.M., 2009. Molecular systematics and population genetics of biological invasions: towards a better understanding of invasive species management. Ann. Appl. Biol. 154, 1–17.

Le Roux, J.J., Wieczorek, A.M., Ramadan, M.M., Tran, C.T., 2006. Resolving the native provenance of invasive fireweed (*Senecio madagascariensis* Poir.) in the Hawaiian Islands as inferred from phylogenetic analysis. Divers. Distrib. 12, 694–702.

Le Roux, J.J., Wieczorek, A.M., Wright, M.G., Tran, C.T., 2007. Super-genotype, global monoclonality defies the odds of nature. PLoS One 2, e590.

Le Roux, J.J., Wieczorek, A.M., Tran, C.T., Vorsino, A.E., 2010. Disentangling the dynamics of invasive fireweed (*Senecio madagascariensis* Poir. species complex) in the Hawaiian Islands. Biol. Invasions 12, 2251–2264.

Le Roux, J.J., Brown, G.K., Byrne, M., Ndlovu, J., Richardson, D.M., Thompson, G.D., Wilson, J.R.U., 2011. Phylogeographic consequences of different introduction histories of invasive Australian *Acacia* species and *Paraserianthes lophantha* (Fabaceae) in South Africa. Divers. Distrib. 17, 861–871.

Le Roux, J.J., Blignaut, M., Gildenhuys, E., Mavengere, N., Berthouly-Salazar, C., 2014. The molecular ecology of biological invasions: what do we know about non-additive genotypic effects and invasion success? Biol. Invasions 16, 997–1001.

Leblois, R., Rousset, F., Tikel, D., Moritz, C., Estoup, A., 2000. Absence of evidence for isolation by distance in an expanding cane toad (*Bufo marinus*) population: an individual-based analysis of microsatellite genotypes. Mol. Ecol. 9, 1905–1909.

Lee, C.E., 2002. Evolutionary genetics of invasive species. Trends Ecol. Evol. 17, 386–391.

Leinonen, T., McCairns, R.J., O'Hara, R.B., Merilä, J., 2013. $Q_{ST}$–$F_{ST}$ comparisons: evolutionary and ecological insights from genomic heterogeneity. Nat. Rev. Genet. 14, 179–190.

Lombaert, E., Guillemaud, T., Cornuet, J.-M., Malausa, T., Facon, B., Estoup, A., 2010. Bridgehead effect in the worldwide invasion of the biocontrol harlequin ladybird. PLoS One 5, e9743.

Lynch, M., Conery, J., Bürger, R., 1995a. Mutation accumulation and the extinction of small populations. Am. Nat. 146, 489–518.

Lynch, M., Conery, J., Bürger, R., 1995b. Mutational meltdowns in sexual populations. Evolution 49, 1067–1080.

Mabin, C.A., Wilson, J.R.U., Le Roux, J.J., Robinson, T.B., 2017. Reassessing the invasion of South African waters by the European shore crab, *Carcinus maenas*. Afr. J. Mar. Sci. 39, 259–267.

Mairal, M., Chown, S.L., Shaw, J., Chala, D., Chau, J.H., Hui, C., Kalwij, J.M., Münzbergová, Z., Jansen van Vuuren, B., Le Roux, J.J., 2021. Human activity strongly influences genetic dynamics of the most widespread invasive plant in the sub-Antarctic. Mol. Ecol., 1–17. https://doi.org/10.1111/mec.16045. In press.

Mank, J.E., Carlson, J.E., Brittingham, M.C., 2004. A century of hybridisation: decreasing genetic distance between American black ducks and mallards. Conserv. Genet. 5, 395–403.

Marchini, G.L., Arredondo, T.M., Cruzan, M.B., 2018. Selective differentiation during the colonization and establishment of a newly invasive species. J. Evol. Biol. 31, 1689–1703.

Maron, J.L., Vila, M., Bommarco, R., Elmendorf, S., Beardsley, P., 2004. Rapid evolution of an invasive plant. Ecol. Monogr. 74, 261–280.

Martin, M.D., Zimmer, E.A., Olsen, M.T., Foote, A.D., Gilbert, M.T.P., Brush, G.S., 2014. Herbarium specimens reveal a historical shift in phylogeographic structure of common ragweed during native range disturbance. Mol. Ecol. 23, 1701–1716.

Maslin, B.R., McDonald, M.W., 2004. *Acacia* search. Evaluation of Acacia as a Woody Crop Option for Southern Australia. RIRDC Report 03/017, Rural Industries Research and Development Corporation, Canberra, Australia.

McKay, J.K., Latta, R.G., 2002. Adaptive population divergence: markers, QTL and traits. Trends Ecol. Evol. 17, 285–291.

Medley, K.A., Jenkins, D.G., Hoffman, E.A., 2015. Human-aided and natural dispersal drive gene flow across the range of an invasive mosquito. Mol. Ecol. 24, 284–295.

Medley, K.A., Westby, K.M., Jenkins, D.G., 2019. Rapid local adaptation to northern winters in the invasive Asian tiger mosquito *Aedes albopictus*: a moving target. J. Appl. Ecol. 56, 2518–2527.

Merila, J., Crnokrak, P., 2001. Comparison of genetic differentiation at marker loci and quantitative traits. J. Evol. Biol. 14, 892–903.

Mesgaran, M.B., Lewis, M.A., Ades, P.K., Donohue, K., Ohadi, S., Li, C., Cousens, R.D., 2016. Hybridization can facilitate species invasions, even without enhancing local adaptation. Proc. Natl. Acad. Sci. U. S. A. 113, 10210–10214.

Messing, R.H., Wright, M.G., 2006. Biological control of invasive species: solution or pollution? Front. Ecol. Environ. 4, 132–140.

Millar, M.A., Byrne, M., Nuberg, I., Sedgley, M., 2008. A rapid PCR-based diagnostic test for the identification of subspecies of *Acacia saligna*. Tree Genet. Genomes 4, 625–635.

Millar, M.A., Byrne, M., O'Sullivan, W., 2011. Defining entities in the *Acacia saligna* (Fabaceae) species complex using a population genetics approach. Aust. J. Bot. 59, 137–148.

Mooney, H.A., Cleland, E.E., 2001. The evolutionary impact of invasive species. Proc. Natl. Acad. Sci. U. S. A. 98, 5446–5451.

Moran, E.V., Alexander, J.M., 2014. Evolutionary responses to global change: lessons from invasive species. Ecol. Lett. 17, 637–649.

Morin, P.A., Luikart, G., Wayne, R.K., SNP Workshop Group, 2004. SNPs in ecology, evolution and conservation. Trends Ecol. Evol. 19, 208–216.

Novak, S.J., Mack, R.N., 1993. Genetic variation in *Bromus tectorum* (Poaceae): comparison between native and introduced populations. Heredity 71, 167–176.

Oh, K.P., Shiels, A.B., Shiels, L., Blondel, D.V., Campbell, K.J., Royden Saah, J., Lloyd, A.L., Thomas, P.Q., Gould, F., Abdo, Z., Godwin, J.R., Piaggio, A.J., 2021. Population genomics of invasive rodents on islands: Genetic consequences of colonization and prospects for localized synthetic gene drive. Evol. Appl. 00, 1–15.

Olden, J.D., Poff, N.L.R., Douglas, M.R., Fausch, K.D., 2004. Ecological and evolutionary consequences of biotic homogenization. Trends Ecol. Evol. 19, 18–24.

Ouborg, N.J., Piquot, Y., van Groenendael, J.M., 1999. Population genetics, molecular markers, and the study of dispersal in plants. J. Ecol. 87, 551–569.

Paterson, I.D., Coetzee, J.A., Weyl, P., Griffith, T.C., Voogt, N., Hill, M.P., 2019. Cryptic species of a water hyacinth biological control agent revealed in South Africa: host specificity, impact, and thermal tolerance. Entomol. Exp. Appl. 167, 682–691.

Phillips, B.L., Brown, G.P., Webb, J.K., Shine, R., 2006. Invasion and the evolution of speed toads. Nature 439, 803.

Phillips, B.L., Brown, G.P., Shine, R., 2010. Evolutionarily accelerated invasions: the rate of dispersal evolves upwards during the range advance of cane toads. J. Evol. Biol. 23, 2595–2601.

Poulin, J., Weller, S.G., Sakai, A.K., 2005. Genetic diversity does not affect the invasiveness of fountain grass (*Pennisetum setaceum*) in Arizona, California and Hawaii. Divers. Distrib. 11, 241–247.

Price, T.D., Sol, D., 2008. Introduction: genetics of colonizing species. Am. Nat. 172, S1–S3.

Pyšek, P., Hulme, P.E., Meyerson, L.A., Smith, G.F., Boatwright, J.S., Crouch, N.R., Figueiredo, E., Foxcroft, L.C., Jarošík, V., Richardson, D.M., Suda, J., Wilson, J.R.U., 2013. Hitting the right target: taxonomic challenges for, and of, plant invasions. AoB PLANTS 5, plt042.

Reznick, D.N., Ghalambor, C.K., 2001. The population ecology of contemporary adaptations: what empirical studies reveal about the conditions that promote adaptive evolution. Genetica 112, 183–198.

Reznick, D.N., Losos, J., Travis, J., 2019. From low to high gear: there has been a paradigm shift in our understanding of evolution. Ecol. Lett. 22, 233–244.

Rhymer, J.M., Williams, M.J., Braun, M.J., 1994. Mitochondrial analysis of gene flow between New Zealand Mallards (*Anas platyrhynchos*) and Grey Ducks (*A. superciliosa*). Auk 111, 970–978.

Rius, M., Bourne, S., Hornsby, H.G., Chapman, M.A., 2015. Applications of next-generation sequencing to the study of biological invasions. Curr. Zool. 61, 488–504.

Rosen, D., 1986. The role of taxonomy in effective biological control programs. Agric. Ecosyst. Environ. 15, 121–129.

Sagata, K., Lester, P.J., 2009. Behavioural plasticity associated with propagule size, resources, and the invasion success of the Argentine ant *Linepithema humile*. J. Appl. Ecol. 46, 19–27.

Sakai, A.K., Allendorf, F.W., Holt, J.S., Lodge, D.M., Molofsky, J., With, K.A., Baughman, S., Cabin, R.J., Cohen, J.E., Ellstrand, N.C., McCauley, D.E., O'Neil, P., Parker, I.M., Thompson, J.N., Weller, S.G., 2001. The population biology of invasive species. Annu. Rev. Ecol. Syst. 32, 305–332.

Schierenbeck, K.A., Ellstrand, N.C., 2009. Hybridization and the evolution of invasiveness in plants and other organisms. Biol. Invasions 11, 1093–1105.

Selechnik, D., Richardson, M.F., Shine, R., Devore, J., Ducatez, S., Rollins, L.A., 2019. Increased adaptive variation despite reduced overall genetic diversity in a rapidly adapting invader. Front. Genet. 10, 12–21.

Sherpa, S., Després, L., 2021. The evolutionary dynamics of biological invasions: a multi-approach perspective. Evol. Appl. https://doi.org/10.1111/eva.13215 (In Press).

Sherpa, S., Blum, M.G.B., Capblancq, T., Cumer, T., Rioux, D., Després, L., 2019. Unravelling the invasion history of the Asian tiger mosquito in Europe. Mol. Ecol. 28, 2360–2377.

Shine, R., Brown, G.P., Phillips, B.L., 2011. An evolutionary process that assembles phenotypes through space rather than through time. Proc. Natl. Acad. Sci. U. S. A. 108, 5708–5711.

Simberloff, D., 2009. The role of propagule pressure in biological invasions. Annu. Rev. Ecol. Evol. Syst. 40, 81–102.

Simón-Porcar, V.I., Silva, J.L., Vallejo-Marín, M., 2021. Rapid local adaptation in both sexual and asexual invasive populations of monkeyflowers (*Mimulus* spp.). Ann. Bot. 127, 655–668.

Slade, R.W., Moritz, C., 1998. Phylogeography of *Bufo marinus* from its natural and introduced ranges. Proc. R. Soc. Lond. B 265, 769–777.

Smith, C.D., Zimin, A., Holt, C., Abouheif, E., Benton, R., Cash, E., Croset, V., Currie, C. R., Elhaik, E., Elsik, C.G., Fave, M.J., Fernandes, V., Gadau, J., Gibson, J.D., Graur, D., Grubbs, K.J., Hagen, D.E., Helmkampf, M., Holley, J.A., Hu, H., Tsutsui, N.D., 2011. Draft genome of the globally widespread and invasive Argentine ant (*Linepithema humile*). Proc. Natl. Acad. Sci. U. S. A. 108, 5673–5678.

Smith, L., Cristofaro, M., Bon, M.C., De Biase, A., Petanović, R., Vidović, B., 2018. The importance of cryptic species and subspecific populations in classic biological control of weeds: a North American perspective. BioControl 63, 417–425.

Stephens, K., Measey, J., Reynolds, C., Le Roux, J.J., 2020. Occurrence and extent of hybridisation between the invasive Mallard Duck and native Yellow-billed Duck in South Africa. Biol. Invasions 22, 693–707.

Stepien, C.A., Tumeo, M.A., 2006. Invasion genetics of Ponto-Caspian gobies in the Great Lakes: a 'cryptic' species, absence of founder effects, and comparative risk analysis. Biol. Invasions 8, 61–78.

Stuart, K.C., Cardilini, A.P.A., Cassey, P., Richardson, M.F., Sherwin, W.B., Rollins, L.A., Sherman, C.H.D., 2021. Signatures of selection in a recent invasion reveal adaptive divergence in a highly vagile invasive species. Mol. Ecol. 30, 1419–1434.

Suarez, A.V., Holway, D.A., Case, T.J., 2001. Patterns of spread in biological invasions dominated by long-distance jump dispersal: insights from Argentine ants. Proc. Natl. Acad. Sci. U. S. A. 98, 1095–1100.

Sun, Y., Roderick, G.K., 2019. Rapid evolution of invasive traits facilitates the invasion of common ragweed, *Ambrosia artemisiifolia*. J. Ecol. 107, 2673–2687.

Tatem, A.J., Hay, S.I., Rogers, D.J., 2006. Global traffic and disease vector dispersal. Proc. Natl. Acad. Sci. U. S. A. 103, 6242–6247.

te Beest, M., Le Roux, J.J., Richardson, D.M., Brysting, A.K., Suda, J., Kubesova, M., Pysek, P., 2012. The more the better? The role of polyploidy in facilitating plant invasions. Ann. Bot. 109, 19–45.

Thompson, G.D., Robertson, M.P., Webber, B.L., Richardson, D.M., Le Roux, J.J., Wilson, J.R.U., 2011. Predicting the subspecific identity of invasive species using distribution models: *Acacia saligna* as an example. Divers. Distrib. 17, 1001–1014.

Thompson, G.D., Bellstedt, D.U., Byrne, M., Millar, M.A., Richardson, D.M., Wilson, J.R.U., Le Roux, J.J., 2012. Cultivation shapes genetic novelty in a globally important invader. Mol. Ecol. 21, 3187–3199.

Thompson, G.D., Bellstedt, D.U., Richardson, D.M., Wilson, J.R.U., Le Roux, J.J., 2015. A tree well-travelled: global genetic structure of the invasive tree *Acacia saligna*. J. Biogeogr. 42, 305–314.

Travis, J.M.J., Dytham, C., 2002. Dispersal evolution during invasions. Evol. Ecol. Res. 4, 1119–1129.

Travis, J.M.J., Mönkemöller, T., Burton, J.O.J., Best, A., Johst, K., 2007. Deleterious mutations can surf to high densities on the wave front of an expanding population. Mol. Biol. Evol. 24, 2334–2343.

Tsutsui, N.D., Suarez, A.V., Holway, D.A., Case, T.J., 2000. Reduced genetic variation and the success of an invasive species. Proc. Natl. Acad. Sci. U. S. A. 97, 5948–5953.

Tsutsui, N.D., Suarez, A.V., Holway, D.A., Case, T.J., 2001. Relationship among native and introduced populations of the Argentine ant (*Linepithema humile*) and the source of introduced populations. Mol. Ecol. 10, 2151–2161.

Uesugi, A., Baker, D.J., de Silva, N., Nurkowski, K., Hodgins, K.A., 2020. A lack of genetically compatible mates constrains the spread of an invasive weed. New Phytol. 226, 1864–1872.

Valade, R., Kenis, M., Hernandez-Lopez, A., Augustin, S., Mari-Mena, N., Magnoux, E., Rougerie, R., Lakatos, F., Roques, A., Lopez-Vaamonde, C., 2009. Mitochondrial and microsatellite DNA markers reveal a Balkan origin for the highly invasive horse-chestnut leaf miner *Cameraria ohridella* (Lepidoptera, Gracillariidae). Mol. Ecol. 18, 3458–3470.

van Boheemen, L.A., Lombaert, E., Nurkowski, K.A., Gauffre, B., Rieseberg, L.H., Hodgins, K.A., 2017. Multiple introductions, admixture and bridgehead invasion characterize the introduction history of *Ambrosia artemisiifolia* in Europe and Australia. Mol. Ecol. 26, 5421–5434.

van Wilgen, B.W., Forsyth, G.G., Le Maitre, D.C., Wannenburgh, A., Kotzé, J.D.F., Van Den Berg, E., Henderson, L., 2012. An assessment of the effectiveness of a large, national-scale invasive alien plant control strategy in South Africa. Biol. Conserv. 148, 28–38.

Vicente, S., Máguas, C., Richardson, D.M., Trindade, H., Wilson, J.R.U., Le Roux, J.J., 2021. Highly diverse and highly successful: invasive Australian acacias have not experienced genetic bottlenecks globally. Ann. Bot. https://doi.org/10.1093/aob/mcab053 (In Press).

Von Holle, B., Simberloff, D., 2005. Ecological resistance to biological invasion overwhelmed by propagule pressure. Ecology 86, 3212–3218.

Whitlock, M.C., 2008. Evolutionary inference from QST. Mol. Ecol. 17, 1885–1896.

Wilkins, J.S., 2011. Philosophically speaking, how many species concepts are there? Zootaxa 1765, 58–60.

Wright, S., 1965. The interpretation of population structure by F-statistics with special regard to systems of mating. Evolution 19, 395–420.

Wu, K.K., Jain, S.K., 1978. Genetic and plastic responses in geographic differentiation of *Bromus rubens* populations. Can. J. Bot. 56, 873–879.

Wu, N., Zhang, S., Li, X., Cao, Y., Liu, X., Wang, Q., Liu, Q., Liu, H., Hu, X., Zhou, X.J., James, A.A., Zhang, Z., Huang, Y., Zhan, S., 2019. Fall webworm genomes yield insights into rapid adaptation of invasive species. Nat. Ecol. Evol. 3, 105–115.

Wurm, Y., Wang, J., Riba-Grognuz, O., Corona, M., Nygaard, S., Hunt, B.G., Ingram, K.K., Falquet, L., Nipitwattanaphon, M., Gotzek, D., Dijkstra, M.B., Oettler, J., Comtesse, F., Shih, C.J., Wu, W.J., Yang, C.C., Thomas, J., Beaudoing, E., Pradervand, S., Flegel, V., Keller, L., 2011. The genome of the fire ant *Solenopsis invicta*. Proc. Natl. Acad. Sci. U. S. A. 108, 5679–5684.

Yadav, S., Stow, A.J., Dudaniec, R.Y., 2019. Detection of environmental and morphological adaptation despite high landscape genetic connectivity in a pest grasshopper (*Phaulacridium vittatum*). Mol. Ecol. 28, 3395–3412.

# CHAPTER 9

# Incorporating evolutionary biology into invasive species management

## Contents

9.1 Introduction   190
9.2 Evolutionary approaches to enhance the success of classical biological control   192
9.3 The use of native enemies as biological control agents   195
9.4 Engineering genomes to manage invasive species   199
9.5 Gene drives: Frankenstein species or promising tools for the genetic control of invasive species?   200
9.6 Conclusions   203
References   203

## Abstract

This chapter discusses the use of evolutionary biology in invasive species management. Understanding the historical evolutionary context of host–enemy relationships, i.e. co-evolution, can assist the selection of more efficient biological control agents. Moreover, rapid evolution in biocontrol agents for enhanced exploitation of target hosts can be achieved, and accelerated, through genetic augmentation and artificial selection. The colonisation of invasive hosts by *native* enemies, coupled with the rapid evolutionary responses that often underlie these novel interactions, suggests that evolutionary tools can also be applied to control invasive species using native enemies. The fields of biotechnology and genetic engineering provide promising new ways to manage invasive species. For instance, transgenic organisms engineered to impede invasive performance, such as the production of male-only progeny, have been developed. Gene drive technologies provide researchers with unprecedented opportunities to spread 'control' genes throughout invasive populations. This chapter provides examples of laboratory-based studies to demonstrate the prospects of complete eradication of invasive species using gene drives.

**Keywords:** Artificial selection, Classical biological control, CRISPR, Experimental evolution, Gene drive, Host-enemy interactions, Invasive species control, Neoclassical biological control, Resistance evolution

*Why does the change in our perception of evolution from a historical to contem-*
*porary process matter? The general reason is that treating evolution as a contem-*
*porary process can fundamentally change how we approach any topic in ecology*
*and evolution. It causes us to recast old questions, changes the answers to those*
*questions and enables us to ask new questions that were otherwise inconceivable.*

**Reznick et al. (2019)**

## 9.1 Introduction

Aspects of both historical and contemporary evolution have important
implications for invasive species management. For example, classical biological control aims to reunite invasive species with the co-evolved specialist
enemies that they left behind in their native ranges. Such co-evolutionary
histories are, in part, responsible for the often-spectacular success of biocontrol programs (Smith et al., 2018). The invasive erect prickly pear (*Opuntia*
*stricta*) provides us with one such example. This cactus is one of the most
invasive plants globally (Novoa et al., 2015). In Australia, the species was
first introduced around 1840, quickly becoming widespread, found in over
30 million hectares of land less than a century later (Mann, 1970). As part of a
biocontrol program, the Queensland government sent an expedition abroad
to search for cactus-feeding insects in 1912. Initially, the Argentine cactoblastis moth (*Cactoblastis cactorum*), a specialist feeder on *Opuntia* species,
was imported to Australia but did not survive. A second introduction of
the moth in 1927 was successful and finally turned the tide in the battle
against erect prickly pear. The moth's devastating impact on erect prickly
pear was hailed one of the most significant environmental achievements
in the history of Australia (the Australians even built a cactoblastis memorial
hall in Queensland to celebrate the demise of erect prickly pear!). Less than
10 years after the introduction of the moth, only 1% of erect prickly pear
biomass that was present at the peak of invasion remained in the warmer
tropical and subtropical parts of the country (Freeman, 1992). However,
in the cooler parts of Australia the cactoblastis moth appeared to be less effective. This is because the moth typically undergoes two generations per year
in warmer areas, but only one generation per year in temperate and cooler
regions. In addition to the cactoblastis moth, Australian authorities also
imported a cochineal scale insect, *Dactylopius opuntiae*. Like the moth, this
sap-sucking insect feeds exclusively on *Opuntia* species (Dodd, 1959;
Winston et al., 2014). In Australia, the scale insect had high success in controlling erect prickly pear, including in cooler parts of the country (Hosking
et al., 1994). Noticing this success, South African researchers imported

*D. opuntiae* on multiple occasions from Australia (Volchansky et al., 1999; Winston et al., 2014). Despite the insect's success in Australia, all introductions failed to establish on erect prickly pear in South Africa (Volchansky et al., 1999). Instead, *D. opuntiae* quickly colonised another *Opuntia* species, the common prickly pear (*O. ficus-indica*). It was initially thought that the failure of *D. opuntiae* on erect prickly pear in South Africa was because erect prickly pear plants were genetically distinct from those in Australia. However, it was later discovered that Australian authorities unknowingly imported multiple lineages of *D. opuntiae* from the Americas. It is now believed that the *D. opuntiae* lineage originally sent to South Africa was collected on *O. streptacantha* in Mexico, a species that is more closely related to *O. ficus-indica* than to *O. stricta* (Volchansky et al., 1999). This biotype became known as the *D. opuntiae* 'ficus' lineage. The subsequent introduction of the *D. opuntiae* 'stricta' lineage from Australia in 1997 finally led to the successful control of erect prickly pear in South Africa (Hoffmann, 2003; Rule and Hoffmann, 2018).

Contemporary evolution also holds important considerations for invasive species management. For example, invasive species often rapidly evolve strategies to deal with management approaches such as chemical control (Dunlop et al., 2018). These evolutionary responses are unsurprising, as control methods usually exert strong directional selection. For instance, the aquatic hydrilla (*Hydrilla verticillata*) is considered one of the most aggressive weeds in the United States (Schmitz et al., 1991; Langeland, 1996). The species was estimated to occupy over 55,000 ha of Florida's lakes and rivers in the early 1990s, where it causes significant ecological damage (Colle and Shireman, 1980; Schmitz and Osborne, 1984; van Dijk, 1985; Schmitz et al., 1993). Hydrilla is extensively controlled using fluridone (Fox et al., 1996). However, patches of hydrilla with evolved resistance to this herbicide quickly appeared (Michel et al., 2004). Fluridone resistance in hydrilla is emblematic of the ever-increasing occurrence of biocide resistance globally. Tabashnik et al. (2014) reported more than 11,000 documented cases of evolved biocide resistance in nearly 1000 species of insects, plants, and plant pathogens. Much of this evolution has been recent. For example, Gould et al. (2018) reported more than 7700 cases of evolved pesticide resistance, involving over 550 insect species, most of which were documented after 1960.

The examples discussed before suggest that evolutionary biology can no longer be ignored when developing and implementing management strategies for invasive species. Our scientific understanding of how evolution works, from genes to phenotypes to populations, together with our capacity to control the genetic make-up of organisms and the selection pressures they experience, is broadly appreciated. Yet, unlike in other sectors such as public health and food production, using this knowledge to provide solutions for environmental problems such as invasive species is largely unrealised (Reznick et al., 2019, but see Kruitwagen et al., 2018). In this chapter we start by outlining evolutionary concepts already in use, such as incorporating the co-evolutionary histories of invasive hosts and their natural enemies, or the use of experimental evolution, to improve the success of biological control programs. We then consider novel approaches by which evolution, through manipulation of selection (i.e. artificial selection) and mutation (e.g. genetic engineering and gene drives), can benefit invasive species management.

## 9.2 Evolutionary approaches to enhance the success of classical biological control

The erect prickly pear example discussed before illustrates the importance of understanding past evolutionary history when searching for, and selecting, effective and host–specific biocontrol agents (Jourdan et al., 2019; Paterson et al., 2019; also see Chapter 8, Section 8.2.2). That is, the success of biocontrol programmes is expected to correlate with the co-evolutionary histories between invasive hosts and their natural enemies (Goolsby et al., 2006; Le Roux and Wieczorek, 2009). Many researchers have therefore advocated that natural enemies be imported from the exact, or at least approximate, native sources of invasive populations in order to maximise the genetic and phenotypic fit between control agents and their target hosts (e.g. Goolsby et al., 2006; Le Roux and Wieczorek, 2009; Smith et al., 2018).

Just like their hosts, imported biocontrol agents also often experience strong founder events, leading to reduced genetic diversity and inbreeding (Hopper et al., 2019; also see Chapter 2, Section 2.2.1). This may reduce their control efficacy. Biocontrol practitioners typically only import natural enemies from part of their overall native range, often without matching the target (i.e. invasive) populations with their source population(s). Agents are then transported to the lab (almost always transoceanic) and tested under quarantine conditions for host specificity, host damage, and potential non–target impacts on native and economically important species. All these steps may result in consecutive founder events, causing a further reduction in genetic diversity (Hopper et al., 2019). Studies linking genetic diversity of founding biocontrol populations to fitness and performance on their target host remain scarce (Fauvergue et al., 2012). One such study involved the parasitoid wasp, *Aphidius ervi*, the biocontrol agent of the invasive pea aphid (*Acyrthosiphon pisum*) in the United States. Hufbauer (2002) wanted to know whether wasps from biocontrol and native (French) populations differed in their ability to overcome parasitism resistance in the aphid. Using reciprocal transplant experiments Hufbauer (2002) found biocontrol wasps less able to overcome parasitism resistance compared with wasps from the native range, suggesting that fitness is impaired in the biocontrol population due to low genetic variation. Another aphid–parasitoid wasp example provides a more concrete link between genetic diversity and the fitness of a biocontrol agent. Fauvergue and Hopper (2009) manipulated propagule pressure (and genetic diversity) of different releases of the biocontrol wasp, *Aphelinus asychis*, the parasitoid of invasive Russian wheat aphid (*Diuraphis noxia*) in the USA. Wasps at all release sites were equally able to mate, suggesting that no Allee effect related to mate availability exists. However, for a given population density, fecundity was always lower in populations founded by low propagule pressure compared with populations founded by high propagule pressure. This observation persisted over three successive generations, suggesting that the genetic diversity of founder biocontrol populations is positively correlated with fitness (Fauvergue and Hopper, 2009).

Evolutionary arms races between host resistance or avoidance, and the exploitative behaviour of enemies, are commonplace. During biological control releases, target hosts may have an unfair advantage in such arms races when control agents have low genetic diversity (e.g. Stastny and Sargent, 2017). Tomasetto et al. (2017) provide an eloquent example of such evolved resistance against biocontrol. These authors studied levels of parasitism by the biocontrol wasp, *Microctonus hyperodae*, in invasive Argentine stem weevils (*Listronotus bonariensis*) in New Zealand. The wasp was initially highly successful in controlling the weevil. However, levels of parasitism dramatically declined over a 5-year period, from 70% during the first 7 years following the wasp's release into New Zealand, to around 25% in subsequent years. Interestingly, despite the wasp having been released to different localities at different times, declines in weevil parasitism consistently occurred 7 years after introduction of the wasp at all localities. Tomasetto et al. (2017) found these declines to be independent of the number of wasps released (i.e. propagule pressure) or of differences in climatic conditions between release sites. This strong temporal signal in levels of parasitism suggests that evolutionary change, and not local environmental conditions, underlies parasitoid resistance. Importantly, the wasp reproduces asexually, and therefore opportunities for evolutionary responses to host resistance are limited by the amount of genetic diversity that was originally introduced. It is conceivable that resistance evolution in the weevil could have been overcome, or at least slowed down, if genetically diverse wasp populations were initially introduced into New Zealand.

The examples discussed before suggest that biological control practitioners must be cognisant of the amount of genetic diversity they introduce (Szűcs et al., 2019a). In particular, the evolutionary capacities of biocontrol agents will benefit from the introduction of large populations (i.e. high propagule pressure with high genetic diversity) or post-release augmentation of genetic diversity. When no co-evolutionary diversification between hosts and enemies is evident in the native range, then admixture between different native biocontrol populations may benefit founder populations in novel environments. As we discussed throughout this book, such admixture often creates novel genotypes and trait combinations, some of which may be better suited to conditions in the new range. For example, Szűcs et al. (2012, 2019b) found admixture between distinct lineages of the biocontrol beetle *Longitarsus jacobaeae* to result in increased lifetime fecundity on its target host, the invasive ragwort, *Jacobaea vulgaris*. Whether admixed genotypes perform differently on non-target species is currently unknown and a topic that deserves further research attention.

We also have evidence where strong founder events had positive effects on the performance of biocontrol agents. For instance, as we discussed in Chapter 2 (see Section 2.2.1), small founding populations that experience moderate genetic bottlenecks are often purged of detrimental alleles, as was the case for the invasive ladybird, *Harmonia axyridis* (Facon et al., 2011). Ironically, while this ladybird is now considered one of the world's most invasive insects, it was initially introduced to many parts of the world as a biological control agent for soft-bodied insect pests such as aphids, scales, and psyllids.

High genetic diversity, and thus evolutionary capacity, also creates opportunities for the use of experimental evolution, via artificial selection, to enhance the efficiency of biocontrol agents (Roderick and Navajas, 2003; Kruitwagen et al., 2018). Artificial selection in biocontrol agents is not a new idea (for a review, see Lirakis and Magalhães, 2019). Because most biocontrol agents are small-bodied insects or microbes with short generation times, they represent tractable systems to use artificial selection to improve host damage and control. For instance, diapause is a phenotype that ceases development, allowing organisms to overwinter or withstand stressful environmental conditions. Diapause in biocontrol agents could hamper their efficacy if their target hosts can survive and reproduce during times when they are dormant. Artificial selection for non-diapause phenotypes has been achieved in mites like *Neoseiulus cucumeris* and *N. barkeri*, the control agents of western flower thrips (*Frankliniella occidentalis*; van Houten et al., 1995). Importantly, in these instances artificial selection did not lead to any obvious trade-offs and non-diapause mites established successfully in small-scale greenhouse experiments, remaining reproductive under conditions that would normally induce diapause (van Houten et al., 1995). However, artificial selection for one trait does often involve trade-offs in others, often with negative affects on fitness. For example, selection for increased tolerance to low temperatures in the parasitic wasp, *Trichogramma pretiosum*, a biocontrol agent for economically important pests like the fall armyworm (*Spodoptera frugiperda*), leads to lower fitness at high temperatures (Carriere and Boivin, 2001).

## 9.3 The use of native enemies as biological control agents

New associations between native weedy species and exotic enemies that lead to suppression of the weed are often referred to as *neoclassical biological control* (Eilenberg et al., 2001). In a broader sense, we will use this term here to include all instances where novel associations, whether between two exotic species or between a native and an exotic species, lead to the suppression of invasive populations (e.g. Hokkanen and Pimentel, 1984, 1989; Carroll et al., 2005). We will discuss how ongoing evolution in native enemies can be utilised and accelerated (via artificial selection) to control invasive species. This approach may offer exciting, but hitherto largely overlooked, opportunities for neoclassical biological control (Carroll, 2011).

As we discussed in Chapter 3, Darwin suggested that exotic species may fare better in environments where closely related relatives occur, because of their preadaptations to local conditions. However, he also thought that exotics may experience more intense competition from closely related natives (Cadotte et al., 2018). These contrasting perspectives suggest that intermediate phylogenetic relatedness between exotic and native species may benefit establishment success and invasion. For example, exotic species that share some phylogenetic history with resident species may be similar enough to some natives to prove phenotypically well matched to conditions in the new range (Petitpierre et al., 2012), but at the same time sufficiently different to others to avoid competition (Catford et al., 2019). These differences in eco-evolutionary experience (Heger et al., 2019; also see Chapter 3) also critically inform classical biological control, as phylogenetic distance between exotic and native species is expected to correlate negatively with the probability of them sharing natural enemies (Divíšek et al., 2018; Davies, 2021). Considering this, exotic species that are phylogenetically vulnerable to attack by resident enemies are less likely to become widespread to the point that they require (bio)control. Alternatively, exotic species that lack close relatives in the new range may experience substantial enemy release and may become widespread. As we will discuss later, between those two extremes, intermediate phylogenetic relatedness between exotic and native species creates opportunities to utilise rapid evolution to improve neoclassical biocontrol success.

The evolutionary underpinnings of new associations between invasive species and resident enemies have been repeatedly shown (e.g. Strauss et al., 2006; Carroll, 2011). This is perhaps unsurprising since exotic species and natives that share derived traits (i.e. that are phylogenetically related) also often share related enemies. For example, large proportions of native plant species in the same families that occur across continents are also often attacked by common enemies (Strauss et al., 2006; Ness et al., 2011; Davies, 2021). Therefore exotic species that co-occur with congeneric or confamilial natives (i.e. intermediate phylogenetic relatedness) might be vulnerable to attack by resident enemies. When these enemies begin to exploit exotic hosts, they also have the potential to adapt to them. Such evolution may become especially prominent when new hosts become widespread and commonly encountered as reliable resources by resident enemies.

The associations between invasive balloon vines (genus *Cardiospermum* in the soapberry family Sapindaceae) and native soapberry bugs (genus *Leptocoris*) that we discussed in Chapter 3 (see Section 3.3) provide examples of native enemies rapidly adapting to invasive host plants that share intermediate phylogenetic relatedness with their historical host plants. As popular ornamental plants, balloon vines have been moved around the world, in some instances leading to the establishment of invasive populations (Gildenhuys et al., 2013, 2015a,b). These invasive vines have been colonised by native specialist soapberry bugs, seed predators of Sapindaceae plants, on numerous occasions (Carroll et al., 2005; Carroll and Loye, 2012; Foster et al., 2019). The larger size of balloon vine fruits (compared to those of native host plants) has resulted in independent instances where bugs evolved elongated proboscides (or 'beaks') following colonisation of balloon vines (Carroll et al., 2005; Andres et al., 2013; Foster et al., 2019).

A textbook example of rapid evolution in soapberry bugs on invasive balloon vines comes from Australia. The annual balloon vine (*C. halicacabum*) was introduced into Australia between 200 and 400 years ago to northern parts of the Northern Territory (Bean, 2007). Here, two native Sapindaceae species, the whitewood tree (*Atalaya hemiglauca*) and the false currant (*Allophylus cobbe*), are the natural hosts of the soapberry bug, *Leptocoris tagalicus* (Carroll et al., 2005; S.P. Carroll, personal communication). The bug has two subspecies, subsp. *vulgaris* and subsp. *tagalicus* (Gross, 1960; Andres et al., 2013). Subspecies *vulgaris* is smaller bodied and inhabits desert and savanna habitats, whereas subsp. *tagalicus* is larger bodied and found in eastern and northern wet and dry rainforests (S.P. Carroll, personal communication). Following the introduction of the annual balloon vine, the two subspecies came into contact, hybridised, and established a highly differentiated biotype on it: hybrid 'halicacabum' bugs (Andres et al., 2013). This new biotype remained small bodied, but its beak length increased by up to 38% (Andres et al., 2013; Fig. 9.1), resulting in a much higher beak-to-body length ratio than in either parental subspecies. These allometries are comparable to those of efficient Neotropical *Jadera* soapberry bugs (which achieve a 95% seed kill rate; Carroll and Loye, 2006), the seed predators from which annual balloon vine was liberated from upon introduction into Australia.

Since the 1960s–1970s, large areas along Australia's east coast became invaded by the perennial balloon vine, *C. grandiflorum*. This very large-fruited vine was colonised by *L. tagalicus* subsp. *tagalicus*, again in sympatry with the bug's locally native host, the woolly rambutan tree (*Alectryon tomentosus*). Museum specimens collected from eastern Australia between the 1920s and 2000 revealed that the bug's beak became elongated, and by 2004 beaks of field-collected 'grandiflorum' bugs were on average 10% longer than those of bugs on neighbouring woolly rambutan trees (Carroll et al., 2005; Fig. 9.1). This rapid evolution almost doubled the experimental kill rate of perennial balloon vine seeds by the bug (Carroll et al., 2005). It is clear that the intermediate phylogenetic relatedness between balloon vines and Australian Sapindaceae species facilitated the novel associations between these vines and native soapberry bugs. The evolutionary lability observed in wild *L. tagalicus* populations also suggests that introgressive hybridisation and artificial selection in this insect may further optimise trait matching between balloon vines and bugs. Such experimental evolution may be key to unlocking the potential of native insects as efficient neoclassical biocontrol agents.

New associations between native and invasive species may also lead to unique and unforeseen management challenges. For example, in the United States various native pierid butterflies in the genus *Pieris* oviposit on invasive garlic mustard (*Alliaria petiolata*), despite not being able to complete development on it (Chew, 1980; Courant et al., 1994; Porter, 1994). Similarly,

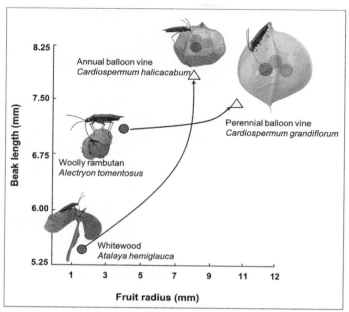

**Fig. 9.1** In Australia, the native *Leptocoris tagalicus* soapberry bug has independently colonised two invasive balloon vine species. Hybrids between the bug's two subspecies (subsp. *tagalicus* and subsp. *vulgaris*) colonised the annual balloon vine (*Cardiospermum halicacabum*; introduced over 200 years ago), in sympatry with one of their native host plants, the whitewood tree (*Atalaya hemiglauca*). More recently, *L. tagalicus* subsp. *tagalicus* has colonised perennial balloon vine (*C. grandiflorum*; introduced around 40 years ago), again in sympatry with its locally native host, the woolly rambutan (*Alectryon tomentosus*). Elongated proboscides (beaks) rapidly evolved in both 'grandiflorum' and 'halicacabum' bugs (*yellow (light grey in the printed version) triangles*) and are significantly longer than those of bugs on neighbouring native host plants (*green (dark grey in the printed version) circles*). These evolutionary changes were driven by directional selection stemming from the size of balloon vine fruits. (*Photo credit: Fruit and bug illustrations: Scott Carroll. Graph redrawn from data reported in Andres et al. (2013).*)

monarch butterflies (*Danaus plexippus*) lay up to a quarter of their eggs on invasive black swallowwort (*Vincetoxicum nigrum*), even in the presence of their native milkweed hosts (Tewksbury et al., 2002). Monarch butterfly larvae are also unable to complete development on their new host plant. Examples like these, where native species show a preference for invasive hosts, despite having negative impacts on their fitness, have been referred to as 'evolutionary traps' (Schlaepfer et al., 2005). Severe evolutionary traps also mean that trapped native species may suffer significant population losses when their invasive hosts are controlled.

## 9.4 Engineering genomes to manage invasive species

Genetic engineering (GE) took off in the early 1970s after geneticist Stanley Cohen and biochemist Herbert Boyer described, for the first time, how to 'nick' DNA of one organism and 'paste' DNA from another into these nicks (Cohen et al., 1973). Along with their colleagues, they also showed that pasted DNA can be functional when put back into a living bacterium. This opened the door to an extraordinary effort to create genetically modified (GM) organisms, whereby traits of one organism are transferred to another. One of the best-known examples of GE is the isolation of the first pest resistance genes from the bacterium *Bacillus thuringiensis*. This bacterium is a common resident of soils and naturally produces so-called Bt toxins against a wide range of insects. By randomly inserting Bt genes into the genomes of plants, crops like corn or cotton produce the insect-killing proteins of *B. thuringiensis*, making them resistant to many pests. Genetic engineering also promises novel ways to minimise invasion risk. For example, horticultural species could be engineered to have slower growth rates or non-flowering phenotypes (Anderson et al., 2006). However, as we will discuss later, these approaches have mostly been tested under laboratory conditions only.

One way in which GE can be applied to control invasive species is through the genetic distortion of sex ratios (Harvey-Samuel et al., 2017). Given that the number and productivity of females determine future population growth in most sexually reproducing species, male-biased sex ratios are expected to suppress, or even crash, populations. The disruption of female sex determination genes can lead to offspring that are female but phenotypically, or entirely, male. Therefore releasing GM individuals with these attributes into invasive populations can cause male-biased sex ratios over successive generations. However, the suppression of populations through male-biased sex ratios will require several releases of GM individuals, as selection against defective female sex determination systems will be strong. Galizi et al. (2014) developed a line of GM malaria mosquitos (*Anopheles gambiae*) that causes distortion of sex ratios in favour of males. The transgenes they employed involved an enzyme that targets and destroys a specific region of the *A. gambiae* X chromosome. Similar to humans, female mosquitoes are homogametic (i.e. XX) and males heterogametic (i.e. XY). Cleverly, these researchers restricted the enzyme's activity to meiosis, the cell division phase responsible for the production of gametes. Therefore the genetic modification does not affect fertility of mosquitoes, but only prevents functional X chromosomes from being passed on to offspring, resulting in male-only progeny. When Galizi et al. (2014) mated these GM mosquitoes with wild-caught ones under laboratory conditions, more than 95% of offspring was male, causing populations to crash after only a few generations. This example illustrates that species genetically engineered to produce male-biased sex ratios may provide effective control of some invasive populations. These approaches are currently being developed and tested in invasive species like common carp (*Cyprinus carpio*) and cane toads (*Rhinella marina*) (Harvey-Samuel et al., 2017).

A lack of implementation of traditional GE approaches to manage invasive species partly reflects the potential risks involved with these technologies. For instance, escape of transgenes into non-target taxa, through hybridisation, has been demonstrated. Ironically, these often involved GM crops and their weedy relatives. For example, Warwick et al. (2008) found hybridisation between GM rapeseed (*Brassica napus*) and weedy field mustard (*B. rapa*) to have led to the stable inheritance of glyphosate resistance transgenes in field mustard populations. These resistance genes persisted in weedy populations under field conditions over 6 years, even in the absence of glyphosate application. Similarly, Zapiola et al. (2008) found that weedy creeping bentgrass (*Agrostis stolonifera*) acquired glyphosate resistance genes via hybridisation with GM creeping bentgrass. In both these examples, transgene escape provided weed populations with immediate glyphosate resistance.

## 9.5 Gene drives: Frankenstein species or promising tools for the genetic control of invasive species?

Driving desirable transgenes to high frequencies in invasive populations may not be straightforward or, sometimes, impossible. For sexually reproducing organisms, Mendelian inheritance means that most alleles, including transgenes, will only have a 50% chance of being passed on from parents to offspring. At the same time, transgenes that lower the performance or reproductive fitness of individuals will experience strong selection, or may be eliminated by genetic drift when present in low frequencies. But, is it possible to cheat the rules of Mendelian inheritance and give transgenes an unfair advantage over other genes? The answer is yes, scientists now have the ability to create 'cheater' alleles, through what is commonly referred to as gene drives, a technology that allows the engineering of genes to be selfish and to favour their own transmission over successive generations.

The discovery of so-called clustered, regularly interspaced short palindromic repeat genes (or CRISPR, Mojica et al., 2000) and their associated proteins (Cas) in the genomes of prokaryotes paved the way for one of the simplest, yet incredibly accurate, gene editing technologies. In prokaryotic genomes, CRISPR genes consist of repeat DNA segments that are separated by short sequences known as 'spacers'. The genetic information encoded by spacers are identical to parts of the genomes of viruses that frequently invade microbes (Doudna and Charpentier, 2014). When invaded, microbes store pieces of genetic information of viruses they successfully fight off in their CRISPR array. These spacers then act as a kind of 'molecular memory' so that when a microbe is reinvaded by the same virus, CRISPR spacer sequences are transcribed into small antisense RNAs (so-called guide- or gRNAs) that act as guides for a gRNA-Cas complex and allow sequence-specific identification and destruction of the invader's genome via the cutting of viral DNA by Cas endonucleases.

The mechanism of CRISPR genes caught the eye of molecular biologists in the early 2010s, who realised it could potentially be used to make precise changes in virtually any piece of DNA. They knew that CRISPR cut DNA at highly specific locations and, together with knowledge of how most eukaryotic organisms typically repair DNA damage, thought that it could be used to delete, replace, or subtly edit genetic information. In general, when DNA is damaged it is repaired by replacement using a template. This is called homology-directed repair, a process whereby one (undamaged) chromosome acts as the template for DNA repair on its damaged homologous chromosome. This enables gene drives to rapidly spread desirable traits throughout a population. The process of constructing gene drives starts by creating transgenic organisms in the laboratory. For this, the wild-type sequence on one chromosome is replaced with a drive allele, containing all CRISPR components (i.e. a guide RNA and Cas endonuclease gene) and a gene (trait) of interest. The transgenic organism can then be released into the wild to interbreed with wild-type individuals. The drive allele will be copied to the homologous wild-type chromosome within cells. First, the CRISPR gRNA guides the Cas endonuclease to the target site on the wild-type chromosome. The endonuclease then cuts the DNA at the target site. The cell then repairs the cut through homology-directed repair, using the drive allele-containing chromosome as template. As a result, the genetic information encoding the guide RNA and Cas endonuclease, and the transgene of interest, is copied 'verbatim' to the homologous wild-type chromosome during repair. During meiosis the drive allele is passed to all gametes and therefore all offspring inherit the transgenic trait, and within a few generations the transgene will become completely fixed in the population.

The potential of gene drives to eradicate, control, or lessen the impacts of invasive species is enormous. For example, the spread of transgenes causing dysfunctional sex determination can crash invasive populations through biased sex ratios, while transgenes that block toxin production may lessen the impacts of invasive species on native species (Tingley et al., 2017; Moro et al., 2018). Alternatively, gene drives may be used to counter rapid evolution in invasive populations, e.g., evolved resistance to biocides could be eliminated by drives that replace resistance alleles with susceptibility alleles (Neve, 2018). To date, the utility of gene drive technologies has been tested only under strict laboratory conditions and not in wild populations. For example, Kyrou et al. (2018) crashed a laboratory population *Anopheles gambiae* malarial mosquitoes by constructing a gene drive that disrupts the *doublesex* fertility gene. Disruption of this gene causes female mosquitoes to stop laying eggs, and within 8–12 generations laboratory populations of GM mosquitoes were unable to lay any eggs (Kyrou et al., 2018).

While laboratory-based studies provide sufficient proof of concept for the utility of gene drives to control or eradicate invasive species, we are probably nowhere near seeing these being released into the wild. A number of unknowns and risks remain. First, gene drives may have poor specificity and may cause mutations in non-target areas of the genome (Webber et al., 2015; Barrett et al., 2019). On the other hand, mutations within the target region of guide RNAs may confer resistance to gene drives (Barrett et al., 2019). Such resistance may be minimised when target regions consist of genes that may be naturally more 'resistant' to mutation. For instance, the *doublesex* gene we discussed before is probably less prone to mutation than many other genes given its vital role in mosquito reproduction.

If invasive species hybridise with closely related native species then there is the obvious risk of gene drives escaping into non-target species, with the same detrimental effects on their populations. We also know that many invasive populations originate from multiple introductions or bridgeheads (e.g. Kolbe et al., 2007; Javal et al., 2019; Vicente et al., 2021). This implies that the spread of gene drives aimed at controlling or eradicating invasive populations in one area may unintentionally spread to other areas, including to the native range of invasive species. In the worst-case scenario this can result in the global extinction of the target species (Webber et al., 2015; Barrett et al., 2019).

A less risky approach might be to engineer gene drives to enhance the performance of biological control agents, eliminating the risk of irreversible extinction of the modified species. Spreading traits such as high tolerance to temperature extremes or pesticides throughout biocontrol agent populations may greatly enhance their efficacy against invasive populations (Gurr and You, 2016). Of course, the risk of biocontrol gene drives dispersing to the native range also exists.

## 9.6 Conclusions

Biological invasions are spatially and temporally dynamic and the evolutionary responses in invasive and native species often rapid. This lability suggests that we can no longer ignore the solutions offered by evolution to manage invasive populations. Classical biocontrol will benefit from a better understanding of the co-evolutionary dynamics between invasive species and their natural enemies, e.g. by incorporating phylogenetic information to infer co-evolution between host and agent species. Biocontrol agents themselves will benefit from genetic augmentation that will enhance their adaptability to target hosts in the new range. A particularly promising tactic may be the application of experimental evolution in biocontrol agents, via artificial selection, to maximise their control efficacy of invasive species. The time is also ripe to explore the potential use of native enemies to control invasive species. The often-strong evolutionary underpinnings of these novel associations suggest that we can borrow evolutionary tools from agriculture, such as introgressive hybridisation and selective breeding, to accelerate ongoing evolution in natives for enhanced invasive species control. Advances in biotechnology, together with evidence from laboratory-based studies, suggest that genetic engineering will provide important ways to manage invasive species in the future. These include the release of genetically modified species, via traditional or gene drive technologies, to control or eradicate invasive populations.

## References

Anderson, N.O., Gomez, N., Galatowitsch, S.M., 2006. A non-invasive crop ideotype to reduce invasive potential. Euphytica 148, 185–202.

Andres, J.A., Thampy, P.R., Mathieson, M.T., Loye, J., Zalucki, M.P., Dingle, H., Carroll, S.P., 2013. Hybridization and adaptation to introduced balloon vines in an Australian soapberry bug. Mol. Ecol. 22, 6116–6130.

Barrett, L.G., Legros, M., Kumaran, N., Glassop, D., Raghu, S., Gardiner, D.M., 2019. Gene drives in plants: opportunities and challenges for weed control and engineered resilience. Proc. R. Soc. Lond. B Biol. Sci. 286, 20191515.

Bean, A.R., 2007. A new system for determining which plant species are indigenous in Australia. Aust. Syst. Bot. 20, 1–43.

Cadotte, M.W., Campbell, S.E., Li, S.-.P., Sodhi, D.S., Mandrak, N.E., 2018. Preadaptation and naturalization of nonnative species: Darwin's two fundamental insights into species invasion. Annu. Rev. Plant Biol. 69, 661–684.

Carriere, Y., Boivin, G., 2001. Constraints on the evolution of thermal sensitivity of foraging in *Trichogramma*: genetic tradeoffs and plasticity in maternal selection. Am. Nat. 157, 570–581.

Carroll, S.P., 2011. Conciliation biology: the eco-evolutionary management of permanently invaded biotic systems. Evol. Appl. 4, 184–199.

Carroll, S.P., Loye, J.E., 2006. Invasion, colonization, and disturbance; historical ecology of the endangered Miami blue butterfly and a guild of subtropical seed-predators in Florida. J. Insect Conserv. 10, 13–27.

Carroll, S.P., Loye, J.E., 2012. Soapberry bug (Hemiptera: Rhopalidae: Serinethinae) native and introduced host plants: biogeographic background of anthropogenic evolution. Ann. Entomol. Soc. Am. 105, 671–684.

Carroll, S.P., Loye, J.E., Dingle, H., Mathieson, M., Famula, T.R., Zalucki, M., 2005. And the beak shall inherit—evolution in response to invasion. Ecol. Lett. 8, 944–951.

Catford, J.A., Smith, A.L., Wragg, P.D., Clark, A.T., Kosmala, M., Cavender-Bares, J., Reich, P.B., Tilman, D., 2019. Traits linked with species invasiveness and community invasibility vary with time, stage and indicator of invasion in a long-term grassland experiment. Ecol. Lett. 22, 593–604.

Chew, F.S., 1980. Larval preferences of *Pieris* caterpillars (Lepidoptera). Oecologia 46, 347–353.

Cohen, S.N., Chang, A.C.Y., Boyer, H.W., Helling, R., 1973. Construction of biologically functional bacterial plasmids in vitro. Proc. Natl. Acad. Sci. U. S. A. 3, 240–244.

Colle, D.E., Shireman, J.V., 1980. Coefficients of condition for large-mouth bass, bluegill, and red ear sunfish in Hydrilla-infested lakes. Trans. Am. Fish. Soc. 109, 521–531.

Courant, A.V., Holbrook, A.E., Van der Reijden, E.D., Chew, F.S., 1994. Native pierine butterfly (Pieridae) adapting to naturalized crucifer. J. Lepid. Soc. 48, 168–170.

Davies, T.J., 2021. Ecophylogenetics redux. Ecol. Lett. 24, 1073–1088.

Divíšek, J., Chytrý, M., Beckage, B., Gotelli, N.J., Lososová, Z., Pyšek, P., Richardson, D.M., Molofsky, J., 2018. Similarity of introduced plant species to native ones facilitates naturalization, but differences enhance invasion success. Nat. Commun. 9, 4631.

Dodd, A.P., 1959. The biological control of prickly pear in Australia. In: Keast, A., Crocker, R.L., Christian, C.S. (Eds.), Biogeography and Ecology in Australia. W. Junk, The Hague, The Netherlands, pp. 565–577.

Doudna, J.A., Charpentier, E., 2014. Genome editing. The new frontier of genome engineering with CRISPR-Cas9. Science 346, 1258096.

Dunlop, E.S., McLaughlin, R., Adams, J.V., Jones, M.L., Birceanu, O., Christie, M.R., Criger, L.A., Hinderer, J.L.M., Hollingworth, R.M., Johnson, N.S., Lantz, S.R., Li, W., Miller, J.R., Morrison, B.J., Mota-Sanchez, D., Muir, A.M., Sepulveda, M.S., Steeves, T.B., Walter, L., Westman, E., Wirgin, I., Wilkie, M.P., 2018. Rapid evolution meets invasive species control: the potential for pesticide resistance in sea lamprey. Can. J. Fish. Aquat. Sci. 75, 152–168.

Eilenberg, J., Hajek, A., Lomer, C., 2001. Suggestions for unifying the terminology in biological control. BioControl 46, 387–400.

Facon, B., Hufbauer, R.A., Tayeh, A., Loiseau, A., Lombaert, E., Vitalis, R., Guillemaud, T., Lundgren, J.G., Estoup, A., 2011. Inbreeding depression is purged in the invasive insect *Harmonia axyridis*. Curr. Biol. 21, 424–427.

Fauvergue, X., Hopper, K.R., 2009. French wasps in the new world: experimental biological control introductions reveal a demographic Allee effect. Popul. Ecol. 51, 385–397.

Fauvergue, X., Vercken, E., Malausa, T., Hufbauer, R.A., 2012. The biology of small, introduced populations, with special reference to biological control. Evol. Appl. 5, 424–443.

Foster, J.D., Ellis, A.G., Foxcroft, L.C., Carroll, S.P., Le Roux, J.J., 2019. The potential evolutionary impact of invasive balloon vines on native soapberry bugs in South Africa. NeoBiota 49, 19–35.

Fox, A.M., Haller, W.T., Shilling, D.G., 1996. Hydrilla control with split treatments of fluridone in Lake Harris, Florida. Hydrobiologia 340, 235–239.

Freeman, D.B., 1992. Prickly pear menace in eastern Australia 1880–1940. Geogr. Rev. 82, 413–429.

Galizi, R., Doyle, L.A., Menichelli, M., Bernardini, F., Deredec, A., Burt, A., Stoddard, B.L., Windbichler, N., Crisanti, A., 2014. A synthetic sex ratio distortion system for the control of the human malaria mosquito. Nat. Commun. 5, 3977.

Gildenhuys, E., Ellis, A.G., Carroll, S., Le Roux, J.J., 2013. The ecology, biogeography, history and future of two globally important weeds: Cardiospermum halicacabum Linn. and C. grandiflorum Sw. NeoBiota 19, 45–65.

Gildenhuys, E., Ellis, A.G., Carrol, S.P., Le Roux, J.J., 2015a. Combining known native range distributions and phylogeny to resolve biogeographic uncertainties of balloon vines (Cardiospermum, Sapindaceae). Divers. Distrib. 21, 163–174.

Gildenhuys, E., Ellis, A.G., Carroll, S., Le Roux, J.J., 2015b. From the Neotropics to the Namib: evidence for rapid ecological divergence following extreme long-distance dispersal. Bot. J. Linn. Soc. 179, 477–486.

Goolsby, J.A., De Barro, P.J., Makinson, J.R., Pemberton, R.W., Hartley, D.M., Frohlich, D.R., 2006. Matching the origin of an invasive weed for selection of a herbivore haplotype for a biological control programme. Mol. Ecol. 15, 287–297.

Gould, F., Brown, Z.S., Kuzma, J., 2018. Wicked evolution: can we address the sociobiological dilemma of pesticide resistance? Science 360, 728–732.

Gross, G.F., 1960. A revision of the genus Leptocoris Hahn (Heteroptera: Coreidae: Rhopalinae) from the Indo-Pacific and Australian regions. Rec. Aust. Mus. 13, 403–451.

Gurr, M.G., You, M., 2016. Conservation biological control of pests in the molecular era: new opportunities to address old constraints. Front. Plant Sci. 6, 1255.

Harvey-Samuel, T., Ant, T., Alphey, L., 2017. Towards the genetic control of invasive species. Biol. Invasions 19, 1683–1703.

Heger, T., Bernard-Verdier, M., Gessler, A., Greenwood, A.D., Grossart, H.P., Hilker, M., Keinath, S., Kowarik, I., Kueffer, C., Marquard, E., Müller, J., Niemeier, S., Onandia, G., Petermann, J.S., Rillig, M.C., Rödel, M.-O., Saul, W.-C., Schittko, C., Tockner, K., Joshi, J., Jeschke, J.M., 2019. Towards an integrative, eco-evolutionary understanding of ecological novelty: studying and communicating interlinked effects of global change. Bioscience 69, 888–899.

Hoffmann, J.H., 2003. 107 Biotypes, hybrids and biological control: lessons from cochineal insects on Opuntia weeds. In: Cullen, J.M., Briese, D.T., Kriticos, D.J., Lonsdale, W.M., Morin, L., Scott, J.K. (Eds.), Proceedings of the XI International Symposium on Biological Control of Weeds. CSIRO Entomology, Canberra, pp. 283–286.

Hokkanen, H., Pimentel, D., 1984. New approach for selecting biological control agents. Can. Entomol. 116, 1109–1121.

Hokkanen, H., Pimentel, D., 1989. New associations in biological control: theory and practice. Can. Entomol. 121, 829–840.

Hopper, J.V., McCue, K.F., Pratt, P.D., Duchesne, P., Grosholz, E.D., Hufbauer, R.A., 2019. Into the weeds: matching importation history to genetic consequences and pathways in two widely used biological control agents. Evol. Appl. 12, 773–790.

Hosking, J.R., Sullivan, P.R., Welsby, S.M., 1994. Biological control of Opuntia stricta (Haw.) Haw. var. stricta using Dactylopius opuntiae (Cockerell) in an area of New South Wales, Australia, where Cactoblastis cactorum (Berg) is not a successful biological control agent. Agric. Ecosyst. Environ. 48, 241–255.

Hufbauer, R.A., 2002. Evidence for nonadaptive evolution in parasitoid virulence following a biological control introduction. Ecol. Appl. 12, 66–78.

Javal, M., Lombaert, E., Tsykun, T., Courtin, C., Kerdelhué, C., Prospero, S., Roques, A., Roux, G., 2019. Deciphering the worldwide invasion of the Asian long-horned beetle: a recurrent invasion process from the native area together with a bridgehead effect. Mol. Ecol. 28, 951–967.

Jourdan, M., Thomann, T., Kriticos, D.J., Bon, M.-C., Sheppard, A., Baker, G.H., 2019. Sourcing effective biological control agents of conical snails, *Cochlicella acuta*, in Europe and North Africa for release in southern Australia. Biol. Control 134, 1–14.

Kolbe, J.J., Glor, R.E., Schettino, L.R., Lara, A.C., Larson, A., Losos, J.B., 2007. Multiple sources, admixture, and genetic variation in introduced *Anolis* lizard populations. Conserv. Biol. 21, 1612–1625.

Kruitwagen, A., Beukeboom, L.W., Wertheim, B., 2018. Optimization of native biocontrol agents, with parasitoids of the invasive pest *Drosophila suzukii* as an example. Evol. Appl. 11, 1473–1497.

Kyrou, K., Hammond, A.M., Galizi, R., Kranjc, N., Burt, A., Beaghton, A.K., Nolan, T., Crisanti, A., 2018. A CRISPR–Cas9 gene drive targeting *doublesex* causes complete population suppression in caged *Anopheles gambiae* mosquitoes. Nat. Biotechnol. 36, 1062.

Langeland, K.A., 1996. *Hydrilla verticillata* (L. F.) Royle (Hydrocharitaceae), 'the perfect aquatic weed'. Castanea 61, 293–304.

Le Roux, J., Wieczorek, A.M., 2009. Molecular systematics and population genetics of biological invasions: towards a better understanding of invasive species management. Ann. Appl. Biol. 154, 1–17.

Lirakis, M., Magalhães, S., 2019. Does experimental evolution produce better biological control agents? Entomol. Exp. Appl. 167, 584–597.

Mann, J., 1970. Cacti Naturalised in Australia and their Control. Brisbane Department of Lands, Queensland Government Printer, Queensland.

Michel, A., Arias, R.S., Scheffler, B.E., Duke, S.O., Netherland, M., Dayan, F.E., 2004. Somatic mutation-mediated evolution of herbicide resistance in the nonindigenous invasive plant hydrilla (*Hydrilla verticillata*). Mol. Ecol. 13, 3229–3237.

Mojica, F.J.M., Díez-Villaseñor, C., Soria, E., Juez, G., 2000. Biological significance of a family of regularly spaced repeats in the genomes of Archaea, Bacteria and mitochondria. Mol. Microbiol. 36, 244–246.

Moro, D., Byrne, M., Kennedy, M., Campbell, S., Tizard, M., 2018. Identifying knowledge gaps for gene drive research to control invasive animal species: the next CRISPR step. Glob. Ecol. Conserv. 13, e00363.

Ness, J.H., Rollinson, E.J., Whitney, K.D., 2011. Phylogenetic distance can predict susceptibility to attack by natural enemies. Oikos 120, 1327–1334.

Neve, P., 2018. Gene drive systems: do they have a place in agricultural weed management? Pest Manag. Sci. 74, 2671–2679.

Novoa, A., Le Roux, J.J., Robertson, M.P., Wilson, J.R.U., Richardson, D.M., 2015. Introduced and invasive cactus species: a global review. AoB Plants 7, plu078.

Paterson, I.D., Coetzee, J.A., Weyl, P., Griffith, T.C., Voogt, N., Hill, M.P., 2019. Cryptic species of a water hyacinth biological control agent revealed in South Africa: host specificity, impact, and thermal tolerance. Entomol. Exp. Appl. 167, 682–691.

Petitpierre, B., Kueffer, C., Broennimann, O., Randin, C., Daehler, C., Guisan, A., 2012. Climatic niche shifts are rare among terrestrial plant invaders. Science 335, 1344–1348.

Porter, A., 1994. Implications of introduced garlic mustard (*Alliaria petiolata*) in the habitat of *Pieris virginiensis* (Pieridae). J. Lepid. Soc. 48, 171–172.

Reznick, D.N., Losos, J., Travis, J., 2019. From low to high gear: there has been a paradigm shift in our understanding of evolution. Ecol. Lett. 22, 233–244.

Roderick, G.K., Navajas, M., 2003. Genes in new environments: genetics and evolution in biological control. Nat. Rev. Genet. 4, 889–899.

Rule, N.F., Hoffmann, J., 2018. The performance of *Dactylopius opuntiae* as a biological control agent on two invasive *Opuntia* cactus species in South Africa. Biol. Control 119, 7–11.

Schlaepfer, M.A., Sherman, P.W., Blossey, B., Runge, M.C., 2005. Introduced species as evolutionary traps. Ecol. Lett. 8, 241–246.

Schmitz, D.C., Osborne, J.A., 1984. Zooplankton densities in a *Hydrilla* infested lake. Hydrobiologia 111, 127–132.

Schmitz, D.C., Nelson, B.V., Nail, L.E., Schardt, J.D., 1991. Exotic aquatic plants in Florida: a historical perspective and review of the present aquatic plant regulation program. In: Proceedings of the Symposium on Exotic Pest Plants, 2–3 November, 1988. University of Miami, Miami, FL, pp. 303–326.

Schmitz, D.C., Schardt, J.D., Leslie, A.J., Dray Jr., F.A., Osborne, J.A., Nelson, B.V., 1993. The ecological impact and management history of three invasive alien aquatic plant species in Florida. In: McKnight, B.N. (Ed.), Biology Pollution: The Control and Impact of Invasive Exotic Species. Indiana Academy of Science, Indianapolis, IN, pp. 173–194.

Smith, L., Cristofaro, M., Bon, M.C., De Biase, A., Petanović, R., Vidović, B., 2018. The importance of cryptic species and subspecific populations in classic biological control of weeds: a North American perspective. BioControl 63, 417–425.

Stastny, M., Sargent, R.D., 2017. Evidence for rapid evolutionary change in an invasive plant in response to biological control. J. Evol. Biol. 30, 1042–1052.

Strauss, S.Y., Lau, J.A., Carroll, S.P., 2006. Evolutionary responses of natives to introduced species: what do introductions tell us about natural communities. Ecol. Lett. 9, 354–371.

Szűcs, M., Eigenbrode, S.D., Schwarzländer, M., Schaffner, U., 2012. Hybrid vigor in the biological control agent, *Longitarsus jacobaeae*. Evol. Appl. 5, 489–497.

Szűcs, M., Vercken, E., Bitume, E.V., Hufbauer, R.A., 2019a. The implications of rapid eco-evolutionary processes for biological control—a review. Entomol. Exp. Appl. 167, 598–615.

Szűcs, M., Salerno, P., Teller, B., Schaffner, U., Littlefield, J., Hufbauer, R., 2019b. The effects of agent hybridization on the efficacy of biological control of tansy ragwort at high elevations. Evol. Appl. 12, 470–481.

Tabashnik, B.E., Mota-Sanchez, D., Whalon, M.E., Hollingworth, R.M., Carrière, Y., 2014. Defining terms for proactive management of resistance to Bt crops and pesticides. J. Econ. Entomol. 107, 496–507.

Tewksbury, L., Casagrande, R., Gassmann, A., 2002. Swallow-warts. In: van Driesche, R., Blossey, B., Hoddle, M., Lyon, S., Reardon, R. (Eds.), Biological Control of Invasive Plants in Eastern United States. Forest Health Technology Enterprise Team, Morgantown, WV, pp. 209–216.

Tingley, R., Ward-Fear, G., Schwarzkopf, L., Greenlees, M.J., Phillips, B.L., Brown, G., Clulow, S., Webb, J., Capon, R., Sheppard, A., Strive, T., Tizard, M., Shine, R., 2017. New weapons in the toad toolkit: a review of methods to control and mitigate the biodiversity impacts of invasive cane toads (*Rhinella marina*). Q. Rev. Biol. 92, 123–149.

Tomasetto, F., Tylianakis, J.M., Reale, M., Wratten, S., Goldson, S.L., 2017. Intensified agriculture favors evolved resistance to biological control. Proc. Natl. Acad. Sci. U. S. A. 114, 3885–3890.

van Dijk, G., 1985. *Vallisneria* and its interactions with other species. Aquatics 7, 6–10.

van Houten, Y.M., van Stratum, P., Bruin, J., Veerman, A., 1995. Selection for non-diapause in *Amblyseius cucumeris* and *Amblyseius barkeri* and exploration of the effectiveness of selected strains for thrips control. Entomol. Exp. Appl. 77, 289–295.

Vicente, S., Máguas, C., Richardson, D.M., Trindade, H., Wilson, J.R.U., Le Roux, J.J., 2021. Highly diverse and highly successful: invasive Australian acacias have not experienced genetic bottlenecks globally. Ann. Bot. 128, 149–157.

Volchansky, C.R., Hoffmann, J.H., Zimmermann, H.G., 1999. Host-plant affinities of two biotypes of *Dactylopius opuntiae* (Homoptera: Dactylopiidae): enhanced prospects for biological control of *Opuntia stricta* (Cactaceae) in South Africa. J. Appl. Ecol. 36, 85–91.

Warwick, S.I., Légère, A., Simard, M.-J., James, T., 2008. Do escaped transgenes persist in nature? The case of a herbicide resistance transgene in a weed population of *Brassica rapa*. Mol. Ecol. 17, 1387–1395.

Webber, B.L., Raghu, S., Edwards, O.R., 2015. Opinion: is CRISPR-based gene drive a biocontrol silver bullet or global conservation threat? Proc. Natl. Acad. Sci. U. S. A. 112, 10565–10567.

Winston, R.L., Schwarzländer, M., Hinz, H.L., Day, M.D., Cock, M.J., Julien, M.H., 2014. Biological Control of Weeds: A World Catalogue of Agents and their Target Weeds, fifth ed. USDA Forest Service, Forest Health Technology Enterprise Team, Morgantown, WV.

Zapiola, M.L., Campbell, C.K., Butler, M.D., Mallory-Smith, C.A., 2008. Escape and establishment of transgenic glyphosate-resistant creeping bentgrass *Agrostis stolonifera* in Oregon, USA: a 4-year study. J. Appl. Ecol. 45, 486–494.

# Index

Note: Page numbers followed by *f* indicate figures, *t* indicate tables, and *b* indicate boxes.

## A

Abiotic conditions, changes in, 23–24
    climate and physicochemical conditions,
        85–88, 86*f*
    habitat availability, 82–85
    heterogeneity, 82–85
*Acacia*, 169–170, 170*f*
*Acacia mangium*, 21
*Acacia saligna. See* Port Jackson willow
*Achatina fulica. See* African land snail
*Acridotheres tristis. See* Indian myna
Activated carbon, 141
*Acyrthosiphon pisum. See* Pea aphid
Adaptation
    genetic signatures of
        genomics, 175–177
        neutral genetic makers, 174–175
        trait data, 174–175
        transcriptomics, 175–177
        whole-genome sequencing, 177–178
    local, 23–24, 84–85, 117–118, 121–123,
        167–168
    native animals
        to compete with invasive animals,
            141–142
        to invasive predators, 142–143
    native predators to invasive prey, 143–145
    native species to invasive mutualists,
        145–147
Adaptive flexibility, 64
Admixture, 14–15, 20, 119–121, 165–166,
    194
African land snail, 149–150
*Ageratina adenophora. See* Crofton weed
Agricultural pests, 91–92
*Alectryon tomentosus. See* Woolly rambutan
    tree
Allelochemicals, 140–141, 148–149
*Alliaria petiolata. See* Garlic mustard
*Allophylus cobbe. See* False currant
Allopolyploid cordgrass, 26
Allozymes, 160

*Ambrosia artemisiifolia. See* Common ragweed
American red swamp crayfish, 142–143,
    171–172
Anal pads, 100–101
*Anas platyrhynchos. See* Mallard duck
*Anolis carolinensis. See* Green anole lizard
*Anolis sagrei. See* Brown anole lizard
*Anopheles gambiae*, 199–200, 202
Anthropocene, 1
    biodiversity in, 2
    habitats, 2–3
Aphids, 39, 88–89
*Apis mellifera. See* Honey bee
*Arabidopsis thaliana*
    habitat availability and heterogeneity,
        82–84
    histone modifications and phenotypic
        variation, 61
Arbuscular mycorrhizal fungi (AMF), 89
Argentine ant, 44–45, 177–178
Argentine cactoblastis moth, 190–191
Argentine stem weevil, 193–194
Artificial selection, 25, 191–192, 194–195
Assortative mating, 14–15, 23, 117–118
*Atalaya hemiglauca. See* Whitewood tree

## B

*Bacillus thuringiensis*, 91–92, 199
Balearic lizard, 146–147
Balloon vine, 45–47, 46–47*f*, 196–197
Beach gladiolus, 84
Big-headed ant, 115
Biocontrol agents, 192–198
Biodiversity, in Anthropocene, 2
Biological invasions, 2–3, 12–14, 48–49,
    162–163
    anthropocene habitats, 2–3
    in benign environments, 18
    microevolution
        artificial selection, 23–25
        assortative mating, 23
        founder events, 16–19

Biological invasions *(Continued)*
  genetic drift, 16–19
  hybridization, 19–23
  inbreeding, 16–19
  mutation, 19–23
  natural selection, 23–25
  whole-genome duplication, 19–23
Biotic conditions, changes in
  competition, 90
  food resources, 91–92
  mutualisms, 88–90
Biotic interactions, 36–39, 37–38*t*
Birdsfoot trefoil, 146
Black swallowwort, 197–198
Bladder campion, 20
Blueberry, 150
Bluebunch wheatgrass, 140–141
Blue-eyed Mary, 136
Bluehead sucker fish, 148
*Boiga irregularis. See* Brown tree snake
*Brachypodium sylvaticum. See* Slender false
  brome
Brown anole lizard, 141–142
Brown tree snake, 43
Bt toxins, 91–92, 199
Burr medic, 102–114, 121
*Butomus umbellatus*, 16

**C**
*Cactoblastis cactorum. See* Argentine
  cactoblastis moth
Cane toad, 11–14, 13–14*f*, 42, 48, 117–118
  dispersal, 84
  habitat availability and heterogeneity, 84
  range expansion, 12
  toxins, 143–144
  trait diversification, 176–177
*Cardiospermum corindum*, 47
*Cardiospermum grandiflorum. See* Perennial
  balloon vine
*Cardiospermum halicacabum*, 47
Carolina horsenettle, 17–18
*Carpobrotus chilensis. See* Sea fig
*Carpobrotus edulis. See* Hottentots fig
Cas endonuclease, 201–202
*Catostomus commersonii. See* White sucker fish

*Catostomus discobolus. See* Bluehead sucker
  fish
*Catostomus latipinnis. See* Flannelmouth
  sucker fish
*Caulerpa taxifolia*, 137–138
*Centaurea solstitialis. See* Yellow starthistle
*Centaurea stoebe. See* Spotted knapweed
Chinese red pine, 27
Chinese tallow, 24
Chromatin, 61
Chromatin remodeling, 61
Cichlids, 42–43
*Ciona savignyi*, 66–68, 67*f*
Classical biological control, 190–191
  evolutionary interventions, 192–195
  native enemies as biocontrol agents,
    195–198
Clearweed, 148–149
Climate, 85–88, 102–114
Clustered, regularly interspaced short
    palindromic repeat genes
    (CRISPR), 201–202
*Cneorum tricoccon. See* Spurge olive
*Coccinella septempunctata. See* Seven-spot
    ladybird
Codling moth, 91
Cold tolerance, epigenetic regulation
    of, 68
*Collinsia parviflora. See* Blue-eyed Mary
Colonisation, 169–171
Common keelback snake,
    42, 143–144
Common ragweed, 85–86, 118
Common tree snake, 143–144
Competition, 90
  interspecific, 115
  sperm, 100–101
Competitive interactions, 36–38
Contemporary evolution, 45–48,
    191
Corticosterone, 69
*Coturnix chinensis*, 43
Crayfish, 40–42
Crofton weed, 68
*Cydia pomonella. See* Codling moth
*Cytisus scoparius. See* Scotch broom
Cytochrome P450, 178

## D

*Dactylopius opuntiae*, 190–191
*Danaus plexippus*. *See* Monarch butterfly
Dandelion, 66–68
Darwin's naturalisation hypothesis, 3, 36, 43, 80–81
*Dasyurus hallucatus*. *See* Northern quoll
Demographic processes, 4, 81–82, 163–164, 167
*Dendrelaphis punctulatus*. *See* Common tree snake
*Dendroctonus valens*. *See* Red turpentine beetle
*Dermolepida albohirtum*. *See* Grey-back cane beetle
Diamondback moth, 91–92
*Digitalis purpurea*. *See* Foxglove
DNA methylation, 26, 59–63, 66–68
DNA repair, 201–202
*Drosophila subobscura*. *See* Fruit fly

## E

Earth System, 1–2
Eco-evolutionary experience (EEE), 15–16, 38–39, 138, 140, 143–144
  contemporary evolution, 45–48
  ecological impacts, 45–48
  exotic species
    native competitor interactions, 43–44
    native herbivore/predator/pathogen interactions, 40–42
    native mutualist interactions, 44–45
    native prey/resource interactions, 42–43
  and invasiveness, 40–45
Ecological drift, 26–27
Ecological interaction networks, 39, 49
*The Ecology of Invasions by Animals and Plants*, 3–4
Ectomycorrhizal fungi (EMF), 45
Ectotherms, 85
Edith's checkerspot, 136
EEE. *See* Eco-evolutionary experience (EEE)
*Eichhornia crassipes*, 17
Enemy release hypothesis, 36–38, 40, 44

Engineering genomes, as management tool, 199–200
Environmental drivers, of rapid evolution
  abiotic conditions, changes in
    climate and physicochemical conditions, 85–88, 86f
    habitat availability, 82–85
    heterogeneity, 82–85
  biotic conditions, changes in
    competition, 90
    food resource, 91–92
    mutualisms, 88–90
Epigenetics, 25–26, 69–71
Epigenetic variation, 25–26
  chromatin remodelling, 61
  DNA methylation, 59–61
  ecological consequences, 63–64
  evolutionary consequences, 64
  facilitated, 59
  histone modifications, 61
  mechanisms, 59–63
  non-coding RNAs, 61–63
  obligate, 59
  phenotypes, 56–58
  pure, 59
  range expansions, 68–69
  role in establishment success, 64–68
Erect prickly pear, 190–192
*Euglandina rosea*. *See* Rosy wolf snail
*Euhrychiopsis lecontei*. *See* Milfoil weevil
*Euphydryas editha*. *See* Edith's checkerspot
Eurasian watermilfoil, 40–42
European starling, 172–173, 173f
  habitat availability and heterogeneity, 84
  mitochondrial mutation in, 19–20
Evolutionary impacts
  direct, 140–147
  genetic, 147–148
  indirect, 148–149
Evolutionary traps, 197–198
Evolution of increased competitive ability (EICA), 24, 90, 118–119
Exotic species
  evolution of increased competitive ability (EICA), 24
  inbreeding, 18–19

Exotic species *(Continued)*
  introduction-naturalisation-invasion-
      continuum (INIC), 15–16
  native competitor interactions, 43–44
  native herbivore/predator/pathogen
      interactions, 40–42
  native mutualist interactions, 44–45
  native prey/resource interactions, 42–43
Extinction, 149–151

**F**

False currant, 196–197
Flannelmouth sucker fish, 148
*Floracarus perrepae*, 167–168
Fluridone, 191
Fluridone resistance, in hydrilla, 191
Food resources, 91–92
Foxglove, 114–115
*Frankliniella occidentalis. See* Western flower
      thrips
French's cane beetle, 11–12
Fruit fly, 87–88

**G**

Garlic mustard, 148–149, 197–198
Genetic admixture, 20, 119–121, 161–162
Genetically modified (GM) organisms, 199
Genetic diversity, 119–121, 161, 169–170
Genetic drift, 16–19, 117, 137–138,
      163–164, 176–177
Genetic engineering (GE), 199–200
Genetic markers, neutral, 174–175
Genetic paradox, 121, 169
Genetics
  of colonisation, 169–171
  of dispersal, 171–174
*The Genetics of Colonizing Species*, 4, 6,
      160–164
Genome doubling, in *Spartina × townsendii*,
      69–71
Genomics, 175–177
Genotype x environment interactions
      (GxE), 14–15
Genotypes, 14–15, 21, 145, 168, 171
*Gladiolus gueinzii. See* Beach gladiolus
Great Acceleration, 2–3
Green anole lizard, 141–142

Grey-back cane beetle, 11–12
gRNA-Cas complex, 201

**H**

Habitat
  availability, 82–85
  heterogeneity, 81–85
*Harmonia axyridis. See* Ladybird
*Hemiaspis signata. See* Swamp snake
HGT. *See* Horizontal gene transfer (HGT)
Himalayan balsam, 115–116
Histone modifications, 61
Holobionts, 26–27
Hologenomes, 26–27
Honey bee, 63–64
Honeysuckle, 150, 151*f*
Horizontal gene transfer (HGT), 145–146
Hottentots fig, 20–21
House sparrow, 64–65, 69, 70*f*
Hybridisation, 19–23, 69–71, 166
  genetic impacts via, 147–148
  interspecific, 163–164
Hydrilla, 191
*Hydrilla verticillata. See* Hydrilla
*Hypericum perforatum. See* St. John's wort

**I**

Ibiza wall lizard, 146–147
*Impatiens glandulifera. See* Himalayan balsam
Inbreeding, 16–19
Inbreeding depression, 17–18
Increased susceptibility hypothesis, 40
Indian dwarf mongoose, 100–101
Indian myna, 84
Introduction-naturalisation-invasion-
      continuum (INIC), 15–16
Introgression, 147
Invasion genetics
  colonisation, 169–171
  dispersal, 171–174
  inferring invasion histories, 166–168
  molecular innovations, 162*b*
  taxonomy, 165–166
Isolation by distance (IBD), 172–173
Italian agile frog, 142–143

**J**

*Jadera haematoloma. See* Soapberry bug
Japanese knotweed, 56, 58, 72

**L**

Ladybird, 18–19, 19*f*, 84, 87–88, 194
*Lates niloticus*, 42–43
Legume-rhizobium symbiosis, 145
Legumes, 145
*Lepidiota frenchi. See* French's cane beetle
*Leptocoris tagalicus. See* Soapberry bug
*Leptographium procerum*, 27
Limiting similarity hypothesis, 44
*Linaria vulgaris. See* Yellow toadflax
*Linepithema humile. See* Argentine ant
*Listronotus bonariensis. See* Argentine stem
    weevil
Local adaptation, 23–24, 84–85, 117–118,
    121–123, 167–168
*Lotus corniculatus. See* Birdsfoot trefoil
*Lygodium microphyllum*, 167–168
*Lythrum salicaria. See* Purple loosestrife

**M**

Mallard duck, 147–148, 166
Marbled crayfish, 21–23, 22*f*, 65–66
*Martes martes. See* Pine marten
*Medicago polymorpha. See* Burr medic
Mendelian inheritance, 200
*Microctonus hyperodae*, 193–194
Microevolution, biological invasions
    artificial selection, 23–25
    assortative mating, 23
    founder events, 16–19
    genetic drift, 16–19
    hybridisation, 19–23
    inbreeding, 16–19
    mutation, 19–23
    natural selection, 23–25
    whole-genome duplication, 19–23
Microsatellites, 161
Milfoil weevil, 40–42
*Mimetes cucullatus. See* Red-crested pagoda
Missed mutualisms hypothesis, 36–38, 45
Mitochondrial DNA (mtDNA), 161
Modern Evolutionary Synthesis, 64
Molecular memory, 58

Monarch butterfly, 197–198
Mutation, 19–23
Mutualism, 44–45, 88–90, 145–147
Mycorrhization, 88–89
*Myriophyllum spicatum. See* Eurasian
    watermilfoil
Myrmecochory, 88–89

**N**

Naïve prey hypothesis, 42–43
National Centre for Biotechnology
    Information (NCBI), 177
Native animals, adaptation
    to compete with invasive animals,
      141–142
    to invasive predators, 142–143
Native enemies, as biocontrol agents,
    195–198
Native plants, adapting to compete with
    invasive plants, 140–141
Native predators, adaptation to invasive
    prey, 143–145
Native species
    adaptation to invasive mutualists,
      145–147
    eco-evolutionary experience, 41*f*, 42–43
    evolutionary responses, 23–24
    phylogenetic relatedness, 38–39
    selection on, 138, 139*f*
Natural selection, 23–25
*Natural Selection in the Wild*, 80
New associations hypothesis, 40
Next-generation sequencing (NGS), 162,
    164–165
*Nicotiana glauca. See* Tree tobacco
Nodulation genes, 145
Non-coding RNAs (ncRNA), 61–63
Northern quoll, 144–145
Novel weapons hypothesis, 36–38, 43

**O**

*Opuntia stricta. See* Erect prickly pear

**P**

Parasitism, 192–194
Parasitoids, 39, 92
*Passer domesticus. See* House sparrow

Pea aphid, 192–193
Perennial balloon vine, 197
*Phaulacridium vittatum*, 175–176
*Pheidole megacephala. See* Big-headed ant
Phenotypic plasticity, 56–58, 68, 86–88, 164–165
Phenotypic variation, 19–20, 25–27, 56–58, 64
Phylogenetic relatedness, 36–39, 49, 196
Phylogeography, 161
Phytochemicals, 140–141
Pierid butterfly, 197–198
*Pilea pumila. See* Clearweed
Pine marten, 146–147
Pine tree, 88–89
*Pinus tabuliformis. See* Chinese red pine
*Plantago lanceolata. See* Ribwort plantain
Plasticity, 86–87
*Plutella xylostella. See* Diamondback moth
*Podarcis lilfordi. See* Balearic lizard
*Podarcis pityusensis. See* Ibiza wall lizard
Pollination, 88–89
Polyploidization, 21, 69–71
Port Jackson willow, 44–45, 165–166, 171
Preadaptation, 4, 15–16, 38–39, 138, 195–196
Predation, 43, 80, 142–143
*Procambarus clarkii. See* American red swamp crayfish
*Procambarus fallax. See* Slough crayfish
*Procambarus virginalis. See* Marbled crayfish
Propagule pressure, 167, 169–171
*Pseudechis porphyriacus. See* Red-bellied black snake
*Pseudoroegneria spicata. See* Bluebunch wheatgrass
Purple loosestrife, 86*f*, 90

**Q**

Quantitative trait variation, 123

**R**

*Rana latastei. See* Italian agile frog
Random dispersal, 26–27

Rapid evolution, 101–102, 103–113*t*
    abiotic conditions, changes in
        climate and physicochemical conditions, 85–88, 86*f*
        habitat availability, 82–85
        heterogeneity, 82–85
    biotic conditions, changes in
        competition, 90
        food resource, 91–92
        mutualisms, 88–90
    causes of, 102–118
        abiotic factors, 102–114
        biotic factors, 114–117
        stochastic factors, 117–118
    experimental design, 122–123
    introduction history, 119–121
    traits, 118–119
Rapid range expansions, 68–69, 70*f*.
    *See also* Epigenetic variation
Reciprocal transplant experiment, 122–124, 148–149, 167–168
Red-bellied black snake, 143–144
Red-crested pagoda, 44–45
Red turpentine beetle, 27
Reid's paradox, 172–173
Resource-enemy release hypothesis, 40–42
*Rhinella marina. See* Cane toad
Rhizobia, 145–146
Ribwort plantain, 136–137
Rosy wolf snail, 149–150

**S**

St. John's wort, 84–85, 89, 167
*Sapium sebiferum. See* Chinese tallow
Scotch broom, 146
Sea fig, 20–21
Seven-spot ladybird, 87–88
Sex determination, 199–200
Shifting defence hypothesis, 40
*Silene vulgaris. See* Bladder campion
Single nucleotide polymorphisms (SNPs), 175–176
Slender false brome, 174–175
Slough crayfish, 21–23
Smooth cordgrass, 18, 62–63, 150
Snowberry, 150
Soapberry bug, 138, 196–197, 198*f*

*Solanum carolinense*. *See* Carolina horsenettle
*Spartina alterniflora*. *See* Smooth cordgrass
*Spartina anglica*, 69–71. *See also* Allopolyploid cordgrass
Spatial sorting, 14–15, 23, 68, 82–84, 117–118
Speciation, 149–151
Sperm competition, 100–101
Spotted knapweed, 140–141
Spurge olive, 146–147
*Sturnus vulgaris*. *See* European starling
Sunbird, 88–89
Swamp snake, 143–144
Symbiotic genes, 146

**T**

*Taraxacum officinale*. *See* Dandelion
Toll-like receptor 4 gene (*TLR4*), 69, 72
Transcriptomics, 175–177
Transgenes, 200, 202
Transposon islands, 60–61
Transposons, 60–61
Tree tobacco, 88–89
*Trichogramma pretiosum*, 194–195

Tripartite symbiosis, 27
*Tropidonophis mairii*. *See* Common keelback snake

**U**

*Urva auropunctata*. *See* Indian dwarf mongoose

**V**

*Vincetoxicum nigrum*. *See* Black swallowwort

**W**

Western flower thrips, 194–195
White sucker fish, 148
Whitewood tree, 196–197
Whole-genome duplication, 19–23, 69. *See also* Polyploidization
Whole-genome sequencing, 177–178
Woolly rambutan tree, 197

**Y**

Yellow starthistle, 168–169
Yellow toadflax, 63

| | | |
|---|---|---|
| 4.18 | The Scene Contained | 95 |
| 4.19 | Perpetrator in the Doorway | 96 |
| 4.20 | Three Moods of Mom | 98 |
| 5.1 | Solid Coping Foundation | 105 |
| 5.2 | The Strong Tower Holds Up Through Heavy Rains | 106 |
| 5.3 | Creating Glimpses and Watching Them Disappear | 107 |
| 5.4 | A Broken Heart in the Kinetic Sand | 108 |
| 5.5 | A Broken Heart on Canvas | 109 |
| 5.6 | The Owls Can't Come Out | 111 |
| 5.7 | The Mighty Battle | 112 |
| 5.8 | Two Guards Come to Help. The First Humans in the Animal World | 113 |
| 5.9 | Manta Ray | 115 |
| 5.10 | Full of Flies | 117 |
| 6.1 | Fire-Breathing Nun Symbolizing Anger | 120 |
| 6.2 | Sam's Somatic Experience of Anger | 121 |
| 6.3 | Sam's Elaboration of Self-Harm | 122 |
| 6.4 | Linnea's Expression of Anger | 123 |
| 6.5 | Unbridled Fear | 125 |
| 6.6 | The Real Josh | 127 |
| 6.7 | Worried Hand Puppet | 128 |
| 6.8 | Enhancing Emotional Connections | 130 |
| 6.9 | Left Out | 135 |
| 6.10 | Sydney's Snake Coiling | 140 |
| 6.11 | Hopeless | 142 |
| 6.12 | Hopelessness Held | 143 |
| 7.1 | Trust and Risk in Action | 148 |
| 7.2 | They Don't Make 'Em Like They Used To!!! | 150 |
| 7.3 | Risk | 151 |
| 7.4 | Still Risking | 152 |
| 7.5 | Safely Across | 153 |
| 7.6 | Rock Pancake With Chocolate Chips and Bananas: Delicious! | 156 |
| 7.7 | The Raging and the Still Water | 172 |
| 7.8 | Late Bloomer and Early Bloomer | 174 |
| 8.1 | The Love Is All Over Their Faces | 187 |
| 8.2 | Fake Teeth | 189 |
| 8.3 | Changing Face and Voice App | 190 |
| 9.1 | Recognizing Numbness | 202 |
| 9.2 | Individual Growth | 203 |
| 9.3 | The Five-Year-Old Harley | 204 |
| 9.4 | Jenny Accepts Harley as a Safe Part | 205 |
| 9.5 | Just Another Body | 206 |
| 9.6 | A Human Being Front View | 208 |
| 9.7 | Take It or Leave It | 209 |
| 9.8 | Encoding Transmission of Positive Attributes | 216 |

9.9    The Baby in Mother's Womb Activates the Client           218
9.10   Self-Object Covered in Handcuffs and Weapons             221
9.11   Hard to Forget                                           222
9.12   Unintegrated Sense of Self                               223
9.13   First Cover of Sally's Life Book                         225
9.14   Sally's Internal Story About Her Biological Mom          226
9.15   Family of Origin                                         227
9.16   Internalized Nuclear Family                              227
9.17   A Day at the Beach, Emerging Boundaries                  228
9.18   Future Circle Sandtray                                   230

# Foreword

*Eliana Gil*

From my point of view, some authors inspire us with academic theories and scientific facts while others guide us with practical, clinical applications that bring theories to life. Paris Goodyear-Brown is unique in her ability to do both: she presents a conceptual foundation for building a trauma-informed practice, and then she gives us specific examples that illustrate how theories inform (and anchor) the creation of developmentally appropriate, inviting, and innovative interventions. Thus, this book reflects a truly integrated model that both informs and inspires those of us for whom working with childhood trauma is our passion.

This book affirms that we've come a long way. In the early and mid-seventies, there were several courageous professionals working hard to bring attention to a number of different topics: child maltreatment, child sexual abuse, traumatic impact, attachment, and domestic violence, among others. We truly stand on the shoulders of greats. And in these early efforts, the topics emerged separately and reflected work that had been ongoing for years. But they were occurring discretely, without a lot of dialogue and information-sharing. We are fortunate indeed that these topics have emerged as shared interests. As professionals working with children and families, we have all gained a great deal by being exposed to these previously separate topics, as well as the field of neuroscience, which has provided us with yet another shared context.

This book is a snapshot of what trauma and attachment specialists hold dear at this moment in time. Goodyear-Brown challenges us to shift paradigms in favor of treating the client or family as uniquely individual. Her model is a motif, a palette, not a rigid protocol. She defies the idea of "either/or," inviting professionals to be more or less active, to lead and to follow, along a continuum. This requires more from clinicians. It requires them to make assessment an ongoing process and to be willing and able to shift approaches based on the child client's changing needs. There is an obvious privileging of play therapy as a treatment of choice; however, as a truly integrated model, it is flexible and respectful of other evidence- and practice-based interventions that promote trauma-focused goals. These other approaches fit easily into this model, finally setting aside the separatism that can exist among those who support one approach to the exclusion of others.

The model is user-friendly; it proposes that the child leads the way and clinicians follow the child's needs at all times. It provides easily applied "components" that respond to specific trauma-related symptoms or concerns. It is guided by the healing power of relationships and safety. It incorporates prevention, anticipating children's needs, and being prepared to help with a range of behavioral, social, psychological, and relational concerns.

Goodyear-Brown walks the walk and talks the talk. She has built a child and family treatment center, Nurture House, and it is an incredible example of how we can incorporate what we know into consistent, warm, and empathic practice. This book describes how she puts her knowledge into practice, both in program development and in the practice of child and family therapy. There is ample evidence of all her greatest current influences, including Theraplay, polyvagal theory, neurosequential treatment, Circle of Security, and, of course, trauma-focused play.

I congratulate Paris for pulling together all the threads of knowledge, for adding her own creativity and conceptual genius, and for leading with the heart. The interactions she demonstrates with her child clients include respectful touch, co-regulating relational opportunities, respect and dignity, necessary structure and limit-setting, coherent narrative building, and kindness. This is a must-read for anyone hoping to be of true service to others.

Coincidentally, I reviewed this book during the same time frame that the US government had a policy of separating (immigrant) children from their families at the southern border. The news coverage has been appropriately focused on this unprecedented policy, and many child welfare and health professionals have made clear and consistent statements about the potential traumatic impact of these separations on young children and their parents. Reading this exquisitely thoughtful and sensitive book at this time made the author's voice even more compelling, more relevant, and more necessary.

# Preface

As I reflect on the words of Eliana Gil's Foreword, I am humbled, challenged, and inspired. I feel this particular combination of emotions more and more frequently these days as I watch our field maturing, as I watch our professional community grappling with the numerous disheartening aspects of our world, and as I watch the warrior spirits of my colleagues as they hold the hard stories of traumatized children while simultaneously holding out hope for healing.

The complexity of tiny humans who have somatically, emotionally, cognitively, and relationally encoded trauma responses requires mental health professionals who work with children to be lifelong learners. I have been taking a deep dive into what it means to titrate a child's approach to trauma content, to appreciate the myriad ways children invite us into their play and their pain, and the power of the one to heal the other. This volume marks the debut of TraumaPlay™, the simplified name of what has been formerly called Flexibly Sequential Play Therapy (FSPT). The evolution from FSPT to Trauma-Play™ mirrors the evolution in my own thinking, my deepening delight in the unlikely power of pairing play with trauma to leach the emotional toxicity out of hard things. These two words paired so closely together are likely to give pause to those who have not yet seen the magic and the mystery that happen as play mitigates the child's approach to trauma content. Just as trauma comes in a bewildering number of forms, so does play. This text will offer a new tool—the Play Therapist's Palette—that articulates in graphic form many of these modes of play in an easily accessible format. Chapters will expand on the nuance of applying these tools to trauma work with children and families. Join me as we explore the power of play together.

# Acknowledgments

A huge thank you, Eliana, for your pioneering work in articulating posttraumatic play. I remember the first time I heard you speak 20-some years ago: your heart, knowledge, and passion painted a path for my growth. Your voice has continued to be the most influential in my conceptualization of trauma work with children and families, and I am honored to have you add your voice to this book.

I am indebted to so many others who have influenced my thinking and shaped my approach. These influences include Charlie Schaefer, Garry Landreth, Sue Bratton, Heidi Kaduson, David Crenshaw, Phyllis Booth, Evangeline Munns, Joyce Mills, Linda Homeyer, Rise van Fleet, Athena Drewes, and Rick Gaskill, to name a few. As the field of interpersonal neurobiology has expanded, I have gained depth of understanding and augmented articulation from the work of Dan Siegel, Bruce Perry, Bessel van der Kolk, Alan Schore, Stephen Porges, Bonnie Badenoch and Theresa Kestly.

Thanks to my play therapy tribe. There are too many to name, but you know who you are! A special thanks to my play therapy women's writing retreat group—Clair, Sueann, Jess, Angie, and Holly—for carving out time and space for writing and laughing.

Thanks to the team that has helped me build Nurture House—Linnea, Kristen, Eleah, Elizabeth, Amy, Eric, Jenny, Kate, Jess, Shelby, Bethany, and Stacey, to name a few. They have grown with me, encouraged me, and held down the fort when I am off gallivanting.

Lastly, my deepest gratitude for my best friend and partner in all things, Forrest. Thank you for taking up the slack again and again with grace and strength—as a parent and as tech support—and for being the love of my life and my biggest encourager. Thank you, Sam, Madison, and Nicholas, my sweet ones, for bringing me hugs, snacks, silly songs, encouraging notes, and for giving me belly laugh breaks from the writing.

# 1  Titration in Trauma Work
## Using the Play Therapist's Palette

This project evolved as a response to repeated requests from clinicians who are familiar with the Flexibly Sequential Play Therapy (FSPT) trauma model, outlined in *Play Therapy with Traumatized Children* (Goodyear-Brown, 2010), to more fully explore the nuance of how we help children move toward and away from trauma content to create the story of what happened. The overarching goal of the work is to leach the emotional toxicity out of the child's experience so that it can be integrated into a healthy sense of self. This model, formerly known as FSPT, has been renamed TraumaPlay™ in order to give more clarity to clinicians and clients alike. TraumaPlay™ is components-based and allows for a variety of interventions to be placed along a continuum of treatment, depending on which treatment goal is actively being addressed. Goals in the early phases of treatment include building safety and security, addressing and augmenting coping, soothing the physiology, enhancing emotional literacy, and helping parents be better partners in regulation while offering additional caregiver support to help them hold the hard stories of the children in their care. The middle phases of treatment provide some form of play-based gradual exposure that may include the continuum of disclosure, experiential mastery play, and/or trauma narrative work. The final phase of treatment is helping the child and family make positive meaning of the post trauma self.

When to invite children to go deeper, when to respect their defenses, when to acknowledge their retreat, when to celebrate that they have come to the end of what they can approach right now, and how to support their exposure work are questions with which both beginning and seasoned clinicians wrestle. An attuned trauma therapist is deciding, sometimes on a moment-to-moment basis, when to invite children out of stuck places, when and how much to push up against avoidance symptoms, when to witness their posttraumatic play (Gil, 2017), and when to support the need to simply rest. This volume is meant to expand significantly on the nuance of these questions while offering a new tool for thinking about the mitigators of a client's approach to hard things.

If you work with traumatized children regularly, my guess is that you have already found yourself in a playroom environment with a child where you have been unsure when to offer invitations to further explore the traumatic event and when to allow the play to do its own healing work. As I train people both here and abroad, the struggle I hear clinicians wrestling with revolves around the question of when, how often, and how directly do I invite children to explore the trauma content in order to bring more coherence and when do I create a space of respite, mindfulness, and the simple and powerfully healing enjoyment of play? If we neglect the avoidance symptoms of posttraumatic reactions, we inadvertently collude with the trauma itself, communicating to the child that it is indeed too big and scary to be approached. If we push a child too hard or too fast, we can cause iatrogenic effects, flood them with anxiety, and shut down their healing process.

Another way this question gets put to me in training sessions is, when do you lead and when do you follow? The question is critically important and always excites me because the clinician asking is wanting to follow the child's need rather than becoming dogmatic about either following the child's lead or directing the therapeutic interactions in the room. A third way this question is asked is as follows: when does the therapist invite the child/family into new kinds of interaction and when does the therapist wait to be invited? Moreover, how do we extend moments of therapeutic work or help open and close circles of communication? Sometimes this question is not specific to the traumatic event but may be asked in relation to other goals of treatment, ones that would be considered symptomatic of the trauma itself, such as self-regulation abilities, hyperarousal symptoms, and a child's ability to connect with others.

My response to this question of when to lead and when to follow, when to be more or less directive, is unpopular in our current protocolized culture: it depends. Beginning clinicians really do not like this answer, as the clearly defined steps of a protocol are soothing to them. As we grow, we become more comfortable with ambiguity and with trusting the process, but it remains important, even for seasoned clinicians, to ask this same question of when to lead and when to follow in a present-minded way in each and every session. One of my goals in supervision is to help clinicians become more comfortable with not having a definitive answer for what is needed per protocol so they can hone their ability to follow the child's need from moment to moment. My goal is to help clinicians nurture the nuance in trauma work with children and families.

## Nurturing the Nuance

Nuance is necessary—arguably in all forms of therapy and most certainly when using play therapy—with traumatized children and families. I have learned this lesson by going too slowly with a client system—not offering enough invitation to change or not providing enough strengthening of the muscles that hold hard things. I have also failed in the opposite extreme by attempting to take a child too quickly toward hard things. My concern about treatment becoming "cookie cutter-ish" applies to my model as much as to any other model of treatment. The flow of the treatment goals codified in TraumaPlay™ is pictured in Figure 1.1).

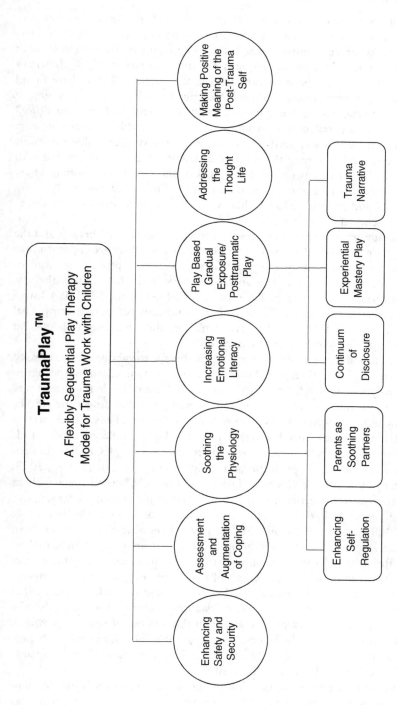

*Figure 1.1* Flowchart of TraumaPlay™, Formerly Known as FSPT

As clinicians have used the model, they have brought back questions and insights that have informed this deeper delineation of factors that I am offering here. As clinicians move through the components of TraumaPlay™, each individual goal of the model requires a nuanced titration of the clinician's therapeutic self to see positive movement in the system. What we have found over and over again in supervision is that nuance is necessary every step of the way. In other words, the questions of "how much?" and "in what dose?" become constant reflections, helping clinicians hone this approach in ways that meet the individual needs of the client in front of them. Along the way, we are constantly aware of the titration of many aspects of engagement with the child and/or caregivers. The first goal of the model, establishing safety and security, may take minutes or months depending on the unique dynamics of the case and the interaction of these with the unique dynamics of the therapist's presence and play space.

A continuum of directiveness exists within the field of play therapy, and the consensus is growing that there are times when a nondirective approach may be most beneficial and other times when a directive approach may lead to greater therapeutic growth. It is an exciting time in the field of play therapy. Several forms of play therapy have just been added to the SAMHSA list of evidence-based treatments. These include child-centered play therapy, filial therapy, and CPRT (all less directive in nature) and Theraplay and Adlerian play therapy (both more directive in nature).

With the addition of certain forms of play therapy to various evidence-based treatment lists, play therapy is becoming more recognized and respected in the larger therapy world. Although more research is needed in the area of play therapy, a vast amount of experimental research exists that proves the efficacy of many different forms of this therapy. Many of these studies were essential in the attainment of evidence-based therapy status. For example, numerous studies spanning over seven decades show the efficacy of using CCPT for children with various presenting problems and of differing ages (Bratton et al., 2013; Cochran, Nordling, & Cochran, 2010). Additionally, several meta-analyses show the general effectiveness of various types of play therapy interventions for children and adolescents (Bratton, Ray, Rhine, & Jones, 2005; Bratton & Ray, 2000; Leblanc & Ritchie, 2001). Beyond showing the effectiveness of play therapy, many studies also proved the effectiveness of additional clinical factors that most play therapists intrinsically incorporate into their practice. These additional factors, length of treatment and parental involvement, also impact the effectiveness of play therapy. Though research does show that play therapy is effective in short-term instances, it also demonstrates that the effectiveness of play therapy increases with the number of sessions provided. Numerous studies also show that when a parent is fully immersed in therapy and has ample opportunity to practice new skills, the chances for therapeutic success are increased (Bratton et al., 2005).

Unlike studies performed on adult populations, which usually exhibit similar results, play therapy research provisionally shows that some forms of play therapy are more effective with certain populations than others (Bratton et al., 2005). As mentioned in this chapter, play therapy exists on a continuum, and at times a client needs both directive and nondirective approaches in the same session. This recent research adds even more importance to the clinical task of assessing our client's needs and responding in an effective manner.

## Core Agents of Change in Play Therapy

Those of you who are familiar with Charlie Schaefer's *The Therapeutic Powers of Play* (Schaefer & Drewes, 2014) will embrace the idea that play provides therapeutic value of its own, that play can be the bridge to attachment enhancement, and that play can lead to increased solutioning abilities, foster useful exploration of roles (that may later be embraced or discarded) through a judgment-free role play environment, etc. These therapeutic powers of play are referenced here and broken down into broader categories of therapeutic goals (see Figure 1.2).

If you are new to play therapy as an overarching approach, it is worth familiarizing yourself with these, as these "powers" are foundational to the process,

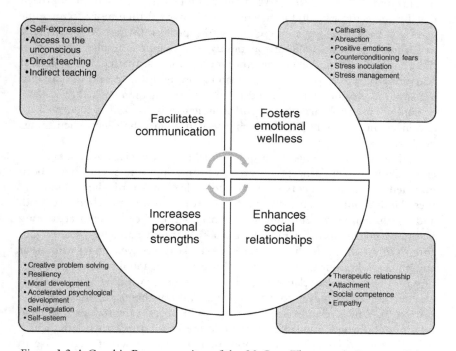

*Figure 1.2* A Graphic Representation of the 20 Core Therapeutic Powers of Play

Parson, J. (2017). Puppet Play Therapy-Integrating Theory, Evidence and Action (ITEA) presented at International Play Therapy Study Group. Champneys at Forest Mere, England. June 18. Adapted from Schaefer, C. E., & Drewes, A. A. (2014). *The therapeutic powers of play: 20 core agents of change* (2nd ed.). Hoboken, NJ: Wiley Publishing.

and only in a play therapy environment where these powers have been given freedom to work can clinicians begin to ask themselves these questions of nuance. Charlie has quoted Gordon Paul's question to me many times over the years: "What treatment, by whom, is most effective for this individual with that specific problem, under which set of circumstances, and how does it come about?" (Paul, 1967, p. 111). Gordon Paul talked about the set of change mechanisms as "active ingredients," and Schaefer and Drewes (2014) talk about them as "core change agents."

Developing a treatment plan for a traumatized child requires us to identify the areas in which a child's window of tolerance for the stress involved in basic life tasks has been compromised. In which area is the child's window of tolerance most compromised? Is the child's window of tolerance for acknowledging or holding big feelings the most compromised? Is the child's window of tolerance for imperfection or mess the most compromised? Is the child's ability to tolerate unknowns compromised? If so, which approach will help to expand the child's window of tolerance in a baby-stepping way that does not create iatrogenic effects by tripping the child's neurophysiology abruptly into hyperarousal or hypoarousal/collapse? This set of concerns is always in the background as I am working with a family. Once I have chosen the approach that I think is most helpful in this moment of work, I then move to my mitigators. I have found that these mitigators protect therapists—particularly new clinicians—from holding too tightly to a dogmatic implementation of the TraumaPlay™ model, or any other model, for that matter. The mantra at Nurture House is to prepare for a session thoroughly, have a plan going into the session, and then let it go and be in the present moment with the child or family. The whole array of potential mitigators is "tucked in their pocket" and gives them options for continuing to dance with the client when things get hard in session.

During the assessment phase of TraumaPlay™, clinicians are looking for clues about which aspects of growth-promoting interaction have been impeded by trauma. Clients feel uncomfortable or awkward when attempting these kinds of interactions. We call this the scary stuff for the child, the dyad, and/or the family system, and we begin targeting these growth areas long before we engage in trauma narrative work. I recently had a first session with parents in which the mother was unable to make direct eye contact with me at any point during our time together. This experience taught me that connection on this particular relational level was outside her window of tolerance; it was uncomfortable—even foreign—for her and was therefore, on some level, scary stuff. Which relational risks are uncomfortable for the traumatized child or family and which mitigators (named on the Play Therapist's Palette, which will be explored later) of playful approach are most inherently attractive to that same child or family are two curiosities that are always held by clinicians during the assessment phase of TraumaPlay™.

*Titration* is a weird word. With its origins in hard science, the definition took me a while to absorb. But now that I have, I think it serves as a powerful way of framing the work we do with traumatized children and their families. Titration has to do with taking a beginning solution (a mixture of things) and adding bits of something else—sometimes called the change agent—in different amounts until you arrive at a new solution. A simplified version of titration happens when my children are making a Shirley Temple for a fancy event. They begin with a generous portion of ginger ale and then titrate the dose of grenadine by slowly adding drops of the bright red liquid until they have achieved the perfect sparkly, pinkish red color they are looking for. Add too much and you end up with a beverage that is too sweet to drink; add too little and it doesn't seem very different from regular ginger ale. If the verb *titrate* refers to continually measuring and adjusting the balance of something, then I would suggest that play therapists are constantly engaged in titration whenever we are interacting therapeutically with a traumatized child and their caregivers.

For traumatized children, the beginning solution they bring to play therapy includes their unique set of experiences and resiliencies; their early relational ruptures, neglect, and maltreatment experiences; whatever traumas have occurred in their young lives and the way these may be manifested in their bodies by episodes of hyperarousal or hypoarousal; their beliefs about themselves; their "go to" emotions; and the ways they have learned to cope. Their beginning solution also includes their beliefs about how to navigate the world and how they negotiate getting their needs met in daily interactions. In order for change to occur, these children often need new experiences—doses of various aspects of therapeutic interaction—to help rewire the brain and body for healthier interactions with the world. In trauma work, the play therapist is the change agent: what we add and how much of it we add matters.

Although the concept of titration has long been understood and applied in the medical community—even among psychiatrists who work to titrate psychotropic medications to their most beneficial dose—there is little application of this term to behavioral health. The discussions that have occurred thus far seem to have more to do with optimal overall time frames for the implementation of treatment protocols—i.e., is 14 weeks enough time to see the change being hoped for, or is 18 weeks needed? The conversation can be much deeper than this. One benefit of the evidence-based practice (EBP) movement is that interventions have become more honed and targeted for specific populations; another is that therapists are being held to a higher level of accountability to be able to defend why they are doing what they are doing in therapy. With wide dissemination of this practice approach, adherence has become a constant topic for reflection, and literature abounds regarding fidelity to the protocol. Fidelity can be defined as "the degree to which a program is implemented following the program model, i.e. a set of well-defined procedures for such intervention" (da Silva, Fernandes, Lovisi, & Conover, 2014). Many of the new clinicians I help train come into this field already anxious about their performance—anxious about "getting therapy right"—and trainees may cling to a stepped protocol like it is a life vest.

The fidelity checklists provided in some of these protocols can shift our focus from a child-centric view to a self-centric view. The questions can move from "what does the child in front of me need right now?" to "am I implementing this part of the protocol well right now?" The "Am I?" questions lead to a therapist-centered orientation. It is a slippery slope for a clinician, moving from asking clinical questions that are child-responsive to asking questions that are focused on the performance of the therapist. There is a danger that we may become overly rigid in implementation of a protocol (regardless of which specific protocol you are using), promoting the mandates of the protocol above the needs of the unique child in front of us. A core value embraced by the TraumaPlay™ model is that we are following the child's need at all times, and although all treatment goals are evidence-informed and arranged in a clinically sound sequence, the time spent in pursuit of each goal and the various models and mitigators that will be implemented along the way may shift based on the needs manifested in the attachment relationship on a moment-to-moment basis. This requires bringing as much of your fully grounded, curious, and compassionate self to each session as possible, making a plan, and then being willing to let that plan go if it does not serve the needs of the attachment relationship and ultimately the need of the child in the moment.

This text is meant to offer a paradigm shift for therapists, enabling them to begin to see themselves as human titrators equipped to offer varying doses of varying kinds of therapeutic interaction to change the experience of the client system they are working in. This idea of titration is an outgrowth of my ever-growing respect for the unique clinical presentation that each client brings us based on all the experiences that have shaped them thus far, each client's window of tolerance for stress, and the idea of optimal arousal zones for therapeutic work.

When we apply the idea of titration to our moment-to-moment interactions with children in the playroom, we open up a world of possibilities. When we explore titration as it applies to the use of various aspects of the self in the play therapy relationship, a rich conversation begins. Our tone of voice; our proxemics; our decision to reflect verbally, remain silent, or offer interpretation or suggestion; whether or not we use touch in any given therapeutic interaction; and how our choice meets the need of the client in front of us are all dimensions of titration worthy of our focus. How we titrate the use of various forms of play with the child needs to be based on our growing clinical understanding of where the child is—developmentally, kinesthetically, emotionally, socially, and cognitively. How we titrate the use of the play materials and the play space to further optimal change at a pace and in a way that are best for each client requires an approach nuanced to the needs of the moment. In the implementation of TraumaPlay™, *fidelity to the model is defined as fidelity to the child*—to the need being expressed and the most appropriate way for you to meet it at each moment within the session, with a secondary goal revolving around expanding their window of tolerance for whatever is hard to face.

So what are we titrating? Put simply, we are titrating the approach to the "scary stuff"—whatever that may be for a particular child at a particular time. It may be that trusting a grown-up to meet needs is the scary stuff for some clients; deciding between the choices offered by a caregiver may be the scary stuff, receiving nurture from a grown-up may be the scary stuff, eye contact may be the scary stuff, positive self-talk may be the scary stuff, and certainly aspects of the trauma itself may be the scary stuff. In earlier writings, I have talked about titration mainly in relation to how children and their caregivers navigate the creation of more coherent narratives of the trauma content itself. The dance toward and away from trauma content that occurs between children, parents, and counselors is certainly a part of this. Many times this shows up as posttraumatic play (Gil, 2012; Gil, 2017; Goodyear-Brown, 2010) or repetitive symbolic play (Campbell & Knoetze, 2010), and case examples of this kind of self-titrated exposure will be given in the chapter entitled Holding the Hard Story: Narrative Nuance.

However, the need to titrate various aspects of therapeutic interaction extends far beyond the trauma story itself. Many other moments in treatment are different expressions of the encoded trauma: a foundational attachment wound that has created an intense challenge to a child's ability to trust, an overtly observable pattern of using too much physical force on their physical environment, or a need for continual reassurance that there is food available. Each of these markers is a different "showing or telling" of the trauma and requires just as much finesse in titration as the moments that are specific to narrative processing. Something as seemingly simple as eye contact may need to be nuanced with great intentionality, maintaining an awareness at all times of the child's current window of tolerance for stress and their zone of proximal development.

Why would a traumatized child have difficulty with eye contact? Some children, while not having words for this discomfort, may feel unmasked by allowing another to "see" them. There is a traditional Zulu greeting in which the greeter says, "I am here to be seen," and the one being greeted says, "I see you." This greeting begins with the two people looking deep into each other's eyes. Whether you are an African bushman or a 12-year-old girl living in urban America, this kind of greeting requires that both parties wish to be seen and are willing to open themselves up to being vulnerable in this way. This is no small task for the children in our care. These children carry such a deep perception of being damaged and of having this damage reflected and amplified in the eyes of another that they avoid eye contact at all costs. These children may also have core beliefs surrounding their "badness" and carry great shame related to their abuse. Eyes are indeed the window to the soul, and these children fear in their deepest selves that holding the gaze of another will result in a being seen that ends in rejection or a validation of their worst fear—that they truly are damaged goods. Other children have experienced such extreme neglect that they simply have had no practice holding the gaze of another, a skill that evolves as a foundational part of the healthy relational exchange between infants and parents in good enough family systems. The neural pathways that train the brain to seek out the nurturing presence of another through eye contact and the neurochemical release that often occurs as a result of a nurturing eye gaze are missing from some children's experiences.

Still other children may have a neurobiologically wired warning system that says eye contact equals danger and perhaps wisely tells them not to make eye contact with grown-ups. Some of my clients were taken out of environments where terrifyingly dysregulated parents would insist, "Look at me when I'm talking to you!" However, as soon as eye contact was made, these children were smacked, pushed, or frozen in place while an enraged parent screamed in their face. In other cases, children had been raised in environments where an alpha dog mentality was modeled by an aggressive parent. In these situations, direct eye contact was interpreted by the aggressive parent as a challenge and always ended in some sort of minimizing, shaming interaction for others. Put in a neurobiological framework, the experience of making eye contact with caregivers did not produce a neuroception of safety. In fact, for many of my clients, direct eye contact with their first caregivers was punitive, scary, and neurologically overwhelming. Instead of supporting the social engagement system, the threatening gaze—often paired with violent words or actions— would spark the immobilization or mobilization system (Porges, 2001).

I have seen all of these presentations in my playroom, and in each of these, the eventual risk the child took to meet my gaze, then hold my gaze, and ultimately the risk taken in allowing me to show the child their preciousness and my delight in them (expressed through my gaze) were part of the set of new experiences that was healing for this child. Healing—but not easy. Respecting the child's window of tolerance for eye contact (how much, how often, and with what mediators) requires us to hold several challenges in mind. The stress experienced by becoming vulnerable and risking eye contact in the first place, the nervous system response as the interpersonal neurobiological process of shared eye contact begins, and the stress of the somatic excitation that is part of this simple action are all critical parts of deciding how you titrate the dose in this instance. Something as fundamental as eye gaze seems like it should come easily, but assuming ease inhibits our ability to respect the dance and ask the most helpful questions.

Nurturing the nuance in this titration starts with asking, "Which play therapy approach best titrates the dose of exposure to extended eye gaze most sensitively and effectively at this point in treatment?" This question is not asked once but is constantly before us as we see how the child/family responds to the approach we have chosen. For example, one child might benefit from diving right in with a course of Theraplay® that extends—very intentionally— the moments of connected eye contact as a form of engagement. For another child, beginning with child-centered play therapy might be most appropriate, as it would allow the child to choose the moments of connection, discovering how it feels to connect with the eyes of the therapist through brief moments of connection, mitigated by kinesthetic engagement with other self-chosen play materials. For a teenager with limited eye gaze, beginning work in the sandtray may offer the safest approach, as therapist and client can together focus joint attention on a separately defined space. The answer regarding which approach is best may be this, this, and this at different points in treatment (I have often ended up utilizing multiple approaches in the course of my work with a traumatized child), but when and how I order them can depend on a variety of factors, not least of which is this guiding question. I conceptualize the initial question as a decision point, a term borrowed from the play therapy dimensions model (Gardner & Yasenik, 2012).

## Mitigators

A respect for the process of titration led me naturally to ask the question, "How can we mitigate the child's approach to the hard thing?" My answer to this became the Play Therapist's Palette. The idea evolved out of lots of supervision sessions in which clinicians were actively engaged in a course of TraumaPlay™ and would say they "hit a wall" or felt stuck with a child or family. Sometimes a therapist will say about a client, "He has gone as far as he can go." In some cases, this may be true, and a season of rest and celebration can be embraced. In other cases, the stuckness may have more to do with a lack of clarity about the scary stuff that needs to be held. A movement from less directive to more directive can simply mean having the therapist share what they know about the scary thing. Sometimes the therapist's fear or hesitation to bring up the scary thing that happened can be contributing to the lack of traction in the work. New clinicians can feel that the words surrounding trauma—suicide, murder, rape—will be too much for the child to hold. It can help to understand that the child is already holding (often somatically and potentially through the unspoken, but still communicated, stories of those around him) the hard thing. Naming their somatically known experience can often bring relief and should not be withheld because of our discomfort. Naming *it*, whatever it is, matter-of-factly can bring permission for deeper work. Therefore, an ongoing part of TraumaPlay™ supervision is helping grow our abilities to contain and name in a safe, grounded way what children need us to hold and name.

Other times the movement to get out of a stuck place may need to occur away from an agenda or direction of any kind. When a child has developed a need to control all aspects of his environment as a response to trauma, the therapist's agenda for checking the next treatment box is likely to cause distance and division. Helping clinicians return to simply being with the child, without an agenda, is often the most powerful use of supervision. In yet other cases, what may be needed to help a client process the trauma is the offering of a different medium. Since so much of our posttraumatic encoding occurs in the right hemisphere, using expressive therapies can open up new pathways for exploration and integration. Inviting a child who feels stuck in a visual art process—or having a client who has been working with sand move to clay—can offer new and deeper expression through work-ing with the trauma from different angles. The same can be true in metaphor work: a child who is not embracing a metaphor for change—say the life cycle of the butterfly—might resonate instead with a metaphor of superheroes and their alter egos. In many cases, it may be true that the client has gone as far as they can go (in a certain direction) on their own. The wonderful hope that emerged from Vygotsky's work and his concept of the zone of proximal devel-opment (Vygotsky, 1987) is that the client can do more with the aid of a big-ger, stronger, wiser, kind grown-up than they can by themselves. This phrase, *bigger, stronger, wiser, kind*, offered by the creators of the Circle of Security (Hoffman, Cooper, Powell, & Benton, 2017), has become a standard mantra at Nurture House, both for our clinicians and for the parents we are coaching.

When a supervisee feels stuck with a client, I often ask them this question: "Which mitigators would make this client's approach to the scary stuff easier?" Some children have very little difficulty giving the top-level, linguistic narra-tive surrounding a traumatic event—they just do it in a way that is removed from affective components and somatic experience, leaving little room for pro-cessing it in relationship with another; the narrative can be delivered almost as a book report. These children may be aided by mitigators that allow for deeper emotional expression through attuned exploration and reflection (the attachment relationship) or by deeper somatic experiencing (Payne, Levine, & Crane-Godreau, 2015), often aided by interaction with nature or kinesthetic grounding.

## The Play Therapist's Palette

The most recent evolution in my thoughts about how to make these mitiga-tors accessible to others is imagining them on an artist's palette. The pal-ette reflects my experience of the play therapy process as a creative process in which the traumatized child, the caregivers, and the therapist paint a canvas of connection and healing together while respecting best practice treatment goals in trauma treatment (Briere & Scott, 2006; Cloitre et al., 2011). Out of the need to quantify our offerings to help clinicians make quick decisions, I have developed a palette of nine potential mitigators in a child's approach to trauma. Each chapter in the book expands on the offerings represented on the palette, and each expands on one or more of these mitigators, citing the theo-retical underpinnings and giving many case examples to inform the practical application of the Play Therapist's Palette. You will notice, as you look at the icon below, that the central offering on the palette is the attachment relation-ship. Implicit in the seminal role of the attachment relationship is that we are providing delight in the child at all times, especially when holding hard things or setting boundaries. We become both a secure base and a safe haven in the playroom. Arranged in no particular order around the attachment relationship are the following mitigators: metaphor, nature, kinesthetic grounding, need meeting, humor, containment, novelty, and touch (see Figure 1.3).

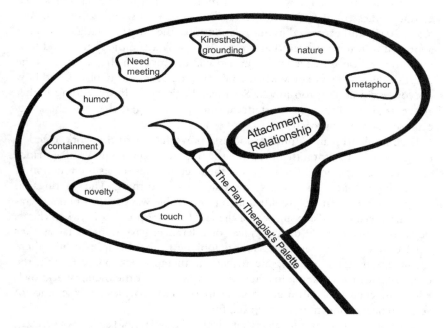

*Figure 1.3* The Play Therapist's Palette

My artist friends have described to me the excitement that comes with a freshly filled palette and a blank canvas. As they approach a blank canvas, they feel that anything is possible. I feel similarly about each new family I meet. How will the trauma recovery process be navigated? How will we expand the window of tolerance? What will I offer to help co-regulate their approach to the hard stuff? No two children are alike and no two trauma recovery processes are the same. There may be parts of the palette that do not get used at all in one clinical relationship and others that may be mixed together or used in large amounts. Either way, the clinician can hold all the possibilities and all the potential avenues for navigation close enough to have them at the ready.

The deeper I have dug into this idea, the more convinced I am that nuanced titration of therapeutic interaction has vast application to therapeutic work. For an integrative play therapist, the foundation of work includes (1) the belief that play is the natural language of children and (2) a respect for the power of play to enhance attachments, heal trauma, create new solutions, build new skills, and help develop new understandings of oneself and the world. We begin by creating a fully equipped play space; developing a safe, nurturing relationship with the child; reflecting their thoughts/feelings/behaviors during the play content; and creating a holding environment for difficult content. From there we can choose from our palette of mitigators.

As we are understanding more and more about how trauma gets stored in our bodies (Ogden, Minton, & Pain, 2006; Rothschild, 2000; van der Kolk, 2015) and expressed in our relationships, the range of mitigators in the approach to trauma content has grown. Who will create the change and how much mitigation is needed are foundational questions in this work, as is the question, how do you know when you've reached the resting place? And how do you avoid adding too much? Scientists begin very slowly, adding higher concentrations of the new substance as they go. I suggest that play therapists can use a similar approach.

Each mitigator can be understood as a powerful ally in the work, offering categories of interaction that promote the neuroception of safety and enhance feelings of competence. Each of these therapeutic allies serves the function of mediating a child's (or parent's) approach to a hard thing. The offerings on the palette are meant to provide a menu of possible ways to go when a therapist hits a relational risk point with a client. These risk points arise when clients begin to move outside their window of tolerance, have gone as far as they believe they can on their own, and are looking to the therapist to help them regulate. Co-regulation may mean that the therapist acknowledges that the client is ready to rest and joins them in doing so or that the therapist acknowledges the client's need for additional support and provides an experience of competence, connection, or plain old fun.

It was in my first painting class that I learned how to hold a palette correctly. I kept dropping it and finally just set it down beside me. When a classmate took pity on me and showed me how to hold it, I could not help but hear the parallel for how a secure base is created through my attachment relationship with the child in my care. The classmate asked me if I wanted to sit or stand to work on my painting, and I chose to stand. She was pleased and said that standing while balancing the palette gives you the most flexibility and mobility, both in creating the work of art and in observing it from multiple angles during the creative process. Then she said, "Use your forearm as a base for the palette and insert your thumb through the hole to help anchor the palette. Then wrap your fingers gently around the other side or let the palette gently rest on your fingers." The first time I tried it, I was concerned about dropping it, so I had a death grip on the palette. She laughed and said, "Relax. You will get cramps in your fingers if you hold on too tight." I relaxed my grip, and she said, "Good. Just remember, you may need to provide a little more support when you put your brush in the paint." It was a constant balancing act between providing enough support and not breaking the palette or my fingers, but I found that resting the palette on my fingers was more comfortable than gripping it tightly.

I feel the same way about these mitigating tools in trauma work. I offer the palette of expressive choices and mitigators for the approach to hard things, but I offer them lightly, supporting the child's choice of all the options available to them. Each time we choose a path together, I have to reassess how much support is needed for the child to explore. My hope is that this text will bring a new awareness of all the pathways available to us when helping children expand their expressive vocabulary. In an art studio, an art student has the freedom to create anything they can imagine, as long as the materials offered are extensive enough to support the work. "I saw the angel in the marble and I carved until I set him free." I reflect on this quote and hear the parallel to play therapy and trauma recovery: if we offer a wide enough range of expression and mitigators for the approach to hard things, we can find the real child inside the hard shell of trauma. I want children to have all the resources we can give them for the journey toward healing. My hope is that, whatever model of play therapy you are using, you will finish this text feeling more deeply equipped with a palette of powerful mitigators that can help you approach the hard moments in your work with traumatized children and families.

A TraumaPlay™ practitioner is operating on multiple levels of awareness simultaneously. The clinician must be aware of where they are on the continuum of treatment, how fully treatment goals have been met, and which goals may need to be returned to, all while remaining as present as possible to discern the child's current window of tolerance for stress. In supervision, I am often helping therapists recognize which parts of the play therapy experience are bringing joy and regulation to the child and how the therapist's use of self is scaffolding a child's approach to the hard things.

## Therapeutic Use of Self

Therapeutic Use of Self in TraumaPlay™ requires the clinician to be comfortable with and always growing in three roles: the role of the witness, the role of the nurturer, and the role of the safe boss. When I am training others in TraumaPlay™, we will map out the therapeutic use of self at various points in a session around a triangle. Each point of the triangle represents one of the roles we may fill in a TraumaPlay™ session: witness, nurturer, safe boss. In the center of the triangle is the child, and all movements by the clinician around this triangle are in direct response to the needs of the child at the current time. Each child in our care requires a unique titration of each role. Some may need more of the witness (seeing) than the safe boss (structuring), whereas others may need large doses of delight from the nurturer before any trauma content is approached (see Figure 1.4).

## The Witness

Being a witness to a child's play therapy process involves hearing, seeing, and holding the child's communications, both verbal and nonverbal. We hold without judgment and, with as much presence as possible, practice staying with the child from moment to moment. Being a witness means communicating through our words, actions, and neurobiological presence that "I see what you are showing me, and you can show me more." We communicate

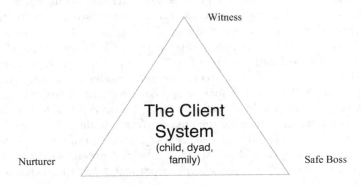

*Figure 1.4* The Triangle of Therapist Roles in TraumaPlay™

this in big and small ways throughout sessions. When a three-year-old puts on a puppet play in which the police officer puppet says, "Put the handcuffs on, you fucker" (as happened in one of my recent sessions), we need to be able to reflect this language without backing up from it. Avoiding the reflection of the word *fucker* can communicate loudly that we are uncomfortable with this language, compromising the child's ability to use us as container and reflector.

All children learn how to stay close to their attachment figures by reading their cues. As soon as we enter a therapeutic relationship with a child, we become a potential attachment figure. In order to provide a corrective emotional experience for the child as an attachment figure, we, as therapists, must stretch ourselves to become large enough containers for whatever hard stuff the child or family system needs us to hold. I must be able to contain difficult images, words, stories, and somatic expressions and reflect them without allowing any personal discomfort to keep me from these roles. When a child's play behavior makes the therapist uncomfortable, it then makes the child uncomfortable. In some cases, the child will curb or even remove the uncomfortable play pattern from their play in order to protect the therapist and maintain the relationship. The three-year-old may understand that the word *fucker* makes *us* feel uncomfortable and so won't play in that way again. In effect, they are being told by us to mask part of their experience from us.

The idea that miscuing begins when a child's feeling or behavior "makes us uncomfortable" was conceived by the creators of the Circle of Security Project to describe the pattern in which children become what they need to be to stay close to their caregiver. So, if we want to get the most authentic version of the child's self in the playroom, we only have to make sure that nothing makes us uncomfortable. Do you hear the laughter stemming from the obvious impossibility of such an ask? In reality, remaining a bigger, stronger, wiser, kind container for our clients means continually working on our own triggers, aggressively pursuing an understanding of countertransference issues as they arise, and intentionally growing our ability to be filled with difficult things and then emptied again. This begs the question, "Where do we put all that we hold?"

Another important aspect of TraumaPlay™ supervision is supporting clinicians in doing their own self-care work, acknowledging compassion fatigue, and transforming heaviness into letting go of what can be let go while holding with gratitude whatever relational gifts or new nuggets of wisdom or insight have been gained. In this way, TraumaPlay™ supervision often becomes a parallel process for the work being done in the client system. From the moment we begin establishing safety and security—the first goal of TraumaPlay™—the child may begin to give us trauma content. The number of times children have created a viscerally disturbing image or engaged in a play pattern that seems to be immediately recreating the situations or experiences in which they have been unsafe supports the idea that we must be ready to embrace our role as witness at any point along the way. The witness communicates, "I see what you are showing me, and you can show me more."

## The Nurturer

Many of the families I see in the wake of trauma are having difficulty with either the giving or the receiving of nurture. Some of our clients come from lonely beginnings. They did not receive nurturing touch to help wire their brains toward connection and regulation. They did not have a continuous stream of reciprocal exchanges of delight with a caregiver. They did not have anyone telling them they are amazing and deserve to exist simply because they were created. These children did not have someone trumpeting their ability to become whatever they chose to become. So, for many of these children, a more direct recapitulation of parent/child interactions may be needed, with attention given to meeting earlier developmental needs that may be hiding inside chronologically well-developed bodies. Delight can be communicated in many ways by TraumaPlay™ practitioners and often includes meeting basic needs, offering nurturing touch, and appreciating a child's sparkling eyes or sense of humor.

## The Safe Boss

The therapist as safe boss provides structure while sharing power with the child. Many of the children we see, as well as many of the parents who care for them, have not had safe boss behavior modeled for them. One of the most rewarding parts of my work is helping parents and teachers breathe in this phrase and grow into safer bosses for the children in their care. I am constantly engaging in parallel process work, and as I model safe boss behavior, the parent begins to internalize this model themselves. As the parent begins to believe their new self-talk, i.e., "I am the safe boss," it changes the way they approach parenting. Safe bosses are people in authority and under authority. A *boss* is the person in charge, the person who has the power to effect change, the person who makes final decisions and holds your future in their hands. If you have been around children for any length of time, you have probably heard one say, "You are not the boss of me." Little ones often fantasize about being the boss, but they are not equipped, and when they have more authority than their parents or teachers—when they actually get to feel like the boss—they end up feeling unsafe.

Children do not have the capacity to be fully in charge. They first have authority over their own mind, will, and emotions, and their growing edge is to bring these more and more under their own control. Theoretically, a safe boss knows more about whatever they are the boss of than the people they are leading and will use this knowledge to equip, encourage, and grow others. Safe bosses do not shame those under their authority. They also set safe boundaries, quickly and efficiently reshape harmful interactions, and maintain appreciation for the person they are leading at all times. Safe boss behavior may involve setting limits, giving explanations, or even doing didactic teaching or skill building.

The safe boss role is one that respects the parallel process dynamics of the therapeutic relationship. The therapist models the dimensions of healthy leadership that can become part of the family system's way of relating. The picture below, taken at Camp Nurture, shows one of our young campers at a time when her body let us know she needed to rest. One of her buddies offered a pillow and lap for the little one to rest on. The child's back-up-buddy got behind the holder and began to braid her hair, providing nurture to the one providing nurture to the child. It was not until long after Camp Nurture was over that I found the picture below (see Figure 1.5). I find it a beautiful representation of the parallel process dynamics we have been exploring. As we, as therapists, hold the holder, the traumatized child is also held.

*Figure 1.5* Holding the Holder

We talk a lot at Nurture House about opening and closing circles of communication. Any opening and closing of a circle of communication between therapist and child is an opportunity to titrate a dose of something therapeutic. Something as simple as saying, "See you later alligator," and the child responding, "In a while crocodile," can be considered an opened and closed circle of communication (Purvis, Cross, & Sunshine, 2007). Any time a child verbally interacts with us and we respond in a way that is reflective of their language, a communication loop has been opened and closed. Even in the more difficult moments of therapy, when a limit has to be set, completing the circle of communication is critical. If a child has thrown something at you, immediate feedback of some kind—whether it is Landreth's ACT limit-setting model, i.e., "The block is not for throwing at me; you can throw it into the block bin or onto the cushy chair," or "Whoa! That made a loud noise when it hit the floor!" or "My stomach clenched when you threw that, as I was worried it might break"—is necessary. There are multiple options for response, depending on your model of intervention, but the circle of communication, opened by the child when he threw the object, must be responded to in order for the circle of communication to be closed and therefore to provide meaningful feedback for the child.

It is important to talk briefly here about how children learn, particularly children who have compromised neurophysiology due to complex trauma. Behaviorally based therapy approaches generally support the ignoring of most negative behaviors. I believe that ignoring these negative behaviors when they begin is the beginning of the end, in the sense that the behaviors will generally continue to ramp up. The child's body language, facial expressions, physical actions, and expressed words (even if the words are curse words) are giving us valuable information about the current regulation of the child—and their underlying needs. When we see negative behavior is when we need to pay more attention, not less. Traumatized children need caregivers who stay very close to them—close enough to sense the changes in reptilian brain stem regulation that matter—and respond quickly, providing valuable feedback that can aid in regulation.

## Conclusion

It is my deep desire that this discussion of titration will expand your sense of freedom in dancing toward and away from the trauma content while at the same time trusting the process and the relationship with the client. It is also my hope that by offering this new tool, the Play Therapist's Palette, the many options available to play therapists (as mitigators in the dance) will feel easily accessible. When I think about the phrase *taming the trauma*, what sometimes comes to mind first is a wild animal and a wild animal tamer wielding a whip as they steer the animal into a cage. The idea of caging the trauma is shortsighted, as it compartmentalizes the trauma within boundaries it will most likely outgrow. In addition, the tamer may one day forget to lock the cage, only to have the trauma come bounding out, more resentful than ever and on the attack. Taming, in the context of this book, has to do with using the process of titration and the Play Therapist's Palette to help clients enhance felt safety, get needs met more effectively, and leach the emotional toxicity out of the trauma in a way that brings greater internal coherence to the trauma survivor.

# References

Bratton, S. C., Ceballos, P., Sheely-Moore, A., Meany-Walen, K., Pronchenko, Y., & Jones, L. (2013). Head start early mental health intervention: Effects of child-centered play therapy on disruptive behaviors. *International Journal of Play Therapy, 22,* 28–42.

Bratton, S. C., & Ray, D. (2000). What research shows about play therapy. *International Journal of Play Therapy, 9,* 47–88.

Bratton, S. C., Ray, D., Rhine, T., & Jones, L. (2005). The efficacy of play therapy with children: A meta-analytic review of treatment outcomes. *Professional Psychology: Research and Practice, 36,* 376–390.

Briere, J., & Scott, C. (2006). *Principles of trauma therapy: A guide to symptoms, evaluation, and treatment.* Thousand Oaks, CA: Sage Publications.

Campbell, M. M., & Knoetze, J. J. (2010). Repetitive symbolic play as a therapeutic process in child-centered play therapy. *International Journal of Play Therapy, 19,* 222–234.

Cloitre, M., Courtois, C. A., Charuvastra, A., Carapezza, R., Stolbach, B. C., & Green, B. L. (2011). Treatment of complex PTSD: Results of the ISTSS expert clinician survey on best practices. *Journal of Traumatic Stress, 6*(24), 615–627.

Cochran, N. H., Nordling, W. J., & Cochran, J. L. (2010). *Child-centered play therapy: A practical guide to developing therapeutic relationships with children.* Hoboken, NJ: John Wiley & Sons.

da Silva, T., Fernandes, C., Lovisi, G. M., & Conover, S. (2014). Developing an instrument for assessing fidelity to the intervention in the critical time intervention-task shifting (CTI TS): Preliminary report. *Archives of Psychiatry and Psychotherapy, 16*(1), 55–62. Retrieved from https://doi.org/10.12740/APP/23279

Gardner, K., & Yasenik, L. (2012). *Play therapy dimensions model: A decision making guide for integrative play therapists.* London: Jessica Kingsley Publishing.

Gil, E. (2012). *The healing power of play: Working with abused children.* New York, NY: Guilford Press.

Gil, E. (2017). *Posttraumatic play in children: What clinicians need to know.* New York, NY: Guilford Press.

Goodyear-Brown, P. (2010). *Play therapy with traumatized children.* Hoboken, NJ: John Wiley & Sons.

Hoffman, K., Cooper, G., Powell, B., & Benton, C. (2017). *Raising a secure child: How circle of security parenting can help you nurture your child's attachment, emotional resilience, and freedom to explore.* New York, NY: Guilford Press.

LeBlanc, M., & Ritchie, M. (2001). A meta-analysis of play therapy outcomes. *Counseling Psychology Quarterly, 14,* 149–163.

Ogden, P., Minton, K., & Pain, C. (2006). *Trauma and the body: A sensorimotor approach to psychotherapy.* New York, NY: W. W. Norton & Co.

Paul, G. L. (1967). Strategy of outcome research in psychotherapy. *Journal of Consulting Psychology, 31*(2), 109.

Payne, P., Levine, P. A., & Crane-Godreau, M. A. (2015). Somatic experiencing: Using interoception and proprioception as core elements of trauma therapy. *Frontiers in Psychology, 6,* 93.

Porges, S. W. (2001). The polyvagal theory: Phylogenetic substrates of a social nervous system. *International Journal of Psychophysiology, 42*(2), 123–146.

Purvis, K. B., Cross, D. R., & Sunshine, W. L. (2007). *Connected child* (p. 264). New York, NY: McGraw-Hill Professional Publishing.

Rothschild, B. (2000). *The body remembers: The psychophysiology of trauma & trauma treatment.* Los Angeles, CA: W. W. Norton & Co.

Schaefer, C. E., & Drewes, A. A. (2014). *Therapeutic power of play: 20 Core agents of change* (2nd ed.). Hoboken, NJ: John Wiley & Sons.

Van der Kolk, B. A. (2015). *The body keeps the score: Brain, mind, and body in the healing of trauma.* New York, NY: Penguin Books.

Vygotsky, L. (1987). Zone of proximal development. *Mind in Society: The Development of Higher Psychological Processes, 5291,* 157.

# 2 The Neurobiology of Trauma and Play

## Understanding the Playroom as a Neurochemical Boxing Ring

Any discussion of how the Play Therapist's Palette helps mitigate the approach to trauma content must begin with unpacking the neurobiology of trauma and the neurobiology of play. As it turns out, play pretty powerfully meets our neurobiological needs for trauma recovery (Badenoch, 2008; Gaskill & Perry, 2012; Gaskill & Perry, 2014; Hong & Mason, 2016; Kestly, 2015; Stewart, Field, & Echterling, 2016). Clinicians who treat traumatized children must familiarize themselves with brain structure, function, growth pathways, and potential injuries to the developing brain. Additionally, clinicians who develop a healthy respect for the neurochemical boxing match that often occurs between stress-related neurochemicals and those released through pleasurable, competency-building experiences in play have the best chance of confidently helping children dance toward and away from the trauma content, trusting the process. We are marvelously made and are meant to be able to co-regulate one another. The more respect we develop for both the power and the limits of neurobiological resonance, the more adept we will be at intentionally crafting experiences and playful interactions that will offer a path toward healing.

It is well understood now that experience shapes the brain and that neural circuitry is use-dependent (Kay, 2009; Siegel, 2001, 2012). The adage "neurons that fire together wire together" (Hebb, 1949) gives scientific explanation to the ways in which interpersonal patterns of relating are shaped. I share this idea with parents by having them imagine that the brain is like a field of tall grass. If you walk a path once, the grass will bounce back, but as you walk the path over and over again, the grass gets pressed down, the path becomes clear, and it becomes easier to get from point A to point B. This laying down of neural pathways can provide a clear path through the weeds and incredible resilience in our development—or the path can become riddled with muddy ruts that are difficult to get out of, even when we are desperate to find another way to interact with the world. If you have a caregiver who comes to soothe you every time you are upset, your brain develops connections that begin to anticipate soothing, and eventually you learn to self-soothe. If you have a caregiver who is continually frustrated and short-tempered with you, your mirror neurons become activated in similar ways, and neural pathways are shaped to more and more quickly override higher brain regions with lower brain reactivity. In other words, you may become short-tempered.

Most of the traumatized children we treat have dysregulation of the brain stem and require co-regulation work and the expansion of their windows of tolerance for stress. Play provides a natural medium for the occurrence of up- and downregulation, and as such is an important component of holistic, healthy development in children (Erickson, 1963; Ray, 2011). Neglected and maltreated children have usually been raised in environments where they have been deprived of play and its natural benefits in shaping our nervous systems. Stuart Brown, founder of the National Institute for Play, unpacks the many benefits that play provides, both in the animal world and in humans. These include opening up our creative potential, providing social training in things like empathy and cooperation, allowing us to try on roles without any long-term consequences, building strength and mastery, and experiencing pleasure. He also discusses the ways in which play deprivation can lead to deficits in connecting with others. These deficits run the gamut from a lack of empathy and ability to problem-solve with others to extreme sociopathy and outcomes such as rape and murder (Brown, 2009).

When a developing mind is subjected to prolonged or overwhelming stressors, use-dependent neural systems can be altered (Perry, 2000). We understand now the multiple ways in which in utero threats, such as alcohol or drug use, malnutrition, and excessive cortisol reactions in the mother's body, can derail healthy brain development. We understand now how neglect, maltreatment, and chronic trauma can negatively impact a child's window of tolerance, creating states of both hypoarousal and hyperarousal (Ogden, Minton, & Pain, 2006). Children who did not have a supportive attachment figure to help mitigate stress for them early in life tend to move outside their optimal arousal window much more frequently than those who had thousands of repetitions of nurturing care early in life.

The groundbreaking work around the Adverse Childhood Experiences (ACEs) study has led to an understanding that early traumatic events have long-term consequences on everything from physical health to addictive behaviors to the kinds of intimate relationships you will have as an adult. This longitudinal study began as a collaboration between Kaiser's Health Appraisal Center and the Centers for Disease Control and Prevention. After visits to the Health Appraisal Center, surveys with detailed questions regarding a host of adverse childhood experiences were mailed out. Three categories were abuse-related (emotional, physical, and sexual), and the other seven were related to household dysfunction (parental separation/divorce, domestic violence, substance abuse, crime, and mental illness). In one phase of the study, researchers looked at the relationship between exposure to ACEs and lifelong exposure to drug use. They found that each ACE increased the risk of drug use earlier in life two- to fourfold (Dube et al., 2003). This study also examined four successive birth cohorts dating back to 1900 and found that the relationship between ACE scores and early onset of drug use crossed all four cohorts. People who have more than four ACEs are twice as likely to be smokers, seven times as likely to be alcoholics, twice as likely to have heart disease or cancer, six times as likely to be sexually active before age 15, and—the most disturbing statistic—12 times as likely to attempt suicide. ACEs predict the ten leading causes of adult death and disabilities. ACES also increase our likelihood of being obese and having impaired cognitive capacities—and therefore school and work performance—often leading to higher rates of poverty (Felitti et al., 1998; Anda et al., 2006). In addition, ACEs increase the likelihood of psychotropic drug use as an adult and are predictive of mental illness (Anda et al., 2007). It is now understood that untreated ACEs can lead to changes at a cellular level, in essence creating intergenerational trauma as genetic code is passed down to one's children and one's children's children.

## Developmental Trauma Disorder

Out of our understanding of the pervasive neurophysiological differences found in people who experience trauma has come an appreciation for the devastation that trauma can wreak in every area of a child's development (D'Andrea, Ford, Stolboach, Spinazzola, & van der Kolk, 2012). Bessel van der Kolk's body of work and his offering of developmental trauma disorder as a more holistic conceptualization of the clinical picture of children with complex trauma histories have shaped our field. He completed a study that compared adult survivors of childhood trauma with adults with acute trauma and adults with domestic violence in their adult histories, and he found that there was a very different presentation of symptoms for adults who had experienced chronic trauma as children than for adults who had experienced a single traumatic event equipped with the resources of adulthood.

The three main areas that van der Kolk identifies as being impacted by developmental trauma are attentional abilities, affect regulation, and how the individual navigates relationships. People who experience chronic trauma in childhood are not able to attend deeply to tasks for long periods of time and are easily distracted by things, unable to filter out extraneous information in order to focus on what matters at the moment. They are also less able than others to regulate their affect, becoming too reactive and too intense and having too high highs and too low lows. He makes the point that as this clinical presentation gets treated for affective arousal, these children are often diagnosed with bipolar disorder and given antipsychotic medication that shuts down the dopamine system, which is important for engagement and motivation. In our generation, more and more children are being treated with psychotropics that numb the excitation response. This numbing (arguably dissociative response) does not allow children to do the work of expanding their windows of tolerance for distress, as distress may not be authentically felt and therefore has no hope of eventually being tolerated.

## Neuroplasticity, Hope, and Balance

Neuroplasticity, the lifelong potential for neural change (Mundkur, 2005; Neumeister, Henry, & Krystal, 2007), supports the idea of a growing edge for both caregivers and children in learning new response patterns and implementing them. We know that synaptic strengthening is experience-dependent, and since play is inherently fun, it is rich soil for the planting of new interactions between parents and kids. Foster and adopted children get an opportunity to create new neural pathways in relationship to their caregivers as new interactional patterns are practiced. In these ways, models such as filial therapy, child-parent relationship therapy (CPRT) (Bratton, Landreth, Kellam, & Blackard, 2006), and Theraplay (Booth, & Jernberg, 2010) are in vivo methods for helping dyads practice new ways of interacting. As parents and children play together, new cognitions and emotional perceptions of earlier events can occur (Siegel, 2012).

I find that one of the hardest balances to strike in providing trauma-informed care for families is unpacking for parents the potential injuries to their child's neural development while continuing to wave the banner of neuroplasticity and hope for change. The brain is a marvelous, magical guidance system, and to diminish the lifelong capacity for the growth of new synapses based on new experiences would be a disservice to our clients. Alternatively, asking a child to do something they simply are not able to do yet (and might not be able to do at all) based on their current brain development would be wildly unfair. Helping parents set the bar appropriately for their traumatized children in each area of development is an important part of our job.

I was speaking recently at a conference for foster parents, adoptive parents, and kinship care providers. An adoptive mother came up to me afterward and explained, "I have a nine-year-old child adopted from Uganda. She has significant cognitive delays, temper tantrums, and isn't learning our rules. I am just wondering if I should keep believing she can change, or if I should accept her limitations and resign myself to the fact that this may be "as good as it gets." My answer was, "Which one will help you remain most regulated, kind, and connected with her?" This was, of course, just a jumping off point for much more detailed conversations she might have with her local therapist around which developmental arenas (including the attachment bond and ultimate growth in social relatedness) can be impacted and which behaviors/symptoms/ways of relating to the world may need to be accepted as the child's current best self.

I could see that my question as response to her question was unexpected, but it seemed what this mom most needed—time to reflect on which stance helps her be the most grounded mother she can be and to help her understand her power as a potentially different relator than anyone else in this child's world. I asked her to list the top two behaviors that are hard for her to tolerate, and after some processing, she was able to separate them. For the relational behavior, we came up with some strategies for reengaging her daughter. For the other behavior, which we agreed had a sensory defensive source, she ended up saying, "You know, I'm really not sure she can help that—and if I just admit that, I'll stop fighting with her about it." Yes!!! I see parents who overestimate the speed with which a symptom "should be" extinguished and set a bar that is higher than the child can successfully reach, creating resentments in both the parent and the child and widening the gulf between them. I also see parents who believe that the damage is done and have disengaged. In these cases, I am often attempting to reopen the caregiver's compassion well so they will continue to open and close circles of communication with their child toward the growth of new neural circuits.

Lev Vygotsky's zone of proximal development has been a key concept in shaping my thoughts about the Play Therapist's Palette. In fact, the intersection of Vygotksy's work on how children learn and Perry's neurosequential model of therapeutics (Perry, 2006; Perry & Dobson, 2013) has sparked a new question that can aid us in treatment planning with traumatized children: "What is this child's neural growing edge?" We talk often in supervision about the "growing edge" for our clients. The growing edge is the boundary between what a child can do on his or her own and what a child can do with the help of a safe boss. Asking this question as we begin treatment planning helps us avoid a scenario in which we are pushing child clients beyond their current capacities. Children who have underdeveloped brain regions (the reptilian brain stem, diencephalon, and limbic system being ones that often take a hit) benefit from targeted interventions that pair the presence of a safe boss with an enriched environment that will stimulate new growth in those underdeveloped areas. The neurosequential model of therapeutics has informed an appreciation for targeting interventions to the specific areas of the brain that are lacking in optimal development. Targeting intervention to the neurophysiological needs of the child informs my choice of which mitigators from the Play Therapist's Palette to use with each child.

## The Brain and the Other

Play therapists have long understood the attachment relationship between the therapist and the child to be the most critical part of the understood change mechanisms in play therapy (Axline, 1947; Landreth, 2012). Pioneers in the field of interpersonal neurobiology (IPNB), such as Dan Siegel and Bruce Perry, have given us a new language, grounded in neuroscience, to support the importance of interpersonal regulation and growth. Perry and Pate (1994) write, "Simply using cognitive and verbal interventions will not alter the parts of the brain mediating trauma . . . the changing element of therapy is the 'relationship,' . . . not the words of therapy." Play remains one of the most powerful and palatable ways to help a child grow in regulation with a co-regulating grown-up, and it offers these experiences without the necessity of linguistic involvement. The nine dimensions of neural integration offered through IPNB (consciousness, bilateral, vertical, memory, narrative, state, interpersonal, temporal, and transpirational, or identity integration) can be positively impacted by play therapy (Wheeler & Dillman Taylor, 2016). The consciousness dimension is often tackled by play therapists almost intuitively as we become more mindfully present in the playroom and offers a reflective presence for children to become more aware of themselves. In the bilateral dimension, as we give words to the feelings we see children expressing, we are facilitating communication between hemispheres (Badenoch, 2008). Play therapists need to understand the unique role that play holds in accessing and remapping lower brain regions. Play therapy may be not only a helpful medium but also the most effective medium for inviting certain areas of brain development.

## Bottom-Up Brain Development

Put simply, the brain develops from the bottom up, and that hierarchical development must be respected if we are to heal from trauma. The hierarchical and systematic development of the brain begins in utero. The brain develops first through genetic scaffolding and very quickly is influenced by environmental factors, including in utero threats to development. The idea of bottom-up brain development was first posited by Jackson (1958). The brain of an infant starts out with a consuming preoccupation with survival. The reptilian brain stem is in charge, and whenever that baby is hungry, cold, wet, etc., a lot of noise is made in order to get the help of a safe boss to bring the infant back to a state of regulation—satiation, warmth, dryness. However, as infants in a satisfactory caregiving system grow and get thousands of repetitions of this co-regulation by parents, their reptilian brain stem begins to pay less attention to moment-to-moment basic need meeting, and more time is spent in relationship-building and learning. If you think of the energies spent by the brain as a pyramid, with the most energy spent at the bottom and the least at the top, the pyramid looks inverted for a typical brain that has had lots of experiences in which safety was a given. The typical brain has only the tip of the pyramid given over to survival responses, the next layer given to regulation, the next to social and emotional involvement, and the largest part of the pyramid, the base, to cognition.

When a brain has experienced trauma, and particularly if the trauma has been chronic, the pyramid is reversed. The base of the pyramid, the largest part by far, is consumed with survival, and the next largest is concerned with regulation. Very little room is left for social and emotional engagement, and only the smallest tip of the pyramid remains given over to cognition. Once we understand this, the paradigm shift is easy. Historically, our treatments for trauma have revolved around cognitively geared therapies. Van der Kolk (2015) makes the argument that somatically regulating interventions, such as mindfulness, trauma-informed yoga, and interaction with nature, is significantly more effective in reducing trauma symptoms than cognitively geared treatments.

When helping caregivers shift their paradigms, I usually begin with parents by showing them the elegantly simple depiction of MacLean's triune brain, consisting of the reptilian brain stem, the limbic brain, and the neocortex, or thinking brain (MacLean, 1990). There is a beautiful spiral staircase made of stone on Vanderbilt's campus. Whenever I see it, I am reminded about the sequential bottom-up nature of the brain's development. I have taken a picture of the spiraling stairs and will sometimes give parents a copy that we label with the functions of each part of the triune brain (from respiration to cognition, etc.). During parent support sessions, I ask them to put an asterisk next to which parts of the child's brain the parent was targeting during a moment of co-regulation or discipline from the week prior.

I often share Dan Siegel's hand model of the brain (Siegel, 2010) with parents early on and talk about how stressful experiences can cause us to flip our lids (Siegel, 2011). My favorite practical application of the triune brain comes from Becky Bailey's conscious discipline model (Bailey, 2001, 2015), in which she attaches guiding questions to each part of the triune brain to make it even more accessible to caregivers. I explain each brain region's "job" and pair it with Bailey's questions. I explain the functions of the reptilian brain stem in regulating heart rate, respiration, body temperature, and sleep (Perry, 2006). While the brain stem must be active for an infant to survive, it continues to grow through the repetition of regulating interactions with caregivers over the first year of life (Perry, 2006). This most foundational level of the brain is always asking the question: am I safe? The limbic brain is represented by the palm of the hand, and the enfolded thumb represents the amygdala. Emotional responses and the fight, flight, and freeze responses originate in this midbrain region. The limbic brain is responsible for emotional regulation and is always asking the question: am I loved? The cortex, or the thinking brain, is represented by the fingers wrapped over the thumb enfolded in the palm of the hand. Various fingers and knuckles represent specific regions of the cortex, but as a whole, the cortex is responsible for executive functioning, goal-driven behavior, and cognition and is always asking the question, what can I learn from this? I talk about it as the journey from regulation to reason. I share the bottom-up language of brain development with them. I help them understand that until that child's brain has answered the questions "am I safe?" and "am I loved?" with a resounding yes, the neocortex is not empowered to ask questions related to learning. It is helpful also to have an understanding of the evolution and anatomy of emotion and emotional circuits in the brain (LeDoux, 1996; Panksepp & Biven, 2012; Panksepp, 1998) in understanding both bottom-up implications for the brain's interpretation of the environment and the potential for top-down regulatory control.

When I present this information to 30-year veteran day care owners and school teachers in the inner city, they usually say something like, "Amen, sister!" Many of these dedicated professionals, who have been in the trenches for years, have their life experiences validated by this expanded language of bottom-up brain development. In less than 15 minutes, they are given permission and perhaps even a mandate to provide regulation and nurture (connection) before trying to teach ABCs and 123s. I was training school teachers in a highly violent urban area of Nashville recently. At the break, a teacher approached me and explained that two kindergartners in her class had come to school the day before saying they had stepped over blood on the sidewalk on the way to school. For these two kindergartners, the answer to the question of whether they are safe may be "absolutely not" or "only for a few hours." Many teachers who work in the inner city have communicated to me the sense that they are starting over at ground zero each morning, and by the time they have rebuilt a child's sense of safety and connection, the final bell of the school day is ringing. How can these children learn if we cannot help their limbic brains to calm and their lower brain regions to regulate their bodies properly? More and more schools that serve at-risk children are moving to trauma-informed models of teaching that spend time with connection and regulation in the classroom before attempting to teach children their ABCs and 123s. The research is showing that when these underlying needs for regulation and connection are met, learning proceeds more spontaneously and more quickly (Walkley & Cox, 2013; Perry & Daniels, 2016).

## The Amygdala Alarm

The amygdala is, among other things, the seat of somatosensory memory as it relates to heightened emotional experience (Davies, 2002; Goleman, 2006; Hughes & Baylin, 2012; Panksepp & Biven, 2012). That's kind of a mouthful, but it basically means that your most joyful memories—the day you got married, the moment your newborn was placed in your arms, the day you graduated from your graduate program—may have a crystal clear sense memory attached to them, i.e., the taste of the wedding cake, the smell of your baby's head, the contrast of the black graduation cap against the blue sky. If that is true of our joyful memories, it is doubly true of the things that terrify us. We are wired to encode danger signals from the environment very deeply and to respond to them very quickly in order to survive. If a fist is coming at you and it punches you, the next time a fist comes at you, you hope to move out of the way, right?

The thing is, the amygdala is a pretty sloppy processor and can have functional impairments related to the anxiety that is being carried (Strawn et al., 2014). Its associative function pairs sights, sounds, and smells with the traumatic experience and is meant to be an early warning system, but sometimes it generalizes too broadly (Goleman, 2006). The way I normally unpack this for the families in my practice is by bringing out an extra-large military figure and telling a story. I say, "Let's pretend this guy is in Iraq, and his job over there is to take care of the tanks. So he gets up in the morning and is washing the windshield of the tank, when all of a sudden he hears gunshots—bang, bang! He drops to the ground, his heart is racing, his body is shaking, but he is safe. After a few minutes, the gunshots subside. He slowly gets up, checks himself out, and he's not hurt. His heart rate and breathing go back to normal, and he goes back to his job. The next day, he and a buddy are getting the mud off the big treads of the tank tires, and again there are gunshots—bang, bang! He drops to the ground, and so does his buddy, although his buddy doesn't drop quite as fast and gets a little bit hurt (he just needs a Band-Aid), but our guy is safe. This happens day after day in Iraq until our guy's tour of duty is over. He is back in the states, he's been home for maybe five months, and he is at the mall doing some Christmas shopping. He is laden down with packages, and as he's walking out to his car, he hears a car door slam—bang! What's he going to do? He's going to drop to the ground."

Kids as young as five will tell me the guy is going to drop to the ground, and while we don't say to children, "Your brain becomes habituated to the trauma trigger," this story illustrates the idiosyncratic, quirky ways we respond to trauma. In fact, one of my favorite definitions of *trauma* is the normal reactions of normal people to events that for them are unusual or abnormal (Parkinson, 1993). This explanation de-pathologizes our response to trauma and gives permission for children to use the playroom in any way needed to express and heal from the trauma.

## Implicit and Explicit Memory

Another way to talk about how trauma gets stored is by describing explicit and implicit memory systems (Siegel & Hartzell, 2003). Explicit memories require autobiographical memory to be available to us. Autobiographical memory relies on the hippocampus, which is highly underdeveloped in infants and young children. In utero injuries and early trauma are stored in implicit memory, which is basically associative memory. As infants, we have thousands of interactions with our primary caregiver, and while we do not consciously remember each of these, we make associations between the smell of mom's hair or her motherese and pleasurable feelings of satiation. If mom was nurturing, these smells and sounds will trigger a full body sense of wellness. If the smell of alcohol on dad's breath was associated with yelling, throwing things, or mom being hurt, the smell of alcohol may trigger a felt sense of danger even if you have no conscious memory of a negative event. The most insidious thing about implicit memories is that you do not have a sense of remembering something, as you are taken over with the physical embodiment of a somatic memory.

When I am training others, I often give this example: imagine you are a nine-month-old baby, and for the first nine months of your life, you were colicky, difficult to soothe, and had a mother who was struggling with postpartum depression. She would leave you in your crib for long periods of time, but when she got to the point where she could not stand your screaming anymore, she would storm into the room, pick you up, and with clenched teeth scream, "Stop crying!!!" The good news is when you reached nine months of age your mother went to see her doctor and got the help she needed to regulate. Since autobiographical memory does not usually develop until around age two, from the time you can remember, she was a terrific mother, providing delight and regulation and meeting your needs from that point forward. Fast-forward 40 years, and you are married, and it is your job to put out the trash on Tuesday mornings. Your wife comes home and says in frustration, with her teeth clenched, "Why didn't you take out the trash?" You are awash in fear and feel as if you have been attacked. You may have no awareness that what you are experiencing is an implicit memory, but it is impacting your relationship in the here and now. I encourage both children and their parents to pay attention to any moments in daily life when they tend to significantly overreact or greatly underreact to a situation happening in real time, as these can give us hints to unintegrated implicit memories that may need exploration. Preverbal implicit memories can be restored and consolidated through play (Badenoch & Kestly, 2015). I have had fascinating cases in which I sit with a tearful mother who tells me in great detail about the traumatic experience of her child's birth. She verifies that she has never told the child the story in the graphic detail she just told it to me, but the child then comes into the playroom and acts out almost identically in play what the mother had described, seeming to be playing out for me the somatic encoding of their birth experience.

Another example of a child communicating his implicit memories of abuse through posttraumatic play was given to me by Danny. Danny was four years old at the time he came to therapy. He had been sexually abused by his mother's boyfriend. When he first entered my playroom, he went directly over to the brand new, extra-large dinosaur I had recently added to my collection. The dinosaur was made in two pieces, and he unscrewed the front of the dinosaur from his tail. The creature now had a gaping hole in his backside. Danny went over to the play kitchen and got the eggbeaters and began to shove them rhythmically into the hole in the backside of the dinosaur, making sexual grunting noises as he did so. I do not believe this behavior was a conscious telling, but rather an implicit reenactment. During a subsequent session, he began to scribble on a large sheet of paper. I asked him to tell me about his drawing. He said, "This is what Daddy did," and he began to jam the marker into the paper, making grunting noises again. Both of these expressions were examples of penetration play, the first an expression of implicit memory and the second a more conscious attempt to communicate the hurtful acts to me.

## Nurture House: An Unexpected Neurochemical Boxing Ring

Everything about Nurture House is based on our most current understanding of the intersection of the neurobiology of trauma and the neurobiology of play. From the way the space is designed to the tools of play offered within to the play therapist's ways of relating, all aspects of the environment are crafted to increase the release of healing neurochemicals in doses that will be effective for individual children who are recovering from trauma. At any given time, as you come into Nurture House, you will hear laughter and see high fives, and you will see children eating, drinking, jumping, and playing—playing for their lives. The secret drama unfolding beneath all of this is an epic neurochemical boxing match in which stress hormones and feel good hormones are vying for primacy, and play gives the joy hormones a leg up. I would encourage you to visit www.nurturehouse.org and watch the introductory video or spend some time with the gallery of images. Chapters 3 and 4 will give you a sense of the trauma-informed choices for the design of the physical space and how the offerings at Nurture House answer those lower brain region questions "am I safe?" and "am I loved?" with resounding yeses.

The hypothalamic-pituitary-adrenal (HPA) axis is intimately involved in the stress response and is a mediating pathway for the release of the stress hormone cortisol. People who have experienced persistent or overwhelming trauma have large amounts of cortisol coursing through their bloodstreams. Cortisol is the stress hormone that in small amounts might motivate us to get out of bed to get to a meeting on time, but when an overwhelmingly terrifying event occurs, it is released in massive amounts and becomes toxic to our bodies and minds. The hippocampus, a brain structure important in memory and emotion, reads cortisol levels and can help modulate them (Gilkerson, 1998). The real problem is that continually high levels of cortisol can lead to cell death and are implicated in hippocampal volume loss in traumatized children and adults (Schmidt, 2007). Cortisol is hard to miss: it looks like the figure pictured below (see Figure 2.1).

This child (OK, he is my son) has just experienced a massive cortisol dump due to this shiny, freaky hat that someone (OK, me) put on his head, thinking it would make for an adorable Halloween costume. The hat was short-lived, but during the time it was on his head, his normal bright-eyed, inquisitive, playful nature was shut down. Cortisol dumps shut down our play system lightning fast. We all experience occasional cortisol dumps, but in children who live under ongoing conditions of threat, cortisol is released daily in almost inverse proportions to the typical population. Most of us wake up with our cortisol at the highest levels it will be all day. This helps us get out of bed, motivates us to face whatever is ahead of us, and slowly lessens throughout the day. Cortisol is at its lowest concentration for most people in the evening, making it easier for us to slip off into sleep. For children who have experienced significant maltreatment or neglect, their cortisol is lowest in the mornings (making it difficult for them to wake up and get motivated for the day) and highest at night (making for knockdown, drag-out bedtime battles in which the child is insisting they are not tired because on one level they are the most alert—or, arguably, hyperaroused—they have been all day). We know that excessive cortisol suppresses our immune system, is implicated in all sorts of gut issues, and can further cause dysfunction of the hippocampus (McEwen & Magarinos, 2004; Shin et al., 2004).

If the above picture represents cortisol, then Figure 2.2 represents its antidote, oxytocin, often called the bonding or calming chemical (Uvnäs-Moberg & Francis, 2003) (see Figure 2.2).

Antidote may be too strong a word, but it makes for an easy way for parents to keep priorities ordered when co-regulating an upset child. Oxytocin is released in the brains of both mom and baby during nursing in order to knit

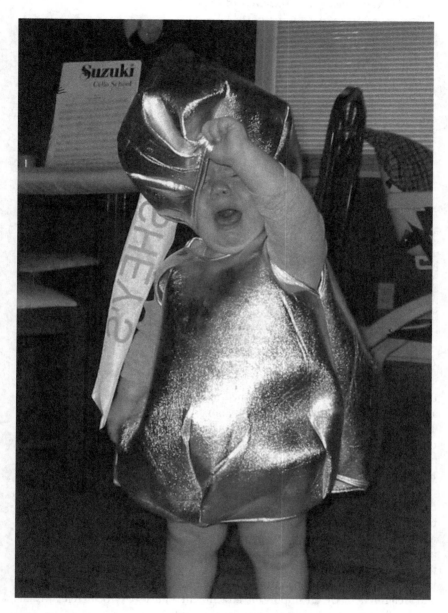

*Figure 2.1* Cortisol Dump

*Figure 2.2* Oxytocin Antidote

them closer together. We want as much of it as we can get on board while we are healing from trauma. Fortunately, nursing is not the only way to get it. This same neurochemical can be stimulated through humor. It can also be released through other forms of nurturing touch. In fact, touch in the form of infant massage has led to clear decreases in infant cortisol (Acolet et al., 1993). Finally, oxytocin can be released through play.

When we are in a state of play, engaged in spontaneous, joyful interaction, dopamine, sometimes called the "joy" chemical, is also released in our brains, so play helps us combat terror with joy. Playing comes naturally to most children, and therefore most feel competent as players and receive dopamine releases when they have a mastery experience in play, e.g., getting a ball through a hoop, painting a pleasing picture, climbing to the top of the monkey bars. When the competency experience (dopamine release) is witnessed and delighted in by an important attachment figure (oxytocin release), an opioid effect is provided for the traumatized child.

Play serves many functions, but in terms of expanding our window of tolerance, it may hold powerful keys. Stuart Brown argues in his book *Play: How It Shapes the Brain, Opens the Imagination, and Invigorates the Soul* that play uniquely positions us to be able to handle the unexpected. The unexpected is always accompanied by stress, so if we can use play to expand our ability to tolerate the unexpected, is it not likely that play in a larger sense helps us expand our window of tolerance for stress? For many children who have experienced neglect or maltreatment, their sympathetic and parasympathetic nervous systems are often strangers to one another, meaning they tend to move quickly out of an optimal arousal window (Ogden et al., 2006) into either hyperarousal or hypoarousal. They may spin wildly out of control, punching, hitting, screaming, and kicking, or collapse in on themselves, turtling up and becoming withdrawn, dissociative, or even catatonic. We see these reactions play out in the playroom often in children who engage in mighty sword fights with me, die, and, after a period of time experiencing what that feels like, come back to life and seem to be processing, through their play, what Stephen Porges would call the mobilization and immobilization responses of the vagus nerve. I sometimes see myself as a greeter for the dysregulated child, helping the sympathetic and parasympathetic responses say hello to each other and learn to live together in more harmony.

## Play and the Social Engagement System

Many trauma survivors, including the children we serve, carry great shame at what they perceive to be their body's betrayal when a trauma reaction is triggered. It can help children to understand, in developmentally appropriate terms, that our trauma reactions are based on the evolution of our autonomic nervous system. Our nervous system is constantly assessing risk, and a *neuroception of safety*, a phrase coined by Stephen Porges, the creator of polyvagal theory, must be present in order for our social engagement system to do its job. Porges' term *neuroception* was a nod toward honoring the involuntary nature of our responses in the face of threat. It is neither conscious awareness nor cognitive perception, but rather a viscerally felt sense of safety. Understanding that this felt safety is the foundational neuro-scaffolding for change supports the enhancement of safety and security as the first and continual goal of TraumaPlay™. Panksepp and Biven (2012) offer a similar idea around the spontaneous activation of the play circuit that occurs when a child is experiencing safety. Understanding that when this felt sense of safety is missing we may respond in highly specialized and self-protective ways can give our clients permission to accept the polarity of responses they may have experienced when danger signals were present. Porges compares our nervous system to a TSA agent doing a body scan of every person who comes near. This process by which we viscerally assess risk and assure ourselves that we are safe enough to engage socially becomes the primary goal when beginning work with a traumatized child or family system.

Porges' theory is based on research tracking evolutionary shifts in the vertebrate autonomic nervous system (ANS). He explains that as vertebrates evolved into social groups, the aggressive and defensive subsystems of our ANS needed to be downregulated in order for spontaneous social behavior to evolve. One form of spontaneous social behavior is the subject of this book: play. Porges talks about the dueling processes of excitation and inhibition that are always ready to play their part and can be honed in large part through the social environment. The social environment of choice for play therapists is, of course, play. Porges endorses play as a neural exercise that shifts affective states within a safe context (Devereaux, 2017). There are many moments of play that are exciting, and as children become excited, we offer them reflection, matching, and modulation—all supported by the attachment relationship. Collapse, or death-feigning, is a phenomenon that all seasoned therapists has seen played out at different times in the playroom. In play therapy, players (either the therapist or client) literally collapse in defeat or death and withdraw into immobilization as they hide in stillness inside a play tent or puppet theater or use a turtle or snail puppet to play out complete withdrawal from those around them.

Porges' study of the phylogeny of the vagus nerve resulted in the identification of three phylogenetically distinct systems: the social engagement system (a uniquely mammalian system) and two defense systems that hearken back to our vertebrate days—the mobilization system and the immobilization system. The three circuits are organized hierarchically, and when we are in a state of felt safety (including a state of play), our myelinated mammalian vagal pathway is in charge, regulating and keeping our heart beating in a more steady rhythm, modulating the HPA axis, minimizing our fight or flight response, and generally bringing us a sense of wellness and literally allowing our bodies to heal by inhibiting our immune reactions (Porges, 2009).

The most fascinating aspect of polyvagal theory, for those of us who place a core value on the attachment relationship and the co-regulation functions humans provide for one another, is that this mammalian myelinated vagus is intimately connected to the muscles of the face and inner ear. What have we always done intuitively to create safety and connection with our little ones? A soft, loving eye gaze, a delighted facial expression, and prosody that includes a nurturing tone of voice become powerful ways in which to bring another person back to a state of calm. Conversely, an overly intense eye gaze, a disapproving facial expression, or a harsh tone of voice can inadvertently cause a "takeover" of the older vertebrate fight or flight (mobilization) system or the death-feigning (immobilization) system. Misusing our tone, eye contact, or facial expression can cause these defense systems to become activated. In play therapy, an attuned reflection of anger or anxiety can build the child's sense of feeling felt, where as expressing these emotions in a way that seems to genuinely emanate from the therapist may actually move that client further into a state of dysregulation. Co-regulation by play therapists does not mean embodying the anger, anxiety, etc., of the child but modulating it by holding the space and understanding our role in keeping the social engagement system activated, consequently keeping the client within his window of tolerance for continued work.

We see then that excitatory pathways and inhibitory pathways are both available to us, and when the children in our care become dysregulated, we can read their physical cues more effectively and provide the right kind and amount of co-regulation more intentionally. This content is a critical part of the paradigm shift that often needs to occur in the parents and teachers of dysregulated children. When I am working with children who have their mobilization system activated frequently, I will work with caregivers on more preventative measures, exploring the conditions that help this child remain in their optimal arousal zone and remain socially engaged. For many children who come from hard places, their sensory systems have been compromised in ways that wire them to be sensory seeking or sensory defensive, so one of the primary treatment goals for these children becomes helping them learn to regulate their own physiology, their own autonomic responses, helping them develop an internal capacity to create a sense of being safe. We will return to this concept in the chapter on need meeting and sensory integration.

It is critical at these times, when children may be moving from a state of hyperarousal to a state of hypoarousal or collapse, for the play therapist to continue to have a duality of being, remaining an anchor and a secure base as the child explores these polarities. The play therapist remains a holding presence, an anchor, and a reflector in a way that is always providing feedback and helping shape the traumatized child's ANS toward the middle of this arousal continuum. Skilled play therapists are able to enter into role play with children in a way that helps them feel felt and that matches their intensity with a duality that maintains a readiness to modulate the child's arousal level as needed.

## Implications for Play Therapy With Traumatized Children

Much research exists to promote the idea that the most effective way to regulate a child's arousal system, particularly when limbic calming is needed, is within their safe relationship with a regulating adult (Perry, 2009; Schore, 2001; Siegel, 2012; Sroufe, Coffino, & Carlson, 2010).

Porges proposes that the vagus nerve offers unconscious mediation of our autonomic nervous system through our social engagement system. The good news is that we can become co-regulators for one another and use our own powerful presence, accessing the social engagement system, to co-regulate the child, parent, or dyad who is with us. Polyvagal theory becomes a powerful argument for the foundational importance of the play therapist as the main agent of change in the room—we literally co-regulate the child in our care through our tone, gestures, and limbic resonance. As such, there is great hope in the use of the play therapist's self to create a feeling of safety. Our presence and the way we use it become the most powerful predictors of state change in the child in our care.

Porges' definition of play is very much in line with that of play therapists. For it to be useful in terms of practicing neuroregulation, play requires reciprocal and synchronous interactions using the social engagement system as a "regulator" of mobilization behavior (e.g., fight/flight). This definition of play may differ from the world in which play is used to describe interactions between an individual with a toy or computer. "Play with a toy or computer lacks face-to-face interaction and will not 'exercise' the social engagement system as a regulator of the neural circuits that foster fight/flight behaviors" (Porges, 2015, p. 5). Porges states in an earlier work that over time, "a face-heart connection evolved with emergent properties of a social engagement system that would enable social interactions to regulate visceral states" (2009, p. 1). What this means is that each of us is at all times a powerful influence on those around us.

## Feeling Felt and Countertransference

The groundbreaking discovery of mirror neurons informs our understanding that increased neurobiological resonance between a client and a TraumaPlay™ therapist is possible. TraumaPlay™ therapists reflect in an attuned way the feelings and emotional expressions we observe in the child. The keys here are attunement and reflection of what we are perceiving in real time. This allows for clients to feel felt and can lead to powerful corrective experiences for the people in our care. However, TraumaPlay™ therapists always keep the awareness of our own countertransference—that is, all the thoughts, feelings, behaviors, and experience (and its somatic encoding), side by side with our attuned responses. Perfect neurobiological resonance is impossible and believing that we can achieve it would be dangerous. It leaves no room for reflection on the anxiety that may be pervasive for a beginning clinician and could easily be misconstrued as the child's anxiety. The clinician could be perceiving anger as emanating from the child in the playroom when the anger being experienced by the therapist may be a carryover from the fight with a family member the night before. Intrapersonal influences of a therapist's moment-to-moment experience of the child in the playroom may even include the simple visceral overload of having had bad pizza before the session. In a system of belief that our neurobiological presence can be thoroughly felt by another, these dimensions have no room. An amplification of a child's dysregulated presence in the playroom by the therapist who believes they are matching the child may leave the child feeling there is no anchor in the room. All of our interactions must be based on a working hypothesis that we gently test out, while always remaining grounded in an understanding that we may have gotten it wrong and will need to switch gears therapeutically.

## Conclusion

Every time I enter a session with a traumatized child, it helps me to remember the neurochemical boxing match that is happening at all times. This visual imagery takes the very complex ideas surrounding the neurobiology of trauma and the neurobiology of play back to the very simple understanding that my internal calm, my attunement, my smiles, my delight, my offerings of ways out, my containment of hard content, and my continual reading of the client's somatic cues help me lower cortisol and increase oxytocin. As I use all of myself as a safe boss, I become the kind of brain-to-brain regulator Schore (1996) talked about—and so can you.

## References

Acolet, D., Modi, N., Giannakoulopoulos, X., Bond, C., Weg, W., Clow, A., & Glover, V. (1993). Changes in plasma cortisol and catecholamine concentrations in response to massage in preterm infants. *Archives of Disease in Childhood, 68*(1 Spec No), 29–31.

Anda, R. F., Brown, D. W., Felitti, V. J., Bremner, J. D., Dube, S. R., & Giles, W. H. (2007). Adverse childhood experiences and prescribed psychotropic medications in adults. *American Journal of Preventive Medicine, 32*(5), 389–394. Retrieved from https://doi.org/10.1016/j.amepre.2007.01.005

Anda, R. F., Felitti, V. J., Bremner, J. D., Walker, J. D., Whitfield, C., Perry, B. D., . . . Giles, W. H. (2006). The enduring effects of abuse and related adverse experiences in childhood: A convergence of evidence from neurobiology and epidemiology. *European Archives of Psychiatry and Clinical Neuroscience, 256*(3), 174–186. Retrieved from https://doi.org/10.1007/s00406-005-0624-4

Axline, V. (1947). *Play therapy.* New York, NY: Ballantine Books.

Badenoch, B. (2008). *Being a brain-wise therapist: A practical guide to interpersonal neurobiology.* New York, NY: W. W. Norton & Co.

Badenoch, B., & Kestly, T. (2015). Exploring the neuroscience of healing play at every age. In D. Crenshaw & A. Stewart (Eds.), *Play therapy: A comprehensive guide to theory and practice* (pp. 524–538). New York, NY: Guilford Press.

Bailey, B. (2001). *Conscious discipline: 7 Basic skills for brain smart classroom management.* Oviedo, FL: Loving Guidance.

Bailey, R. A. (2015). *Conscious discipline: Building resilient classrooms.* Oviedo, FL: Loving Guidance.

Bratton, S. C., Landreth, G. L., Kellam, T., & Blackard, S. (2006). *Child parent relationship therapy treatment manual: A 10 session filial therapy model for training parents.* London: Routledge.

Booth, P. B., & Jernberg, A. M. (2010). *Theraplay: Helping parents and children build better relationships through attachment-based play* (3rd ed.). San Francisco, CA: Jossey-Bass.

Brown, S. L. (2009). *Play: How it shapes the brain, opens the imagination, and invigorates the soul.* New York, NY: Penguin.

D'Andrea, W. D., Ford, J., Stolboach, B., Spinazzola, J., & van der Kolk, B. A. (2012). Understanding interpersonal trauma in children: Why we need a developmentally appropriate trauma diagnosis. *American Journal of Orthopsychiatry, 82*(2), 187–200.

Davies, M. (2002). A few thoughts about the mind, the brain, and a child with early deprivation. *The Journal of Analytical Psychology, 37*, 421–435.

Devereaux, C. (2017). An interview with Dr. Stephen W. Porges. *American Journal of Dance Therapy, 39*, 27–35. Retrieved from https://doi.org/10.1007/s10465-017-9252-6

Dube, S. R., Felitti, V. J., Dong, M., Chapman, D. P., Giles, W. H., & Anda, R. F. (2003). Childhood abuse, neglect, and household dysfunction and the risk of illicit drug use: The adverse childhood experiences study. *Pediatrics, 111*(3), 564–572.

Erikson, E. (1963). *Children and society.* New York, NY: Norton.

Felitti, V. J., Anda, R. F., Nordenberg, D., Williamson, D. F., Spitz, A. M., Edwards, V., . . . Marks, J. S. (1998). The relationship of adult health status to childhood abuse and household dysfunction to many of the leading causes of death in adults. The Adverse Childhood Experience (ACE) study. *American Journal of Preventative Medicine, 14*(4), 245–258.

Gaskill, R. L., & Perry, B. (2012). Child abuse, traumatic experiences, and their impact on the developing brain. In P. Goodyear-Brown (Ed.), *Handbook of child sexual abuse* (pp. 29–67). Hoboken, NJ: John Wiley & Sons.

Gaskill, R. L., & Perry, B. (2014). The neurobiological power of play: Using the neurosequential model of therapeutics to guide play in the healing process. In C. A. Malchiodi & D. A. Crenshaw (Eds.), *Creative arts and play therapy for attachment problems; creative arts and play therapy for attachment problems* (pp. 178–194). New York, NY: Guilford Press.

Gilkerson, L. (1998). Brain care: Supporting healthy emotional development. *Child Care Information Exchange, 66–68.*

Goleman, D. (2006). *Emotional intelligence: Why it can matter more than IQ.* New York, NY: Bantam.

Hebb, D. (1949). *The organization of behavior.* New York, NY: John Wiley & Sons.

Hong, R., & Mason, C. M. (2016). Becoming a neurobiologically-informed play therapist. *International Journal of Play Therapy, 25*(1), 35–44. Retrieved from https://doi.org/10.1037/pla0000020

Hughes, D., & Baylin, J. (2012). *Brain-based parenting: The neuroscience of caregiving for healthy attachment.* New York, NY: W. W. Norton & Co.

Jackson, J. H. (1958). Evolution and dissolution of the nervous system. In J. J. Taylor (Ed.), *Selected writings of John Hughlings* (pp. 45–118). London: Staples Press.

Kay, J. (2009). Toward a neurobiology of child psychotherapy. *Journal of Loss and Trauma, 14,* 287–303.

Kestly, T. (2015). *The interpersonal neurobiology of play: Brain-building interventions for emotional well-being.* New York, NY: W. W. Norton & Co.

Landreth, G. (2012). *Play therapy: The art of the relationship* (3rd ed.). New York, NY: Routledge.

LeDoux, J. E. (1996). *The emotional brain.* New York, NY: Simon and Schuster.

MacLean, P. D. (1990). *The triune brain in evolution: Role of paleocerebral functions.* New York, NY: Plenum Publishing.

McEwen, B. S., & Magarinos, A. M. (2004). In Gorman J. M. (Ed.), *Does stress damage the brain?* Arlington, VA: American Psychiatric Publishing.

Mundkur, N. (2005). Neuroplasticity in children. *Indian Journal of Pediatrics, 72,* 855–857. Retrieved from https://doi.org/10.1007/BF02731115

Neumeister, A., Henry, S., & Krystal, J. H. (2007). Neurocircuitry and neuroplasticity in PTSD. In M. J. Friedman, T. M. Keane & P. A. Resick (Eds.), *Handbook of PTSD: Science and practice* (pp. 151–165). New York, NY: Guilford Press.

Ogden, P., Minton, K., & Pain, C. (2006). *Trauma and the body: A sensorimotor approach to psychotherapy.* New York, NY: W. W. Norton & Co.

Panksepp, J. (1998). Affective neuroscience: The foundation of human and animal emotion. *Consciousness and Cognition, 14,* 19–69.

Panksepp, J., & Biven, L. (2012). *The archaeology of mind: Neuroevolutionary origins of human emotions.* New York, NY: W. W. Norton & Co.

Parkinson, F. (1993). *Post-trauma stress: Reducing long term effects and hidden emotional damage cause by violence and disaster.* Tucson, AZ: Fisher Books.

Perry, B. D. (2000). Traumatized children: How childhood trauma influences brain development. *Journal of California Alliance for the Mentally Ill, 11*(1), 48–51.

Perry, B. D. (2006). Applying principles of neurodevelopment to clinical work with maltreated and traumatized children: The neurosequential model of therapeutics. In N. B. Webb (Ed.), *Working with traumatized youth in child welfare* (pp. 27–52). New York, NY: Guilford Press.

Perry, B. D. (2009). Examining child maltreatment through a neurodevelopmental lens: Clinical applications of the neurosequential model of therapeutics. *Journal for Loss and Trauma, 12*, 240–255.

Perry, B. D., & Dobson, C. L. (2013). The neurosequential model of therapeutics. In J. D. Ford & C. A. Courtois (Eds.), *Treating complex traumatic stress disorders in children and adolescents: Scientific foundations and therapeutic models* (pp. 249–260). New York, NY: Guilford Press.

Perry, B. D., & Pate, J. E. (1994). Neurodevelopment and the psychobiological roots of post-traumatic stress disorder. In L. F. Koziol & C. E. Stout (Eds.), *The neuropsychology of mental disorders: A practical guide* (pp. 129–146). Springfield, IL: Charles C Thomas Publisher.

Perry, D. L., & Daniels, M. L. (2016). Implementing trauma—Informed practices in the school setting: A pilot study. *School Mental Health, 8*(1), 177–188. Retrieved from https://doi.org/10.1007/s123100169182-3

Porges, S. W. (2009). The polyvagal theory: New insights into adaptive reactions of the autonomic nervous system. *Cleveland Clinic Journal of Medicine, 76*(Suppl 2), S86–S90. Retrieved from https://doi.org/10.3949/ccjm.76.s2.17

Porges, S. W. (2015). Play as neural exercise: Insights from the polyvagal theory. *The Power of Play for Mind Brain Health*, 3–7.

Ray, C. D. (2011). *Advanced play therapy: Essential conditions, knowledge, and skills for child practice.* New York, NY: Routledge.

Schmidt, C. W. (2007). Environmental connections: A deeper look into mental illness. *Environmental Health Perspectives, 115*(8), 404–410.

Schore, A. N. (1996). The experience-dependent maturation of a regulatory system in the orbital prefrontal cortex and the origin of developmental psychopathology. *Development and Psychopathology, 8*(1), 59–87.

Schore, A. N. (2001). The effects of early relational trauma on right brain development, affect regulation and infant mental health. *Infant Mental Health Journal, 22*, 201–269.

Shin, L. M., Shin, P. S., Heckers, S., Krangel, T. S., Macklin, M. L., Orr, S. P., . . . Rauch, S. L. (2004). Hippocampal function in posttraumatic stress disorder. *Hippocampus, 14*(3), 292–300.

Siegel, D.J. (2001). Toward an interpersonal neurobiology of the developing mind: Attachment, "mindsight" and neural integration. *Infant Mental Health Journal, 22(1-2)*, 67–94.

Siegel, D. J. (2010). *Mindsight: The new science of personal transformation.* New York, NY: Bantam.

Siegel, D. J. (2012). *The developing mind: How relationships and the brain interact to shape who we are* (2nd ed.). New York, NY: Guilford Press.

Siegel, D. J., & Bryson, T. P. (2011). *The whole-brain child: 12 revolutionary strategies to nurture your child's developing mind.* New York, NY: Bantam Books.

Siegel, D. J., & Hartzell, M. (2003). *Parenting from the inside out: How a deeper understanding can help you raise children who thrive.* New York, NY: Penguin Books.

Sroufe, L. A., Coffino, B., & Carlson, E. A. (2010). Conceptualizing the role of early experience: Lessons from the Minnesota Longitudinal Study. *Developmental Review, 30,* 36–51.

Stewart, A. L., Field, T. A., & Echterling, L. G. (2016). Neuroscience and the magic of play therapy. *International Journal of Play Therapy, 25*(1), 4.

Strawn, J., Dominick, K., Patino, L., Doyle, C., Picard, L., & Phan, K. (2014). Neurobiology of pediatric anxiety disorders. *Current Behavioral Neuroscience Reports, 1*(3), 154–160.

Uvnäs-Moberg, K., & Francis, R. (2003). *The oxytocin factor: Tapping the hormone of calm, love, and healing.* Cambridge, MA: Da Capo Press.

Van der Kolk, B. A. (2015). *The body keeps the score: Brain, mind, and body in the healing of trauma.* New York, NY: Penguin Books.

Walkley, M., & Cox, T. L. (2013). Building trauma-informed schools and communities. *Children & Schools, 35*(2), 123–126.

Wheeler, N., & Dillman Taylor, D. (2016). Integrating interpersonal neurobiology with play therapy. *International Journal of Play Therapy, 25*(1), 24–34. Retrieved from https://doi.org/10.1037/pla0000018

# 3 Need Meeting to Enhance Regulation
## Feeding, Touch, and Sensory Integration

Our case conceptualizations for traumatized children need to begin with a respect for the importance of meeting basic needs and an understanding of dysregulation as an outgrowth of brain and body trauma storage. Dynamics specific to the trauma content—the story of what happened—cannot be addressed until the physiology has been soothed and the family has learned some tools for regulation. Fundamental forms of regulation embodied in hydration, nutrition, and sensory input (Purvis, Cross, & Sunshine, 2007) are sometimes ignored or minimized, when these forms of need meeting can be powerful vehicles for building trust in nurturing caregivers. This chapter will explore how externalizing symptoms of trauma, including behaviors that caregivers find difficult to manage, can be indicators of unmet needs, and that attempting to "give yes" to these unmet needs is a critical part of enhancing felt safety. We will focus on the ways in which meeting children's physical needs builds felt safety.

Reading cues appropriately and finding nurturing, playful ways to meet these needs throughout trauma work can lead to faster resolution of symptoms. When a traumatized child seems resistant to being cared for directly, they will often choose a self-object of some sort and then watch very carefully how the therapist cares for the self-object. For example, Cindy, a five-year-old girl who has deep attachment injuries from years of neglect, chose a baby doll and gave voice to it, saying, "Wah!!! I'm hungry!!! My tummy hurts!!" Cindy looked at me, waiting for my response. If we zoom out from this interaction for a moment, we realize that the play therapist has choices here. One choice would be to simply reflect the content of her play and validate the baby's voice: "The baby's tummy hurts because he's hungry." Another choice would be to enter the pretend scenario with the baby doll, saying, "Oh no, baby! I'm sorry your tummy hurts! Let me get you a bottle, and we can feed you." A third option would be to say to Cindy, "I'm worried about this baby. I wonder if it would be OK for me to feed her." Cindy might then let me know how she would like me to handle this baby's expressed need. This child may be waiting to see if the therapist will meet the expressed need of the baby doll. As a child watches the therapist nurture the self-object, trust is put in the tank that this grown-up is a safe boss.

Titration of nurturing care experiences respects the child's previous experiences while building trust in the therapeutic relationship. As the child watches how the clinician cares for the identified self-objects or vulnerable creatures in the play, they develop expectations of nurturing care from the therapist. Sometimes, in the midst of providing care for self-objects, a child is able to share a new glimpse of their trauma history with us. Miles is a good example of this. Miles was 11 at the time he came to see me. He had spent the first five years of his life in neglect and maltreatment, enduring sexual abuse by several men who paraded through his mom's home. One day, after he had been seeing me for about three months, Miles came into the playroom and said, "I want to wash the babies." This was a brand new request, as he had never expressed interest in the baby dolls before, so I knew it was important for him.

I got down the tub, which he filled with warm water and dish soap. He got a baby doll for each of us, instructed me to undress my baby doll as he undressed his, and then began to wash the head of his baby doll. I imitated his play, washing the head of the baby doll in my arms. As he washed the head of his baby doll, and while his eyes were focused intently on the circular movements of his hands as he washed his baby's head, he said, "My momma used to shove my face in my baby brother's shit diaper." His eyes remained trained on his baby, and he continued to wash its head. I said, "That sounds like a really important part of your story. Thank you for sharing it with me." I believe so strongly in the glimpses and snapshots that children gift us with along the way—and in the power of holding the glimpse together for a moment—that I keep sticky note pads all around the office. I pulled out a sticky note pad and began to write slowly while I spoke slowly out loud: "My momma used to stick . . ." Miles interrupted me: "Shove!" I looked up from the writing/speaking and said, "Oh, buddy, thank you for correcting me. Shove is different than stick." In all the times I have used this approach to extend the moment of holding the story together, I have never had a child cover their ears and not want to hear it repeated. They all come around behind me and watch over my shoulder as I write. They want to make sure I get it right. This exercise extends the moment of holding just a little, and I am careful not to stick with it beyond the child's window of tolerance for looking at it. It becomes a slightly prolonged moment of exposure work. The communication becomes, "I see what you're showing me, and we can stick together in it as you show me." I firmly believe that Miles was only able to tell me about this terribly humiliating, shaming experience because he was actively washing the self-object. Just the act of taking care in play can mitigate the approach to difficult content.

When a child has experienced early neglect, in utero threats, or early trauma, the brain stem may become dysregulated in ways that manifest later as elevated heart rate, difficulty with prolonged attention, sensory integration issues (sometimes sensory seeking or sensory defensive or a combination of the two) (Kranowitz, 2005), or abnormalities in appetite (Neigh, Gillespie, & Nemeroff, 2009; Perry, 2006; van der Kolk, 2015. The importance of providing integrative sensory experiences for children with differences in their needs for proprioceptive and vestibular input will also be described here and will set the stage for some of the practical interventions discussed in the next chapter.

## Need Meeting: It's Never Too Late for Nurture

The need for nurture never diminishes, although the easily discernible signs that a young person needs nurture may diminish as they grow. Teenagers exist in a very uncomfortable space where their bodies are often more adult than their brains, and even healthy teens can miscue parents that they are independent and not in need of nurture. When teenagers have experienced trauma, it can be even more difficult to recognize their underlying needs for nurture and connection, as they will often respond with a bid for independence that in many teens would be age appropriate but may keep the traumatized teen in a place of isolation and fear. That is why, at Nurture House, we place a high value on being able to use your words to ask for what you need. Even if it is not asked for, I will often offer nurture, assuming it will be received. If it is not received, I will have learned something about the teen in my care. In my experience, when it is received, I see my nurturing engagement open the door to that teenager's real connection with me—and often to underlying vulnerabilities that can be held.

## Regulating Body Temperature

If I am meeting with families at Nurture House and another nurturer has turned up the air conditioning, I will begin to find the cold painfully distracting. Sometimes I use these moments to model being attuned to my somatic experience and say, "I'm feeling really cold, and it is keeping me from being able to concentrate on you. I think I need a blanket." I will grab one (we have two in every room of Nurture House) and wrap it around myself. Then I will inquire as to whether the other person or people are chilly and in need of a blanket. This offers just a moment, a titrated stillness of somatic focus, to check in with themselves about whether they are having an uncomfortable experience of being cold. Providing warmth to someone who is cold has always been an expression of need meeting.

Since one of the systems regulated by the brain stem is body temperature, children who had in utero threats to their developing brain stem (drugs, alcohol, excessive cortisol releases) may have impairments in their ability to regulate their temperature and may have an overactive or underactive feedback loop regarding body temperature. These are the children who are running on the playground, red-faced and sweaty, on a hot, humid summer day and are not being signaled by their brains that they need to cool off. These children may also keep their coats on in a warm room until they are invited to take them off because their bodies are not giving them signals that their body temperature is rising. I will sometimes take stones that look almost identical, place one in the refrigerator and one in the sun outside, and then have the client experiment by holding each of the stones and noticing what communication their hand is sending their brain about the stones.

Nadine, a 16-year-old young woman, was referred to me after years of managing an eating disorder. She had run the gamut of services: residential treatment out West, intensive outpatient programs locally, dietitians, nutritionists, therapists. She had also tried acupuncture and yoga but continued to feel disconnected from the world and to have great difficulty coexisting with family members. When she came to me, she was therapy weary (and wary) and knew all the lingo. She came up to the loft of Nurture House with me, sat down, and began to recite her diagnoses and treatment history. The feeling I had was that she had distanced herself from her own experience—she could play the game of therapy and talk the talk, but she kept the real child locked up tight inside the teenage persona.

At one point, while she was giving a litany of her previous treatment, she shivered absentmindedly. I immediately said, "Oh, your body just let me know you might be cold. Let me get you a blanket." I leapt up fast enough that she had very little time to react, and I grabbed an extra soft cream-colored throw, wrapped it around her, and—as I was leaning in to tuck in the edges—noticed her really cool hair color. I commented on it with delight and briefly touched the edges of her hair, wondering out loud if that hair color might work for me. You would have thought I switched on a lamp. From that moment on, that moment of caretaking, of meeting a basic physical need, one that she most likely would not have mentioned or asked for help in rectifying, she was like a different child. I use the word *child*, as opposed to referring to the beautiful 16-year-old she was chronologically, because in meeting those basic needs for her, I ended up meeting the younger version of her, the "self" buried under all the diagnoses and interventions. I liked what I saw—and eventually so did she.

We are never too old for nurture. Robby taught me this lesson when I was just beginning my work in disaster relief settings. Robby was 16 years old when his family was displaced from their home in the wake of Hurricane Katrina. I was three years out from my graduate training and had been to my first Theraplay training. I was using the model regularly in my work with children in the therapeutic preschool where I worked. I had been through disaster relief training for mental health professionals with the Red Cross a year earlier, and I was called when masses of displaced families arrived in Nashville for emergency services. They asked me if I would come and run a group for the children in the temporary housing shelter. Although I was unsure what to expect, I said yes, and when I showed up in the large parking lot taken over with triage tents and asked where I was to run the group, they pointed me toward the abandoned office building that was to serve as a temporary housing shelter for these families. On the second floor, I set up my supplies as children began to straggle in. Their parents were busy filling out FEMA forms, and I was the only adult available. I ended up with a group of about 13 children ranging in age from three to 16 years old. Robby, at 16, was the oldest youth there and already had the stature and muscles of a grown man. We made introductions, played connecting and regulating games, and generally had a good time. At the end of the group meeting, I introduced the rock-the-blanket exercise, a game I had learned during my Theraplay training. To do this exercise, you simply put a blanket on the ground and have a child lie in the middle of it, and then two people pick up the ends of the blanket and rock the child back and forth while singing. I always sing a song to the tune of Frere Jacques that goes like this: "You are safe here, you are safe here, yes you are, yes you are."

I laid out the blanket, and while I was laying it out, Robby saw the predicament I was in: there was no other equally strong person to hold the other side. He jumped up and said, "I can help!" I was admittedly relieved. Disaster work is particularly challenging because you never know what you are going to get. Best practice standards for group work would discourage the idea of a group this large being run by one adult only. In a clinic setting, we might also frown on a group covering such a wide age range. The phrase from the book series Pinkalicious always guides me in times and places where I am trying to provide clinical care but may not have a best practice environment: "You get what you get and you don't get upset."

We began the rocking game with a little three-year-old named Trina. We rocked her in the blanket while singing, "You are safe here, you are safe here . . ." We made our way up the age chain from three-year-olds to eight-year-olds and finally to the 14-year-old girl, and rocked all of them in the blanket. When we had rocked the last of them, I went to lay down the edges of the blanket on my end. While Robby was laying down the edges of his side of the blanket, he said very quietly, "Can I have a turn?" I caught his eyes and said, "Absolutely. We will make that happen." I finished group with the other children and asked Robby to stay for a moment. I went and got another staff person, and we rocked Robby in the blanket. We are never too old to need rocking, to need holding. Thank you, Robby, for teaching me this profound truth.

## Conversations About Care: *What Do They Need?*

Inviting conversations about caretaking of others can serve two purposes with clients: I can begin to get a sense of the child's core beliefs—do they believe that children, animals, and baby plants should have specialized care? What are the child's *shoulds*? Should vulnerable, young, or small living things be taken care of differently than big things? If so, do they deserve to be taken care of in those ways? These conversations can be difficult, and play therapy can help. Creating environments for miniatures or Beanie Babies—simply offering a child a large shoebox, asking them to choose a Beanie Baby from among several choices, and then asking them to create an environment from lots of creative supplies that will give that creature the environment it needs to thrive—can playfully mitigate an approach to this content. I was first exposed to this intervention while training in Trust-Based Relational Intervention (TBRI) (Purvis, Cross, Dansereau, & Parris, 2013) and used it individually with lots of children but had the most powerful results during Camp Nurture. I brought in a large amount of wet clay and several miniatures from the sandtray room: horses, puppies, dragons, and dolphins. Campers paired up, chose a miniature, and then created the environment that creature needed to grow. The horses needed lots of room but also fences to keep them safe. Out of this came some powerful conversations about boundary setting and safe bosses. The dolphins needed the open ocean, but also other dolphins, in order to thrive. Out of this came powerful conversations about our need for community, friends, and others who understand what we need.

I love to use a version of this intervention in self-care work with our team. I ask each clinician to create a symbol for an anxiety that he or she is carrying. We place all the anxieties into a container, and then each clinician chooses one symbol (not their own) to work with. That clinician creates an environment of soothing for the symbol. It is only after this has been done that the creator of the environment pairs up with the person who created the symbol. Clinicians are deeply touched simply having someone "take care" of their anxiety, and the cross-hemispheric intervention often opens up new possibilities for managing the anxiety in new and soothing ways that the original symbol maker may not have been able to access alone.

Potted plants and healing gardens can open up similar conversations about what these living things need to grow. At Christmas this year, I had a client bring me an indoor bulb planting set. It sat on the kitchen counter for a couple of days, and then Jake, a seven-year-old client who had been adopted at birth but dealt with lots of anxiety and sensory processing issues, asked, "Why do you have a bucket of dirt?" Our conversation is below:

ME: Well, it's supposed to grow a plant, but I haven't opened it yet.

JAKE: Can we open it?

ME: Sure. I've never done one of these, but I think you plant the bulb in the bucket and it grows inside.

JAKE: Not outside? At school we learned plants need soil, water, and sun. How will it get all that inside?

ME: Good question. The directions say to fill the pot three-quarters of the way full with dirt.

JAKE: There has to be some in the bottom. [*He began filling the pot with dirt.*] There's more dirt than I thought [*when he got the level right*].

ME: There is a lot in there. It's like a bed so that there's a place for the roots to grow.

JAKE: Well, it's bigger than I thought [*picking up the big bulb*]. No wonder it needs so much dirt. It needs more room.

  I felt like here we began talking on two levels, although we kept the conversation all about the plant.

ME: Yes, sometimes the bulb has to rest there for a while before the roots even begin to grow.

JAKE: Well, do we just lay it in there [*getting back to the task at hand*]?

ME: Place it on the dirt and loosely cover it with the moss [*reading aloud, quoting the directions*].

Jake carefully placed the bulb on the dirt, picked up the moss, and began placing it on top. In truth, it looked to both of us like there was not enough moss to really cover it. When we had placed all the moss, Jake said, "Where's the rest of it?"

ME: It looks like there should be more, but that's all there is.

JAKE: Well, how will it grow? It is not covered.

ME: Yes, it does feel like it's kind of exposed, but we've followed the directions and used everything that came with the plant. The directions say to add water now.

JAKE: It won't be hidden.

ME: Seems like you are worried about it being uncovered.

JAKE: Well, plants push up, right? This one won't have anything to push through, so maybe it'll grow faster.

The parallel between this statement and some of his current struggles was profound. His fear of whether his family would keep him even if he didn't perform well often led him to cover up the "real child" in lots of big behaviors or to retreat into himself, hiding his big feelings from everyone. He would ping-pong back and forth in a way that often led to big behaviors and dysregulated meltdowns. The realization that the plant might be able to grow faster with less struggle, as it had fewer layers of covering, represented a potentially profound shift in his ability to come out of hiding and struggle less in his environment. We watered the plant together.

JAKE: What's next?

ME: The directions say to put it in indirect sunlight.

JAKE: What does that mean?

ME: I think it means not in bright, hot sunlight but kind of off to the side so it can absorb light without the light being too intense for it.

JAKE: I know where. [*He put his hand in the shaft of light coming through the kitchen window.*] That's too hot. I'll put it just over here.

I was a little anxious about whether the plant would grow, but a green stalk began to appear within a couple of days. When Jake returned for his next appointment, the plant had grown up, surprising both of us with how much it had grown in such a short period of time. He said, "See? I knew it would grow faster 'cause it didn't have so much stuff on top of it" (see Figure 3.1).

## Sensory Savvy Safe Bosses

Children with chronic trauma in their history often have sensory differences that must be respected, and needs arising from these differences must be met for trust to be adequately built with their caregivers. While the differences in the way these children experience the world are more clearly understood than ever, our behavioral science traditions continue to inform how these children

*Figure 3.1* The Plant in Full Bloom

get treated in families and classrooms until a different way is learned. Helping parents and teachers understand the sensory seeking or sensory defensive leanings of their children and helping them meet the needs that arise from these leanings is foundational for the rest of our trauma work at Nurture House. In American culture, we have been taught to see smaller amounts of dysregulation—a child who is tapping their pencil repeatedly, a child who is rocking back and forth in their chair or jiggling their leg up and down—as behaviors to be ignored, the idea being that as you ignore the behavior it moves to extinction. However, for many of the traumatized children we see at Nurture House, their bodies are actually trying to give the child and their caregivers valuable information about what they need. When children come in with sensory dysregulation, the therapist joins the parents and child in being detectives, attempting, with deep curiosity, to figure out what the child's body is saying it needs. You will hear the phrase, "Your body is letting us know that . . ." many times a day at Nurture House.

Billy is a good example of how much connection and shift can occur when a client's bodily needs are being met. He was internationally adopted at age three and diagnosed with autism spectrum disorder by age four. I met him when he had just been asked to leave a public school environment because of "acting out behavior." The teacher in question had been trying to give individualized attention to Billy for most of the year and was exhausted and resentful at the amount of time his needs required. She saw him as defiant and unteachable. When I asked Billy's mom what she thought had been the final straw, she explained that on his final day of school, the teacher had told everyone to line up for recess. However, instead of lining up, Billy hid underneath his desk. The teacher asked him to come out, and Billy shrank further under his desk. She spoke sharply to him, saying that if he did not come out from underneath the desk *right now* he would lose his trip to the library that week. Now, Billy was having such a miserable time at school that the only activity that brought him solace was library time, as he could delve into a book for half an hour and not have to interact with others. The teacher's withholding of this privilege at a time when his amygdala was already armed caused him to upset the desk and run out of the room.

What would have happened if, instead of perceiving the behavior as defiant, the teacher had seen it as a valuable indicator of what his body needed right then? What if she had been able to say, "Oh, buddy, your body is letting me know you need a small, quiet space right now. How clever of you to know what you need." I have recently worked with an amazing principal in town who understood the sensory needs of children like Billy so well that she had cleared out one of the built-in cabinets in her office specifically so that escalated children could curl up in there and regroup before she tried to engage them, regulating the reptilian brain stem before attempting to engage in any cognitive exchange.

One day early on in our sessions together, Billy and his mom were in the kitchen with me. When I had greeted them in the lobby, I had offered Billy a variety of snacks, and he had chosen Cheetos. He had chomped through all his Cheetos, licked his fingers, and begun chewing on his shirt and hands. Billy's mom was sharing a story with me around a recent moment in which she and Billy had had a playful, connected mother-son date. She interrupted herself several times to gently touch his hand and say, "Stop putting your hands in your mouth, Billy." She was pleasant and kind in her tone, and Billy would take his hands out for a moment, but they would migrate back to his mouth quickly. The third time his mom stopped to correct him, I intervened, saying, "I too am noticing that Billy is putting his fingers in his mouth over and over again." I made eye contact with Billy and said, "Buddy, it looks like your body is saying, 'I need to chew on something—I need something in my mouth!' Is that what your body is saying?" Billy moved his eyes over to his mom but nodded. I said, "Well, here at Nurture House we love it when kids know what their bodies need and can show us. I have three options for you: you can suck on a lollipop, you can chew on a gumball, or you can have one of my pieces of chewelry." I proceeded to open a drawer in the kitchen and show his mom several kinds of chewelry. He chose the lollipop. Once he had unwrapped it and put it in his mouth, I said, "Now, since your mouth is busy with what it needs, the rest of your brain can stick with us." Billy's mom laughed out loud and said, "Oh my goodness, I can't tell you the number of times I have asked him to stop putting things in his mouth, but you're saying that maybe he needs to?" I explained what I believed to be some of Billy's sensory seeking behaviors, including his need for almost continual oral gratification. I thought it would be helpful for the family to create stress engines together.

The Alert Program (Williams & Shellenberger, 1996), a tool developed by an occupational therapist, teaches children how to become more attuned to their bodies, equating the child's current state of regulation with how slow or fast their engine is running. A personalized stress engine is created with the child and becomes a concrete tool for reflection on internal states. If you are in the blue zone, your engine is running too slow (this might occur during times of hypoarousal, and children are likely to be sluggish, grumpy, and slow to respond to prompting). The green zone is the just right zone, where you feel regulated and alert for new learning. When you are in the red zone, your engine is revving too fast. I asked Billy if his need to put stuff in his mouth felt more urgent when his engine was running too fast or too slow. His response was *both*, and we talked about how that was probably true for him. So we dug deeper, Billy, his mom, and I, wondering out loud what sorts of oral gratification needs he had when his engine was running too slow. He was able to discern that crunchy or sour things would be helpful in upregulating him at those times. He also began to see that sucking on something or chewing gum could be calming for his central nervous system when he was running too fast. This activity invites children and their parents to investigate what helps in each of the somatic states (see Figure 3.2). On the very first morning of Camp Nurture, we made stress engines for each child. Every two hours we checked in with the whole group about how their engines were running. It was through this repetition of mindful internal focus that the children began verbalizing when they needed to go to Crash and Bump and use their bodies and when they needed quiet, and even when they needed weight to be added to help them regulate (each child had their own weighted blanket, and weighted lap pads, ankle/wrist weights, and weighted vests were also available). Halfway

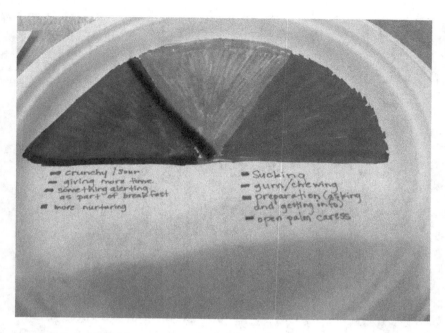

*Figure 3.2* An Example Stress Engine With Articulated Strategies

through the session, Billy's mom joked that she could use one of these for herself, so we made one for her as well. I think we could all use one.

As we began to understand what Billy's body was telling us and began "giving yes" to the underlying needs, he became more connected and was able to take more risks in therapy. He was making great gains in his social relationships but was still having difficulty receiving nurture from his mom. We thought some work around the adoption would be helpful, but Billy became very shut down whenever we began work around this. One day toward the end of a session, in which he had had lots of child-led play, I brought out a book of one-page descriptions of amazing people who were adopted. He said, "I don't want to read it!" I replied, "I hear you. You don't like that book. You can choose what we play." He immediately began stacking three Big Joe chairs with nylon covers that slip around on each other. When stacked on top of one another, they reach a height of at least four feet. After he stacked them, he then scrambled up on top, somehow arranging himself so that he was able to maintain his equilibrium. As soon as he was settled on top, I said, "Wow! You are doing it. You are staying on top!" He grinned really big and said, "OK, give me the book." I gave it to him, and he sat on top and read a whole page out loud. Below is the picture he requested I take while he was reading the book perched on top of his pillow mountain (see Figure 3.3).

So what happened? What helped? I believe the intense exertion needed to wrestle those pillows into place and then scramble up on top of them and the

*Figure 3.3* Kinesthetic Competence

continued engagement of Billy's core that was required to remain balanced up there gave him neurochemical competency surges—pleasurable experiences of mastery that mitigated his approach to the harder content related to adoption. While he was feeling strong, powerful, and in control of his created (if precariously balanced) world, he could choose to delve into content that made him uncomfortable. This is an example of expanding the child's window of tolerance for looking at hard things through mitigation. In this case, the mitigator is somatic mastery, or kinesthetic competence.

## Sensory Static

We had several children at Camp Nurture who had never been able to successfully play with other children without hurting the other children or themselves.

When we met them, these children were isolated and socially impoverished. Each child was assigned a buddy and a "backup buddy," and as these buddies met needs, provided structure and nurture, and gave constant feedback (which required staying within three feet of the child at all times), the children began to change. The inward focus that is required by states of hyperarousal and collapse creates a kind of static that gets between the hurt child and the rest of the world, including potential friendships. When we are in a state of hyperarousal or collapse, we are unable to learn from our experiences.

The first level of feedback always had to do with what their bodies needed to remain regulated. On the first morning of camp, each child was given a backpack and went around the room to stations that each provided a regulation tool. Each child got chewelry, bubble gum, and lollipops. As we know that sucking can provide soothing to our reptilian brain stem, they also received a specific kind of water bottle that requires a stronger suck than others. It seemed that these children had so much sensory static they could not see clearly through it to be curious about the people around them. Peers were not seen as relational resources. As we met underlying needs and used ourselves as the external modem for each child, the static began to clear. We also diffused essential oils at camp. At Nurture House, we value the use of scents as a sensory portal for providing upregulation or downregulation. The olfactory bulb is the only part of the central nervous system that is exposed to the elements, so it is the fastest way in for a change of state. We brew coffee for parents who need both the smell and the caffeine to upregulate, and we keep diffusers in most of the rooms of Nurture House and diffuse essential oils (Johnson, 2015) Setzer, 2009) based on the regulation needs of individual children—and sometimes the regulation needs of a therapist. I tend to hit a downregulated slump in the middle of the afternoon and often diffuse lemongrass oil at that time in order to help upregulate myself so I can remain as fully present and engaged with my clients as possible for the rest of the afternoon. We often make "regulation lollipops" with children as we are helping them appreciate their own up- and downregulation needs. We begin with cotton balls and popsicle sticks and cover the cotton balls with small squares of fabric. Then we invite the client to explore all of our oils and choose the one that is most stimulating to them and the one that is most calming for them. We put drops of each on the lollipops, seal them in separate Ziploc baggies, and have the family practice using them as different regulation needs arise.

On the third morning of camp, Scott, one of our most dysregulated kiddos, asked for one of the "body socks" that had been made by one of our gifted buddies and completely enclosed himself in it. He wrestled around on the floor within a foot or two of his primary buddy for an extended period of time. Eventually, he pulled apart the Velcro sealing the two sides of the sack together at the top, and then spent considerable time and energy wiggling out of the sack. He did not ask for help, would not allow help, and spent great focused and sustained bursts of energy trying to free himself from the sack. His buddy described it to me as a live demonstration of a caterpillar coming out of a chrysalis. It was after this point that he began asking another little boy to play with him. It was as if he noticed the other little boy for the very first time, as if a complete metamorphosis had to occur for him to become aware of the "other." The large, intentional, and constant doses of need meeting by the buddies had to become internalized and reshaped into an ability to connect with others. One of my life mantras is "you can only give what you have received." This little boy had to have enough nurture and need meeting offered to him in high enough doses to allow him to begin to look around. The buddies of both Scott and the little boy he asked to play were still both highly engaged and within three feet of the two boys at all times. There were multiple helps in interactions and multiple redos and compromises needed for the boys to begin socializing, but it was as if the awareness of the "other" was not even initially possible because first one had to cut through all the kinetic static of unmet early needs that circled around Scott.

The ability to develop a friendship requires a capacity for "other-centeredness," which arises from the development of a sense of self as separate from but connected to the "other." It also requires that one have assurance that basic needs will be met. I am reminded of Maslow's classic hierarchy of needs whenever I watch the evolution of children who have experienced attachment trauma. His iconic pyramid of needs placed physiological need meeting as the seminally important foundation for any further growth, followed by security, and then belonging. I believe that capacity for other-centeredness can be grown and involves being able to have a sense of self as secure and cared for.

In many cases, traumatized children do not have basic needs met consistently or are harmed by those who are supposed to care for them. Even when these children wind up with a caregiver who is meeting all their basic physical needs, they may still have the terrifying core question, will I have enough? This is why, at Camp Nurture, we always provided *more than enough*. Because a child's perception that the need has been met is critical to quieting the reptilian brain stem and regulating the limbic brain, having just enough of the needed item is often not sufficient—having backups and backups of backups is sometimes necessary. The fear of not having enough is experienced as an ongoing threat to survival and results in a preoccupation with the self. When they first come to treatment, adoptive parents are often bewildered about why their child "melts down" every time they "don't get their way." We validate and appropriately hold the parents' frustration, begin to offer appropriate psychoeducation, and gently support a shift in paradigm. Parents often use words like "manipulative," "selfish," "spoiled," or "demanding" when they first come to Nurture House and are attempting to paint the picture of their daily lives with their children. It is difficult to meet any need effectively when you feel you are being manipulated or believe what is being demanded is "too much" or unreasonable.

When I am supporting parents, I will explain that the neurophysiology of a traumatized child is such that anxiety, excitement, and aggression all balance on the head of a pin. In the blink of an eye, and through the neural confusion of excitement and anxiety, a want becomes a need. Parents will often feel like they cannot win for losing. If they offer the potential of a trip to the toy store or the friendly possibility of a trip to the park, they may initially be rewarded by the flush of pleasure on their child's face but end up feeling held hostage to the child's demand for the thing or held captive by the child's anxiety about when, where, and how quickly they will get it. These children often need to be taught, with great patience and reflection by a safe boss, the difference between a want and a need. This teaching does not happen with words—or certainly not by words alone—but through the titrated experience of having needs met and exceeded. During the initial titrations of having needs met and exceeded, the parental presence is intentionally paired with the need meeting. We are attempting here to recapitulate the healthy attachment dance in which basic needs are lovingly paired with a safe boss. The relationship between a nursing infant and mother is the most powerful example of this natural association between physical need meeting and the nurturing connection with the "other." The baby's body is snuggled warmly against the mother, often accompanied by a loving eye gaze and playful interaction, while the physical hunger is satiated. Bottle feeding, too, is usually done in such a way that the baby is nestled in the crook of the mom or dad's arm while being fed. As the baby grows and starts eating solid foods, sweet circles of communication surrounding the eating evolve: the baby opens the mouth wide again after swallowing a bite, and the parent, picking up on the nonverbal cue and perhaps even giving voice to it—"you are letting me know you're ready for some more"—gives the baby another spoonful of food, rewarding both of them. When the baby is super hungry, it may be hard to wait, and the baby's window of tolerance is expanded while waiting for mom to spoon out more mashed bananas.

Satiation, or the sense of being truly full, is another way to characterize the neuroception of safety. In the natural feeding cycle of caregivers and infants, satiation, or fullness, is delivered through the relationship with the attachment figure, and eventually the food or the connection in the relationship fills them. I have been repeatedly struck by the profound nature of the trust established in these early feeding routines and the distrust established when a child did not get them. At Nurture House, we use a dyadic assessment that is partially informed by the Marschak Interaction Method (the assessment tool used in Theraplay) (Booth & Jernberg, 2010). One of the tasks directs the adult and child to feed each other. Watching the way adults and children navigate this task is fascinating. In families where a secure attachment exists between parent and child, they will often put food directly into one another's mouths. The parent sometimes finds great joy in this and replicates a game, such as airplane, that used to be played during feeding time during the child's first year. Other parents will elaborate on the task. On numerous occasions, I have watched a parent choose multi-flavored gummies and invite the child to play a game of "guess the flavor." This requires a great deal of trust, as the child closes his eyes, opens his mouth, and allows the parent to place a gummy inside. When the task is navigated in this way, it often makes a powerful statement around the neuroception of safety shared between the parent and child.

In other dyads, and in many of the adoptive families I see, children will begin to make demands almost as soon as the task has been shared out loud. They may simply say, "I want to feed myself!" or they may grab the bag of crackers from the parent or play their own version of the game in which they get close to putting the food in the parent's mouth and then either jerk it away at the last minute or aggressively insert it. These ways of coping with the stress induced by a task that calls for intimacy and vulnerability can tell us a lot about the child's earlier experience and lack of trust that basic needs will be met. On the other hand, it is sometimes the parent who is unable to feed the child. The parent may cope with a task that calls for intimacy that exceeds the parent's current comfort level by encouraging the child to hold the bag themselves, buffering the potentially intimate eye contact that can occur during feeding by "noticing" something else in the room at the same time the feeding is occurring, precluding the extra layer of intimacy felt through eye contact, or by saying, "You don't want to be fed, you're a big boy now!" This idea that the older you are the less likely you are to be comfortable with this intimacy is not supported in my observations of healthy dyads. Many teens, when faced with this feeding task, will laugh and acknowledge the weirdness but then move right along and feed and be fed. Really attuned parents of teens will raise the challenge level for this task by having them toss food into each other's mouths from a distance. Regardless of how they navigate it, securely attached teens and their parents have fun with the task (there is usually a lot of giggling or joking) and do not find it difficult or frightening. An insecure attachment pattern, however, is often evidenced by the parent and child not being able to feed each other with ease and connection.

What starts as excitement about the potential for a new toy or a chosen food or a chosen activity can quickly morph into anxiety around whether they will get to have it. The discomfort of having an out-of-reach potential can trigger the wanting/needing response. This want/need pattern scratches at the thin layer of new neural experiences of safety—their tenuously developing neuroception of safety—as if it were a scab. If we extend the metaphor, what happens when the scab of safety is scratched? The core fear oozes up. In these cases, the question asked by the fear-infused core self is, will there be enough for me? The sandtray you see below was created as a first world by a ten-year-old girl who had been adopted from a very poor country where she had lived in one of the least resourced orphanages. I simply asked her to create a world in the sand (see Figure 3.4).

Notice that the tray is full to its edges with plates, bowls, and cups, and all of them are filled. None of them are empty. This overcompensation for the previous lack of resources is something we see frequently when children are beginning to try to answer the question of "will there be enough for me?" with new answers. Another example of the irrational perception of needing more, even when we have plenty in front of us, may aid our understanding. My fingers are hovering over the keys to delete the word "irrationality" as

*Figure 3.4* Overflowing

I write it because it sounds "judgy," but it only sounds judgy because we, as a society, have judged rationality to be the ultimate goal of experience and have not fully understood and embraced the truth about how the triune brain functions. Within at least the recent American psychological tradition, the subjugation of needs, drives, emotions, and somatic experiencing to the neocortex and the executive functioning system has been, at the minimum, the implied goal of many evidence-based treatments. Respect for the work of lower brain regions and an appreciation for the importance of meeting the needs of lower brain regions are growing in trauma work. Perhaps it remains the overarching goal of mental health to have the lower brain regions deliver input to be interpreted by the higher brain regions, but in order for this to occur, we must agree that irrationality is not a dirty word. In fact, irrationality must be heard, must be explored, must be given a voice in order for regulation to occur. What if instead we reframed *irrationality* as *really good and important information about what our bodies are needing for physical/sensory regulation* (reptilian brain stem/diencephalon) and what our feeling brain needs for emotional balance in order to clear a path to our executive functioning potential.

Whenever I begin to wonder if meeting (and exceeding) the need serves any valuable purpose, I always think of Tasha. Tasha was a five-year-old who had been adopted from India at the age of two. Her orphanage experience was one in which she often had to scramble for rice that was dropped in a heap on the floor in the midst of a circle of hungry children. Her parents had to help her learn how to use utensils when she came home. However, as her buddy and backup buddy met her needs in excess in repetition after repetition, she began to shift.

During the camp day, she was hyperkinetic, sometimes aggressive with other peers, and sometimes territorial. We included several snack times in the schedule at Camp Nurture, and children were always given a yes to getting food or drink between snack times if they simply used their good words to ask their buddies for what they needed. Each child had their own basket of approved snacks (because of allergy constraints), but each basket was heaped with food, and the whole basket was presented at snack time. The only exception to this was if we had learned that it was harder for the client to choose between offered snacks (in which case a narrower field of snacks that was not so overwhelming was offered). One morning, I decided to have a snack with Tasha and her two buddies. Tasha and her two buddies moved over to her snack spot, and her buddy picked up the snack basket and set it down on the floor (snack time was always a picnic affair). Tasha immediately demanded, "More Pirate's Booty!" I got down on both knees in front of her so that I was eye level with her and said, "You want more Pirate's Booty? Sounds like you feel like you *need* more Pirate's Booty." She locked eyes with me and nodded vigorously. I said, "You can ask me for anything, and I will listen." She said, "Can I have more Pirate's Booty?" I replied, "Absolutely!" and went and got two more mini bags of Pirate's Booty. I placed them on top of the one bag of Pirate's Booty that was already displayed on top of her basket of food, and she eyed the basket. Apparently, she decided it was "enough" now and sat down. She opened a bag and had two pieces of the cheesy snack, then gave one bag to her primary buddy and another to her backup buddy and offered me the rest of her bag. It was not about the food. The underlying need was not related to physical hunger. The need was to answer the question, "will there be enough?" with a resounding YES. As soon as she had her loud enough answer, the static cleared and she refocused on relationship, actually sharing the treasure with those who had been nurturing her most closely.

The picture below (see Figure 3.5) includes our offerings of water bottles, juice boxes, and snacks. Children can increase negative behavior when they are dehydrated. Some children's bodies may not communicate well to them that they are thirsty. Therefore, hydration is a core regulation need that we are always supporting at Nurture House. Offering choices of beverages, choices of snacks, and even a choice of which color gumball clients would like sets up opportunities for clients to practice using their voice and having it honored. These choices are also a way for us to share power with clients almost as soon as they enter the building.

We have recently added a third hydration option—a large water dispenser with a giant inverted water jug. I was delighted, just the other day, when I entered the lobby with my empty water bottle to refill it. There was a six-year-old child in the lobby, and he eagerly approached me and asked, "Are you going to fill that up?" I said, "Well, yes I am!" matching his enthusiasm. There is some coordination involved in positioning the bottle correctly to catch the stream of water, but once I had done that and pressed the lever,

*Figure 3.5* Meeting Basic Needs: The Empowerment Principles

this six-year-old stood grinning ear to ear and staring at the giant bottle as it glugged and shot bubbles through the surface of the water. I caught his eye, and he started making the glugging noise the bottle was making out loud, matching the cadence of the actual glug. I joined in for the fun of it, and he and I had a moment of shared experience and connection, simply through his enjoyment of the rhythmic sounds of the water dispenser. I did not know this kid from Adam, but he taught me a new way to enjoy the lobby.

## The Gum Guy

Joey was three years old when I met him. He teetered into the playroom on unbalanced legs with his "milk" backpack on his back. One of his first statements to me was "I do not eat by mouth." Joey has a very rare genetic disorder that combined with anxiety issues, spectrum symptoms, and encephalopathy results in a child who is often dysregulated and then very worried about how this affects those around him. He is fed by a tube and has had multiple life-threatening moments in his short life. His parents have wondered just how deeply the inability to eat affects him. At the very least, it seems like it has contributed to the heightening of all his other senses. If the neuroception of safety could only be attained through physical satiation, Joey would never feel safe because he never feels "full." Over the years, his parents have learned how to make accommodations for him and give him partial "yeses" around food. When he goes out to a restaurant with his parents, he licks a lemon, and he can pour salt into his hand and taste that also.

The very first day he came to see me, he asked for a gumball on his way out. His mom and I helped him get one for himself, and then he requested one for his mom. He understood that he would not be able to eat it, but he took a couple of licks of the gumball and then held it in his hand all the way home in the car. At the end of the next session, we went through this same routine. On his eighth visit, the gumball supply was lower than normal. He became very anxious about this, but having learned that if he uses his words to tell me what he needs, we will really try to make it happen, he walked right into the kitchen (now at the ripe old age of four) and told my intern (who I guess he thought was in charge of the gum) that he needed to buy more gumballs please. For Joey to use his words instead of his body to express a need and to trust that someone would be listening was huge progress. To help give Joey more than enough assurance that we would handle the gumball crisis, I got out a piece of paper and had Joey dictate a note to my office manager, asking her to buy more gumballs. I let him write the letters of his name, and we put the note in "a special place" where Miss Linnea would see it. During our team meeting that week, I noticed that she had the paper on her lap. I said, "Oh yes, that's for Joey. The gumballs are really important to him. Even though he can't eat them, he needs to have them." I was profoundly reminded again of how deep our perceived need can go and how powerfully an attempt to give yes can soothe.

Joey's heightened development of his other senses became clear to me again during a recent session. I walked out to the lobby to get him, and he introduced me to "Cubba," his bunny, who also has a feeding tube. As he was showing it to me, he held it out and said, "Smell it!" I did so, and it smelled like fabric softener to me, and I said, "You wanted me to smell it." Then he pulled Cubba close, inhaled deeply, and said, "He smells like snuggles." I was struck by the profoundly simple statement. His olfactory sense is highly developed, and the smell of Cubba has paired with the physical sensations of snuggling with both his mother and the bunny. The snuggle soup of mom and bunny is enhanced by the release of oxytocin and dopamine and further enhances his neuroception of safety. Cubba, then, can serve as a transitional object and olfactory source of soothing even during times of separation from mom (see Figure 3.6).

*Figure 3.6* Smell the Snuggles

## Touch as Need Meeting

Despite a debate over the use of touch in counseling, many play therapists have come to believe that, when counseling children, touch is essential for various reasons. When mediating factors concerning touch, such as the gender, age, cultural background, and diagnosis of the child, are taken into consideration, touch can be both ethical and beneficial in the therapeutic relationship (Courtney, 2017). Sometimes touch is the simplest and most powerful way to help a child regulate. I want to make it clear that we are not endorsing any punitive uses of touch; however, touch can convey delight. It can provide gentle containment, it can provide sensory input that a client needs, and it can let them know that we are with them, that we are connected in relationship. Touch can be grounding, and touch can be playful. It is concerning to me that so many school systems are disallowing touch, one of the natural pathways for providing care and nurture for children. Kara and the Camp Nurture group are good examples of the power of touch in need meeting.

Kara was a child internationally adopted when she was two. As I was using Theraplay with Kara and her mom, it became clear that this child had some sensory seeking tendencies. One day they were snuggling together in my big cushy chair, and Kara's mom was talking to me while she was unconsciously running her fingers up and down Kara's arm. It was clear to me that this gesture was meant to be nurturing, and I recognized from my own Theraplay work with the mom that she preferred this light touch herself. However, Kara started to twitch. I interrupted her mom and called attention to the interaction. I said, "Hey, mom, I'm sorry to interrupt, but I see something happening that I'm really curious about. Kara, as your mom is stroking your arm, your eyes are twitching." I demonstrated, and she giggled. "It's like your body is saying, 'That feels weird.'" She looked up at her mom shyly, and her mom, a pro at giving Kara voice, said, "It's OK to tell me, honey. I want to help." Kara said, "Well, it feels kind of tickly." We went on to experiment with deeper and deeper caresses of her arm, and when we got to the kind of touch that felt "quite right," it was much deeper than what her mom would have preferred for herself. This new understanding helped Kara's mom be a better sensory co-regulator for her daughter.

The group of children who made up Camp Nurture were so dysregulated, so covered up in static for the first few days of camp, that although we had Nurture Groups, Crash and Bump, and art therapy groups, we did not attempt to read to them, as this would require more of their executive functioning skills, and we felt like we had not really provided enough regulation to the lower brain regions yet. On the morning that I decided to try reading to them, I was still expecting to get three short pages in and then have to say, "Your bodies are letting me know that we need . . ." Then I would offer a regulation tool to extend their window of tolerance and reengage them. It was not necessary. I got through the whole book and was pleasantly surprised. It was not until much later, when I was reviewing the pictures from the day, that I understood why it had been doable. Look at the picture below. What do you notice is happening with each child who is attending to the story? Yes! They each have an anchoring touch from their buddy. At no point did I train everyone to put hands on their camper's back, but they had each attuned to the needs of the camper in their care, and the anchoring touch was need meeting for all of them (see Figure 3.7).

I recently saw a remarkable series of snow globes recently at a local art gallery. Humpty Dumpty was perched precariously on top of the wall, but he was looking down at several broken eggs on the ground in front of him. His predicament struck me as a parallel to the tenuous neurophysiological balancing act that is happening cyclically inside the children we serve. Children with early trauma and consequent compromises to the regulatory functions of the brain stem often have difficulty processing their body's signals. In these children, I often see excitement, aggression, and anxiety all balancing on the head of a neurophysiological pin. These children can benefit from repetition upon repetition of gross motor involvement. They can also benefit from cycles of upregulating and downregulating. Since the excitation of their neuronal systems—even if the excitement comes from the anticipation of a trip to the circus—can often move over into anxiety or aggression, they benefit from expanding their window of tolerance for excitation in play. In play sessions, I will sometimes blast music and dance with a child until we are both out of breath and then move to a downregulating activity such as deep breathing or balancing a peacock feather.

*Figure 3.7* Touch Anchors

## Mindfulness

As shown throughout this chapter, children who have endured early trauma or neglect lack the basic body awareness that is essential for regulation. Many of the case narratives show an integration of mindfulness techniques (lollipop) that meet the child's sensory needs while inviting them into a state of awareness. Mindfulness, which is defined by Kabat-Zinn as "paying attention in a particular way: on purpose, in the present moment, and nonjudgmentally," begins the work of meeting the child's basic needs so that they are more fully present to participate in deeper clinical work (2003, p. 4). Usually mindfulness work is most successful when it is connected to the child's somatic experience (Burdick, 2014). Whether mindfulness is practiced through movement, sight, smell, or taste, becoming aware of one's body is the foundation of health.

Although mindfulness practices have been widely incorporated into treatment with adults, research is recently blossoming in the field of mindfulness practices with children and adolescents. A recent study shows that mindfulness-based cognitive therapy for children (MBCT-C) increases attention and significantly reduces anxiety symptoms (Semple, Lee, Rosa, & Miller, 2010). Furthermore, two recent meta-analyses found support concerning the positive impact of mindfulness practices in a clinical setting for children as young as three through adolescence (Burke, 2010; Zoogman, Goldberg, Hoyt, & Miller, 2015). Though mindfulness is difficult to define as a psychological construct because its practice is broad and takes many forms, numerous child therapists have experienced the positive impact of mindfulness with their clients (Shapiro, Carlson, Astin, & Freedman, 2006). My hope is that research will continue to provide evidence for the success that many of us see in our own practice.

## Gross Motor Involvement for Activating Social Engagement

Ben and his younger brother, Stew, were playing in the living room while their dad worked in his home office. Their mom was at a work event for the evening. After playing for a while, Stew went into the office to ask their dad for a snack. He came back out and told Ben that Daddy was asleep. Ben then went into the office and tried to wake their dad. He could not wake him and called 911. The emergency worker who answered the call talked him through how to do CPR, and Ben, at just six years old, pushed on his dad's chest rhythmically while waiting for the ambulance to arrive. However, he was unable to be revived, even by the emergency workers. Ben had tried to call his mom before calling the police, but she had not heard the phone because of the noise at her work event, and it was not until she pulled into the neighborhood and saw the ambulance that she knew there was an emergency. Ben became quiet and withdrawn after his father's death. He was avoidant in his first therapy experience and avoided talking about the events of that night with his mom. He seemed to have responded to the trauma with first a mobilization response, working diligently to resuscitate his dad, followed quickly by a collapse into immobilization when the emergency responders took over. In the language of polyvagal theory, his body responded with a subdiaphragmatic activation of the immobilization response, bypassing his social engagement systems in a way that made it exceedingly difficult to access the social support of his mother and grandmother, reinforcing his sense of isolation over and over again.

When he came to Nurture House, our first goal had to be helping him reactivate his social engagement system. Building on his natural gift for sports, I took him into the big open area near Nurture House and simply played Frisbee with him. The truth is that I had not dressed properly for playing a baseball type of Frisbee, and this young boy, who had been so hard to reach, began to laugh out loud at me almost immediately as I attempted to haul myself quickly around bases in a long skirt. I did not mind being the butt of his joke if it got him socially engaged. He won, hands down—and I did not give it to him. This full body kinesthetic involvement, combined with the triggering of his apparently naturally competitive bent, provoked what I call (unscientifically) a surge of competency hormones in his body. I got smiles and giggles that I understood later to be some of the first in a long time, a change of state brought about by full body engagement focused on the relationship between us as competitors—a form of social engagement that was meaningful to him within this context. I thought, yes! This is the way in for this young boy. I continued to up the ante, saying each time, "OK, I think this was too easy for you, as you won so quickly. This time, you have to catch the Frisbee (and, to be clear, the Frisbee was a newfangled nylon circle with at least a 12-inch diameter opening in the middle) with your head!" This was not an impossibility. He and I would take turns flicking the Frisbee at each other and then running like mad to get our heads under the opening before the Frisbee landed on the ground. We were mainly unsuccessful, but I laughed harder than I have in a long time and so did he. It was in the midst of this shared laughter that his social engagement system was activated with me. It was not until he trusted me to see all the health, all the typical "kidness" inside of him, and delight in it that he was ready to process the really hard and scary moments of his father's death and his own attempts to resuscitate him.

After several sessions that looked a lot like what was described above (although I learned to dress more appropriately for his sessions), I introduced my domino people. I understood that this child, who was just coming back into resonance with his own body, needed more structure and challenge to any activity that might narrate the events of that night, so I enlisted the help of my domino people and some very small sticky notes. Most of the kids at Nurture House are drawn to the idea of lining up dominoes and watching the whole chain of them fall over with just the touch of their finger. What this allows for clinically is a blow-by-blow breakdown of what the child encoded from the traumatic event. I often save this activity for children who are self-blaming, who believe that some failure of their own caused the devastating outcome that they are now living with. In this case, I had been gifted with a bit of self-disclosure in a previous session. Ben, during one of our many contests of strength or endurance, gifted me with his deeply held belief that he was responsible for his dad's death. If he had been stronger, if he had been able to do the compressions harder, his dad would still be alive. Ben was fascinated with the domino people and began to set them up right away. He went through what happened step by step, starting with walking into his dad's office and thinking he was asleep, shaking him, and realizing he was not breathing. He was able to talk about calling 911 and the compressions he tried to give while his dad was still in the chair. He talked about watching the emergency workers enter their home and take over. And as we went back through all the details, he realized there was nothing he would have done differently.

The kinesthetic engagement was part of what helped him stay with the story long enough to make the cognitive shift away from self-blame. None of his actions brought his dad back to life, but none of them contributed to his death either, and there was a certain peace in coming to this awareness. Interestingly, Ben asked me to take a picture of the line-up of dominoes with the parts of the story close by (see Figure 3.8). I did. Finally, he set them all back up, and he asked me to take a slow-motion video of them being knocked down (honestly, he had to show me how to do that). I told him I thought it was important for his mom and grandmother to support his new discovery that his dad's death was truly not his fault. I asked if it was OK to invite his grandmother into the

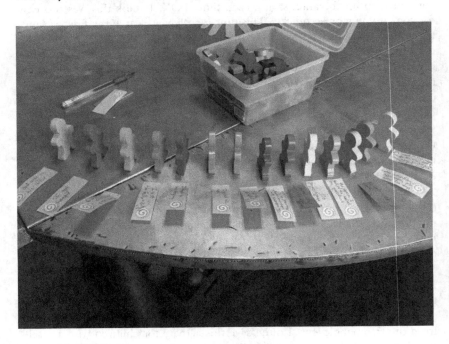

*Figure 3.8* Step by Step

session. She took the video, and for a few minutes he watched the slow, relentless sequence of dominos knocking into each other. As the three of us watched it again together, it was powerful. The sound that had been a tinny "tink, tink, tink" as one domino hit another became a massive echoing bang as each domino hit the next in the slo-mo version. He and I experienced together the power of the initial cardiac event and its relentless cascade toward death. This is not meant to be morbid. Indeed, it is meant to highlight the opposite—the growing sense of freedom from responsibility that my client felt as he watched the inevitable cascade again and again. After several playbacks, Ben sighed deeply in satisfaction and said, "OK. Let's go play outside now." So we did.

## Conclusion

Returning to the beautiful simplicity of need meeting can be the fastest and most effective way to build rapport with the traumatized child and/or his caregivers. In a parallel process dynamic, we train our clinicians (who train the parents who train the children) to check in with their bodies and ask themselves what they need. Particularly when negative behaviors or irritation is present, asking whether the child has had proper hydration, nutrition, or sensory need meeting should be the first line of defense (Purvis et al., 2007). Empowering our bodies and the bodies of those in our care builds safety, establishes us as need meeters, and helps calm the lower brain regions of traumatized children.

## References

Booth, P. B., & Jernberg, A. M. (2010). *Theraplay: Helping parents and children build better relationships through attachment-based play* (3rd ed.). San Francisco, CA: Jossey-Bass.

Burdick, D. (2014) *Mindfulness skills for kids and teens.* Eau Claire, WI: PESI

Burke, C. A. (2010). Mindfulness-based approaches with children and adolescents: A preliminary review of current research in an emergent field. *Journal of Child and Family Studies, 19*(2), 133–144.

Courtney, A. J. (2017). Overview of touch related to professional ethical and clinical practice with children. In J. A. Courtney & N. D. Nolan (Eds.), *Touch in child counseling and play therapy: An ethical guide* (pp. 3–18). New York, NY: Routledge.

Johnson, S. (2015). Evidence-based essential oil therapy: The ultimate guide to the therapeutic and clinical application of essential oils. *Journal of Counseling & Development, 79*(1), 53–60. Retrieved from https://doi.org/10.1002/j.1556-6676.2001.tb01943.x

Kabat-Zinn, J. (2003). Mindfulness-based interventions in context: Past, present, and future. *Clinical Psychology: Science and Practice, 10,* 144–156.

Kranowitz, S. C. (2005). *The out-of-sync child: Recognizing and coping with sensory processing disorder.* New York, NY: The Berkley Publishing Group.

Neigh, G. N., Gillespie, C. F., & Nemeroff, C. B. (2009). The neurobiological toll of child abuse and neglect. *Trauma, Violence, & Abuse, 10,* 389–410.

Perry, B. D. (2006). Applying principles of neurodevelopment to clinical work with maltreated and traumatized children: The neurosequential model of therapeutics. In N. B. Webb (Ed.), *Working with traumatized youth in child welfare* (pp. 27–52). New York, NY: Guilford Press.

Purvis, K. B., Cross, D. R., Dansereau, D., & Parris, S. (2013). Trust-based relational intervention (TBRI): A systemic approach to complex developmental trauma. *Child and Youth Services, 34*(4), 360–386.

Purvis, K. B., Cross, D. R., & Sunshine, W. L. (2007). *Connected child* (p. 264). New York, NY: McGraw-Hill Professional Publishing.

Semple, R., Lee, J., Rosa, D., & Miller, L. (2010). A randomized trial of mindfulness-based cognitive therapy for children: Promoting mindful attention to enhance social-emotional resiliency in children. *Journal of Child and Family Studies, 19*(2), 218–229.

Setzer, W. (2009). Essential oils and anxiolytic aromatherapy. *Natural Product Communications, 4*(9), 1305–1316.

Shapiro, S. L., Carlson, L. E., Astin, J. A., & Freedman, B. (2006). Mechanisms of mindfulness. *Journal of Clinical Psychology, 62,* 373–386.

Van der Kolk, B. A. (2015). *The body keeps the score: Brain, mind, and body in the healing of trauma.* New York: NY: Penguin Books.

Williams, M. S., & Shellenberger, S. (1996). *"How does your engine run?" A leader's guide to the Alert Program for self-regulation.* Albuquerque, NM: TherapyWorks, Inc.

Zoogman, S., Goldberg, S. B., Hoyt, W. T., & Miller, L. (2015). Mindfulness interventions with youth: A meta-analysis. *Mindfulness, 6*(2), 290–302.

# 4 Bigness, Smallness, and Containment
## The Use of Space and Presence in Play Therapy

This chapter will explore the myriad ways we provide containment for children and families as they tame the trauma. A focus on the role of physical space in the child's ability to face trauma content will be discussed as well as the use of containers and containment in chunking trauma work, empowering clients, and reinforcing boundaries. Examples will be given of how the therapist becomes a container and of how boundaries, whether spoken or physical, provide containment. The question of bigness matters in terms of the physical space in which therapy is completed, but it matters even more in terms of the bigness or smallness of the therapist's presence.

A few years ago, at a time when I had been traveling extensively, I came home to find the pantry a mess. I took the kids to The Container Store, and we had a field day choosing containers that would be "just the right size" for the foods that are our family staples. We noticed that some were tall and skinny, some were short and square, and some were large enough to hold quite a lot. The size of the container was chosen based on what would go inside it. Therapists are very much like these containers—and I do not mean individually. I do not mean that each container represents a different therapist. On the contrary, I see each of us as being all the containers at different points in therapy. It is our job to morph into the bigness, smallness, or whatever level of containment is needed by any particular child or family in our care. On certain days, I feel a little like a metamorph, becoming smaller and quieter for one child and large and in charge for another (see Figure 4.1). The only part of the pantry containers that was the same across different sizes and shapes was the lid. Each lid fit the container exactly and would suction to it in a way that did not allow air in to potentially make the food stale. Sticking close to our clients and attuning to their needs allows us to become bigger and smaller in our presence as needed.

## Containment

The idea of containment has long been an important one in the field of therapy. In 1962, Bion talked about parents as the "psychic containers" for their children. Indeed, it was posited that a psychic container was essential for the

*Figure 4.1* What Kind of Container Are You?

healthy development of the child. Infants, particularly, experience waves of intense sensory and emotional experiences that can be overwhelming for them. The parent sorts these experiences, has time to assimilate and consolidate them, and can then give coherence and structure back to the baby to make sense of what has been felt. If the parent is unable to be this psychic container, the child does not have the emotional experience organized and may remain in a heightened state of arousal, attempting to deal with disparate, scary, and sometimes crushing experiences on their own.

Some people are naturally better at containment than others. Some people exude a natural groundedness, a safety that emanates from them and allows others to approach them with difficult things. I think that some of this may have to do with a person's natural internal rhythms, the pace of their communication, their kinetic energy, and their ability to be present. At Nurture House, we begin immediately with new clinicians to hold their big feelings, their anxieties around the work, and in this way model being bigger, stronger, wiser, and kind. Fortunately, no matter what capacity a parent or beginning therapist brings to the work, the mentorship of a bigger, stronger, wiser, kind other can help expand their containment capacity. We have a myriad of psychoeducational strategies and exercises that we employ with both clinicians and parents in training, but none of these is as important as the holding environment created for the mentee by the mentor—the holding environment created by the clinician for the parent.

Being a container also means acknowledging our own pain and our own need for holding along the way. Because we contain so much for the families and children in our care, we must have safe havens of our own. Sometimes our first job is to be a safe place for parents, which means becoming a container for them. In this way, becoming a container is a parallel process. Whenever I sit with a parent, I am cognizant of the truth that we can only give what we have received. Any quality of communication, holding, listening, or responding that I want a parent to have with a child must begin with my modeling that quality with the parent.

## All That We Hold

My Nurture House team went on a staff retreat recently. One of our group-building challenges was to create a sandtray together that characterized what we do and who we are at Nurture House. The symbols that ended up being the center of the tray are pictured below. Added by one or more of my colleagues is the central image of a fiery volcano in active eruption (see Figure 4.2). Surrounding the volcano is the superheroine Elastigirl from the movie *The Incredibles*. She can stretch to reach whatever she needs to get the job done—*and she is flexible enough to bend herself into whatever shape is necessary to save the day.* Elastigirl can put her arms around any situation, no matter how big or scary. When I am in the middle of a difficult session, I will sometimes close my eyes and pull up this image from our joint sandtray and remember that I am meant to be Elastigirl-ish (is that a word?) for the child and/or parents in front of me. As I have reflected on this image, I have understood more deeply that our job is not to keep the eruption from happening but simply to be big enough to contain it—and then begin to invite change.

## The Use of Physical Space

We have long understood that some physical spaces are inviting for us while others are off-putting. Many hospitals are becoming interested in how the arrangement and aesthetics of space enhance or impede healing (Curtis, Gesler, Priebe, & Francis, 2009; Curtis et al., 2013). Gesler, 2003; Gesler, Bell, Curtis, Hubbard, & Francis, 2004). Some places we describe as "warm" or "cozy," others as "bright and spacious," and still others as "overwhelming." Some homes you walk into and feel like you can snuggle right into the couch. In others, there may be plastic on the furniture, and although the home may be "beautifully appointed," it is clear that you are not supposed to touch anything. Large, fluorescently lit chain stores can be immediately overwhelming for people, and certainly are experienced that way by many of my clients with sensory defensive tendencies. For others, they are a beacon of exploration—people are empowered to shop 'til they drop. Some people prefer small spaces and feel best when cozied up in the window nook of an old house with a book and a cup of tea, whereas others may feel claustrophobic, hemmed in by a space that feels too small for them. Part of the work in helping

*Figure 4.2* Wrapping Arms Around the Eruption

families heal from trauma is providing safe space since "safe" may feel different for different people. At Nurture House, we have designed therapeutic spaces that include options for experiences of bigness and smallness, expansiveness and containment. This permission for clients to choose bigness or smallness begins in the lobby area, where we have a larger waiting room with full size adult chairs and tables. Parents and children may decide to camp out here in any number of configurations (child playing at the table and mom on her phone, child and parent building blocks together or playing with puppets, child on mom's lap, child as far away from parent—or vice versa—as one can get in the lobby). Their use of the space provided gives us valuable information and is, in fact, part of our initial assessment of how parent and child function together. Indeed, their connectedness in interaction, their physical proximity to one another in the lobby, and the frequency of their communications with each other can provide a beginning look at how their attachment dance works and can inform the treatment plan.

Understanding the traumatized child's need for containment, for snug spaces, we created a smaller, contained space for kids within the larger room so a child can sit inside the shelter of the enclosure while keeping in visual contact with mom (see Figure 4.3).

There is also a small space set entirely apart from the larger lobby. I often find my clients with anxiety disorders sitting in this area with their parent. I see this part of Nurture House as almost like a "warm-up" area where their senses can "get used to" the smells, sounds, and colors of Nurture House before they have to negotiate more direct interactions with others—even others in the larger lobby. This need for smallness, for close containment, has been shown to us by clients in various ways. The picture below (see Figures 4.4 and 4.5) was created by a boy who had lived in a domestic violence environment for several years and seemed to wish for a return to the enclosure of a womb-like environment. One day early on in treatment, he came to the playroom and began playing out a scenario in which a baby was really scared by what was happening around him. He ran and got the child safe, opened it, carefully arranged the baby (doll) inside, and closed it up tight. The child actually breathed a sigh of relief once the baby was safely enclosed.

## Hiding and Joining

Over the years, I have worked in many different environments, and in each I have had traumatized children who simply needed to hide as part of their joining. Sometimes the hiding is a simple communication that the child does not want to be seen. The child may be showing us the very savvy coping strategy that was used to avoid punishment from the enraged parent. Other times the child is inviting us to find them. They may even be expressing a need to be found. The simple act of hide-and-seek can answer profound questions for traumatized children. I have gone away. I have withdrawn myself. Will anyone notice? Am I worth finding? In the momentary play interaction, the child is saying, "I am unreachable. Can you reach me? Does someone care enough to come after me?" Children who have felt unseen, forgotten, or neglected seem to be asking, will you look for me? will you do the hard work of finding me? On one level, the simple dance of hide-and-seek brings an essence of playful curiosity to the beginning of a session, to the "hello" of a relationship. When I come into the lobby and find a parent (who I know has brought a child with her) alone, I will ask, "Where is Johnny? Did he go to Alaska?" The parent almost always begins to play with me and will offer another place where Johnny may have gone: "I think he flew to Zimbabwe!" I will express my

*Figure 4.3* Small, Snug Area for Children Who Need More Containment for Felt
Safety

disappointment: "I was really looking forward to seeing Johnny today, and I'll
miss him if he's really gone." During this ritual, the child pops out, and the
mom or dad and I can both delight in his return. I often cup his cheek and
check to see if he brought his "beautiful brown eyes with [him] today." With
older kids, I might tussle their hair or offer a high five. In this hello experience,

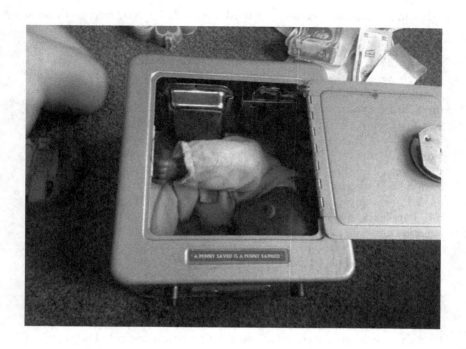

*Figures* 4.4 and 4.5 Baby in the Safe

the child waits for and perhaps even needs the validation/reminder that I want to see him, that the relationship with Johnny matters to me, and that I would miss the connection if he were not present.

Some of this is clear recapitulation of the attachment relationship. When my children were young, the connections between us were neurophysiologically reinforced. I experienced delight with them and they with me. I comforted them and they comforted me. There were exchanges of interaction, from laughter to snuggling, that released oxytocin and dopamine— powerful bonding and joy chemicals—in both our bodies and gave us both a feeling of being "just right" together. When I went to work or went to speak at a conference, there was a period of time during which the dose of connection sustained us, a window of tolerance for time apart that grew incrementally as the children grew, but the stress of being apart would leave us both "jonesing" for a fix. If it sounds like addiction, that's because it is. The pleasure that comes from connection, from the powerful chemical reactions that happen between parents and children, increase the desire and the felt need to stick together. We are not meant to be apart from our young for long stretches of time. Many of my clients from hard places did not have a preferred other and need many reassurances that the relationship is important, that they matter, and that they can have positive power in the relationship—power to provide delight to another. In this hide-and-seek scenario, the child decides when he or she is ready to be seen, which gives them comforting, empowering control. The ability to control the moment at which the child pops out mitigates the vulnerability they may feel in coming to treatment and most likely mitigates the sense of powerlessness felt during whatever experiences of trauma or neglect have brought them to Nurture House. In previous work environments, children have had to hide under a chair because it was the only available space to hide. With this need for cloaking in mind, I positioned my very first play therapy helper underneath the table in the middle of the lobby. Patrick, a giant stuffed dog that I bought when I was attending my very first play therapy conference, has been with me for over 20 years. He sits snuggled under the table unless a child needs him. He is large enough to completely cover a small child and has functioned in that role many times (see Figure 4.6).

Other children benefit from having smaller spaces with clearly identified worlds. Figure 4.7 below shows the inset fireplace of the upstairs loft at Nurture House. It sat empty for a period of time after clients started coming, then it had a wishing jar in it and other prominent items displayed. However, the children mainly overlooked the inset. I was given a fairy door recently, and this sparked an idea to use the space as a little fairy nook. Since we created a total environment in that space, complete with moss, fairies, a mailbox for in and outgoing mail, etc. (see Figure 4.7), children have begun to interact with this world differently. I have one client, in particular, who wants to be a fairy and used the fairy world with his grandmother to play out some interesting power dynamics between himself and his mother, with whom he does not live. For this client, miniaturizing one ecology provided a scaffolding for his story.

*Figure 4.6* A Hiding Friend for Little Ones

*Figure 4.7* Tiny, Controllable Yet Magical World

## Clients' Compartmentalization as Containment

The child may have developed strong and protective containment strategies of his own. Young children who experience something that is overwhelming, terrifying, and beyond their control will "contain" the experiences in a way that leads to dissociative behaviors. They may put the trauma behind such a series of locked doors that getting to them takes time and patience. When multiple instances of abuse have been experienced, it may be necessary for the clinician to offer opportunities for the client to become kinesthetically engaged in chunking the experiences. It can be physically soothing and provide the child with an element of control to quantify the traumatic experiences in some concrete way and then let the child play with the icons. A case example may help to explain. Jack was a six-year-old boy who had been sexually abused at four years of age. After working with him for four months using a combination of child-centered play therapy and the Tackling Touchy Subjects curriculum (Goodyear-Brown, 2013), he had gifted me with multiple glimpses and snapshots of his abuse. As in most work with little ones, we danced toward and away from the trauma together as he gained confidence that we would all survive the telling, and he asked if his mom could tell him the story of what happened.

Jack's mom was a big container and was able to take the pieces of the story that Jack had told us before and create an age appropriate narrative. We had been using a combination of play therapy and eye movement desensitization and reprocessing (EMDR) thus far in treatment. As part of this work, Jack and his mother had identified a song that helped him feel safe and powerful. He sat on his mom's lap, and they sang together "God is bigger than the boogeyman" while I provided bilateral stimulation for Jack. After rehearsing his safety song, Jack's mom began telling the story that he had been telling/showing in snippets throughout the course of therapy.

She was about three minutes into the narrative when Jack's body language began to change, although he did not give any signs of overt distress (he did not tell his mom to stop, cover his ears, shush her, or hop off her lap and effectively distract himself and us from the narrative process). He just appeared to check out. I gently placed my hand on his mom's knee to have her stop telling the story, and I went to get the Slinky. Jack and I had used the Slinky before as a measure of how scared/scary a piece of content was, so he knew what to do with it. I said, "Jack, can you show me how scared you feel right now?" He seemed to come back to himself, hopped off his mom's lap, told her to hold one end of the Slinky, walked all the way across the room to the other wall, and said, "This much." The Slinky and the fear spanned the whole length of the room. I said, "It seemed like you were having a big feeling. I'm glad you showed us how big it is. Seems like even though we've played around with the parts of this story before, putting it all together is just too much right now. So, let's take it apart." Jack looked relieved.

We had a printed copy of the story with us. I got out scissors, and we cut up the parts of the story and put them in an envelope. Jack used the rest of the session to engage in co-regulating play with his mom. In between sessions, I went to The Container Store (one of my favorite places) and bought a set of clear plastic rectangular containers. Each one looked like a pillar a little bit higher than the one before it. They can be stacked inside of each other, much like nesting dolls, but you can see what is inside through the walls of each container. They ranged in size from one to eight inches tall.

When Jack came back for his next session, I said, "I was thinking about you this week and about how it was too much to look at all the scary stuff at once last week. I got these and thought we could play around with which scary thing that happened should go in each, then you can decide if/when we look at each one." Jack became very thoughtful and said, "Should I put the most scary thing in the smallest one or the biggest?" I reflected that in here he could decide. He decided to stack each container within the next. He placed the paper with the least scary experience in the largest container and folded up the most scary experience and put it inside the one-inch container in the very center of the nesting pattern. Jack, his mom, and I spent the next several weeks working through just one paper at a time with a combination of EMDR and play therapy. The Slinky never got too big again, and within six weeks, his mom could tell the whole story again without it being overwhelming. This chunking of experiences accomplishes several therapeutic tasks: it provides literal physical containment and it empowers the client to choose where each experience goes, which requires distancing and reflection on each part in relation to the whole, continuing the desensitization work but with additional levels of structure, containment, distance and choice. Many forms of containment devices for chunking content are available, and I particularly like using nesting dolls in this work.

For many of the families that we see, the number of traumatic events that have occurred may be difficult to quantify and are certainly difficult to pull apart from one another. When a child has a parent who screams at him/her daily or parents who scream at each other daily, it may be difficult for that child to even tease out one incident of yelling from another. When a child has had multiple experiences of sexual abuse with the same person, in the same environment, teasing out discrete traumatic memories, one from another, may be impossible and is not the ultimate goal of treatment. In fact, therapists can create iatrogenic effects by aggressively pursuing "the story" or the details of the story in situations where the memory is stored as a set of bodily sensations and not in conscious or explicit memory. I sometimes combine EMDR with play therapy and find that the two methods work fluidly together. For example, I may enhance the feelings of safety engendered through the creation of a safe place sandtray with bilateral stimulation. I may also enhance attachment relationships or stories around early care and nurture with bilateral stimulation as well.

Recently, I worked with a nine-year-old boy who was having a posttraumatic reaction to a recent car accident. He had developed a set of avoidance symptoms that was going to be best treated with a play-based exposure/response prevention protocol. Several years ago I wrote several therapeutic stories aimed at helping children externalize and then work with their anxiety in order to help them develop a sense of mastery and control over the anxiety. One day, when I went out to get the family in the lobby, Jack had the *Worry Wars* book (Goodyear-Brown, 2011b) on his lap. I could tell that he really wanted to read the story *Daniel the Dragon Slayer*, and when I offered to read it to the whole family, he nodded his head eagerly. I escorted everyone into the front room of Nurture House, a larger room with light walls and high ceilings—plenty of space for mom, brother, client, and myself to read the story. As soon as I began to read, Jack became agitated. He jumped up and ran across the room and got under a blanket. I responded by saying, "Your body is letting me know that this is kind of hard for you, to read this story with me. Seems like this room may be too big for reading. I'm going to let you choose the room in which we read it" (fortunately, I did have the house all to myself at the time).

He said, "Can we go upstairs to the nook?" I said yes, and we went, the child, his mom, and his younger brother. There are two small closets in the loft of Nurture House. Neither has doors, and while one has been outfitted with a nylon circus tent and puppets, the other functions as a small kitchen and dress-up area. Jack dove into the circus tent and then asked for the dragon puppets. Jack and his brother squeezed into the tent (see Figure 4.8), and their mom and I sat right outside the tent while I read the story. If another therapist had come upstairs at that time, she might not have even known we were there. We were all squished into such a tiny part of the room. But this smallness provided Jack with the containment he needed. His neuroception of safety was reinforced by the small space and his ability to decide when he stuck his head out of the "door" to look at the pictures of the dragon, when he used the dragon puppet to speak to us, and when he wanted to stay hidden. He could, without comment or commitment, just listen, which mitigated his approach to the "scary stuff." In his case, the scary stuff was learning how to boss back anxiety. This serves as another example of how play therapists can set up intentionally small, snug spaces as options for helping children regulate as they approach difficult content.

## Offering Titration in Creative Space

In the kitchen of Nurture House is a wall painted with chalkboard paint. On it we have placed several frames of varying sizes and offer paper as well as the chalkboard surface. At first glance, this wall looks like just a cute Pinterest picture—an attractive way to decorate a wall. What we are actually doing is intentionally offering a titration of bigness and smallness for clients. The offering of a chalkboard surface invites people to draw on it—it invites creativity, which is a good thing. Some children may run to the chalkboard wall, ask for chalk, and begin creating large, mural-like art creations on the lower half of the wall. However, some traumatized children can feel overwhelmed by a large blank space but experience increased safety with the boundaries of a smaller frame. Still others might respond best to the roll of paper offered on the wall, as they can choose (control) how much or how little paper they use (see Figure 4.9).

*Figure 4.8* A Smallish Space, Safe Enough to Listen

## Titration in Sand Spaces

Sand has a long history as a therapeutic tool. From tiny Zen sand gardens, often featured in therapists' lobbies, to sandboxes large enough for children to sit in, a child's exposure to sand can be titrated, as can a child's immersion in a world created in the sand. In the sand tray room, we have three kinds of sand: Jurassic sand (that is soft and soothing), white sparkly sand that is

*Figure 4.9* Titration of Space in Art

more granular for sensory seeking clients), and kinesthetic sand. Offering sand of various kinds allows choice to children in meeting their sensory needs in an attuned manor. Bigness and smallness are not just measured by length, width, and diameter—they are also measured by depth. Some children feel most comfortable approaching a shallow tray of sand, one that is only an inch or two deep, whereas others prefer the portable sandtrays we offer that have sand as deep as three and four inches. Especially when children want to engage in burials and resurrections, they often gravitate to the deeper sandtrays. While we are able to offer a sandtray experience in any room of Nurture House, our dedicated sandtray room allows for titration in multiple ways. We have three separate sandtrays in that space. One is a traditionally wooden sandtray with a blue bottom and sides, created in the dimensions of a standard sandtray. In it we keep a snowy-white crystallized sand. It looks sort of magical. In the center of the room is a slightly more shallow but longer and wider sandtray that we keep filled with rich, dusty, orange-colored Jurassic sand. It is the softest sand I have been able to find and is very, very regulating for the sensory systems of the children that we see.

The third sandtray is a portable, extra deep plastic sandtray with kinetic sand in it. Oftentimes our traumatized children will start with the kinetic sand, mainly because it is more controllable, and early in treatment traumatized children often need more control over their mediums. The kinetic sand can be packed and molded, allowing for certain internal states to be manifested in a way not possible with traditional sand. A child's need for bigness or smallness can also be mitigated by the size and kind of sandtray they choose. Since the backyard renovation was completed, children have had their choice of sand enclosures ranging from a boundaried circle of sand that is ten feet in diameter (pretty big) to as small as an octagonal wooden sandtray measuring three inches across (see Figures 4.10 and 4.11).

*Figure 4.10* Our Largest Sandtray Offering

*Figure 4.11* Our Smallest Sandtray Offering

Sand can offer different kinds of sensory experiences based on the consistency of the sand chosen (smooth to rough, dark to light, packable to amorphous), and we can offer titration of expression through the bigness or smallness of the sandtray offered as well as the shape in which it is offered. The traditional rectangular Zen sand garden comes with a rake and allows for the creation of patterns in the sand without a client having to come in contact with the sand itself, a potential mitigation for children with sensory defensive-ness who might avoid physical contact with the sand. So, we find that even the tools offered to clients in the sandtray work can help titrate their approach to letting go of control. When a rake or shovel or smoother sand is offered, clients can spend a great deal of time trying to make the sand "just so." As stated earlier, we recently finished our renovation of the backyard. The project included the construction of a large sand circle in the middle of the space, giving us—and potentially clients—the freedom to do sand work on a full body scale. Already children are having different experiences of "bigness" in this space than they are able to have in a regular size sandtray. Most recently, children have lain right down in the sand to make sand angels, taken off their shoes, asked to have their feet buried in the sand, and used the oversized wooden blocks seen inside the circle below (see Figure 4.10) to build bridges and paths, to divide the space, and to do work related to balance.

## How Children Perceive Their Own Bigness or Smallness

The sandtray is a powerful landscape for helping children communicate their sense of their own bigness or smallness (either in terms of actual physical size or in terms of their personal power) in relationship to the bigness or smallness of others. Below is the family play genogram (Gil, 2014) of a child referred to me for extreme anger outbursts. When I asked him to go and choose toys to be his mother and father, he chose the tiny metal sol-diers. Then I asked him to go and choose a toy to be his brother, and he chose the gummy blue guy. Then I asked him to go and choose a toy to be himself. He chose a disproportionately large figure, the Martian popping doll. While the Martian popping doll communicates volumes all on its own (when you squeeze it, the applied pressure makes the eyes, ears, and nose pop out), effectively mirroring the client's blowups when he gets outside of his window of tolerance, what is most striking here is how much larger and how much more powerful this figure is than the tiny soldiers (see Figure 4.12). This child clearly has an internal schema of himself as bigger than his parents, and, indeed, at the time this child entered treatment, his parents were actively afraid of triggering him, walking on egg shells to avoid his tantrums. His perception of their fear and ineffectiveness in containing him seemed to be scary for him. Notice that his figure does not have any hands or arms—he cannot do for himself, he cannot act upon the environment, he must rely on the capability of others, and yet he deeply doubts others' ability to take care of him. This initial snapshot of his perceptions of his family dynamics was extremely useful in helping the parents understand and support the treatment goals around helping them communicate their role as *bigger, stronger, wiser, kind* safe bosses more effectively, more frequently, and more authentically.

A client's self-perception of physical bigness and smallness can be especially important to understand when working with children and teens who struggle

*Figure 4.12* The Family

with eating disorders. Elli was a teenage client who struggled with anorexia. I invited her to create a sandtray showing how she sees herself and how she wants to see herself (see Figure 4.13).

Elli chose the clearly obese boy pictured below. He is holding a corn dog in one hand and a four-scoop ice cream cone in the other. This was how she saw herself at the time she created the tray. She chose the knight pictured on the right for how she would like to see herself. Notice that the figure is made of metal, the knight's visor is down, and his sword is at the ready. I asked her to give three descriptors for the boy. She chose fat, gross, and laughable. I asked her to give three descriptors for the knight. She chose small, protected, and vigilant. Clearly, neither figure is a fully integrated, healthy sense of self, but the tray gave us both valuable symbols for the work ahead.

## Physical Containers

Containment can be physical. The younger the child is developmentally, the more helpful it is to provide physical containers. As children are concrete thinkers, it can create a sense of mastery for them when they can take difficult content and lock it up somewhere, giving the difficult content solid, visible boundaries. What kinds of difficult content might benefit from containment?

*Figure 4.13* Fat Self and Skinny Self

1. Big feelings that may threaten to overwhelm the child.
2. Visual images, words, or identified somatic experiences related to a discrete traumatic event.
3. Perseverative questions or "confessions" of previously disclosed scary content.
4. Specific instances of traumagenic experience within a scope of trauma work.
5. Perpetrator symbols.

## Big Feelings Containers

Frequently, children who grow up in families with highly conflictual parents can feel overwhelmed by the emotions in the family. Jenny is one such little girl. Jenny's parents sought out treatment for her when she began having heightened emotional outbursts. They readily accepted their role in creating an environment that might foster distress, and they recognized her need for a safe place outside the family system. Jenny's parents sat down with myself and Jenny soon after we met, and I offered the idea of making a container to hold all of her big thoughts and feelings about the things she experiences at home. Jenny's eyes widened, and she clapped her hands, then became quiet very quickly. I said, "It seems like you're thinking about something. You can ask me anything you want." She looked at her parents sideways, from underneath her eyelashes, and said, "Will my parents see it?" I said, "Great question! It sounds like a container will only be helpful to you if you can put any thought or feeling in there without worrying that your parents might see it, that it might hurt their feelings. Mom and dad, can we get your permission to keep the things that go into Jenny's box private?" Both parents agreed and went out of their way to reassure her the box was just for her. I offered Jenny a set of brightly colored note cards to write down anything she felt needed to be "dumped" into or "held" by the "big feelings" box. Jenny would come to session each week and ask for her box. She would write down two or three new cards, go through all her old ones, and then say she was ready for whatever else we might be doing in therapy that day. We purchase basic wooden boxes and vessels of all sizes and shapes. We offer art supplies to decorate the containers. We make sure to have some clear boundary signals among the supplies children can choose from. Below is an example of a Big Feelings Box that is covered in stop signs made of foam. The words "keep out" are also fashioned out of foam stickers (see Figure 4.14).

After a period of time, many of her big feelings were held in the box. We began talking about ways she might use her voice to talk to her parents about some of these big feelings. We agreed that one of her goals could be "using my voice," and that she might begin writing cards for her box that included examples of using her voice. We had a session with her mom and dad that helped everyone understand how to respond when Jenny took the risk to "use her voice" to share her feelings directly with her parents. We created a simple

*Figure 4.14* An Example of a Big Feelings Box

script and role played, having her parents reflect and validate her feelings. Jenny came back the following week and was excited because "it worked!" I asked, "What worked?" She said, "I used my voice, and they listened!"

I talk about this idea of containment regularly at Nurture House and have multiple containment devices for children to choose from. Clinically, having at least one size of a jail is helpful to clients. I have several sizes, including one that is three inches tall and three inches deep—perfect for sandtray work—one that is big enough for full size action figures, and one that is big enough for a whole hand puppet. I also recommend a locking box of some sort, regular and miniature handcuffs for use in containment of perpetrators, and personally created containment devices. We also keep a combination safe in one of our playrooms that displays the combination numbers on the bottom of the safe.

A colleague of mine, Amy Frew, gave me a wonderful example of containment recently. In her own private practice office, she has a safe, and while working with a child who has nighttime anxiety, she offered to contain the worries inside the safe. What follows is her description of the interaction:

> We were playing with the toy safe and talking about how things that were in the safe stayed there until someone decided to open it. I said, "Sometimes kids decide to leave their worries here (in the playroom)." Client started listing the worries she could leave ("I could leave my worry about the dark. I could leave my worry about being alone."). Then she said, "I could leave my toots here!" She walked over to the safe, squatted on top of it, and farted. Smiled and said, "Wait. I have another one to leave!" and farted again.

## Boundaries: Limit-Setting as Caring Containment

Every evidence-based play therapy includes a limit-setting strategy. Limits are vital for children. If done well, giving a limit clearly and calmly when needed establishes a safe boundary while providing freedom within the boundary. I am struck again and again by how scared children can feel when they believe they are actually bigger than the adult who is with them, that they can create harm without having a safe boss helping them regulate. Randy was a ten-year-old child who had been removed from his biological home after five years of intense abuse and neglect. He had intense fits of aggression, and when he moved outside of his window of tolerance, he took his internal chaos and acted it out all around him. Then he would feel worse for having destroyed an environment and would spiral into shame. He carried so much shame, it was difficult for him to play or talk at all about his anger outbursts. One day he asked to borrow my iPad. At Nurture House, we use technology in therapy only selectively and with clear clinical ends in mind (Goodyear & Gott, 2019). I wondered if, in this case, the iPad would serve as a distancing device that would allow us to approach his dysregulation through visual imagery. I opened up an application that allows you to draw colored lines and shapes on a black background, sort of like an Etch A Sketch. He sat and silently drew for

*Figure 4.15* Chaos Spinning

a while. When he was done, I simply asked him if he could give it a title, like a book. He labeled it *Chaos Spinning* (see Figure 4.15).

He then began a new drawing, and when he was done, he again showed it to me. When I asked him to title his second creation, he decided to call it *Chaos with Boundaries*. We sat silently looking at each picture in turn together. Then I asked, "Which one feels better to you?" and he immediately gestured to the *Chaos with Boundaries* creation (see Figure 4.16).

Randy recognized on some level his internal disorganization and his need for a safe boss who could organize his experience and help reassure him he would not be allowed to be powerful enough to bring destruction to his environment. Randy's adoptive mother was learning to provide high structure and high nurture for Randy, and he was beginning to thrive with the calmly enforced boundaries and the close physical proximity his mom kept in order to be able to intervene quickly if he began to become disorganized. He was not, in that session, ready to acknowledge the real-life limits his mom had begun putting in place, so we stayed within the metaphors of the art. However, he did ask to show his mom both of these images. We invited her into the session, and she was clearly struck by the images. At the end of the session, Randy ran ahead as I was walking them out, and his mom said, "It will help me to think of myself as providing those smooth lines for him."

When someone asks for something from someone else, the response becomes the spoken boundary. *No* is a clear demarcation line, whereas *yes* is permission. In all innocence and trust in the better nature of humanity, most of us are born believing that when we say no to other people, the *no* will be respected. When

*Figure 4.16* Chaos With Boundaries

a verbal *no* is not respected, it can call into question our right to set the boundary in the first place. Children are set up for harm when their voices are disrespected. It is part of the safe boss job of caregivers to enforce with the children in their care the boundaries that lead to internalized care of others. Indeed, when boundary violations between two children go unaddressed by the caregiver on duty, both can be set up for unhealthy sexual boundaries (Gil & Shaw, 2013; Goodyear-Brown, 2011a; Goodyear-Brown & Frew, 2015).

The example I give when I am teaching is of two children and their depressed mother. Karson, age five, and Cindy, age three, are siblings. One day Cindy is playing with a shiny red Matchbox car when her brother grabs it out of her hands without asking. She screams, "No, give it back!" Karson ignores her or, worse yet, laughs at her, and Cindy cries and tells their mom, who is lying on the couch watching TV and does not respond in any effective way. This happens over and over, without Cindy's boundaries being reinforced by the safe boss on duty. What are the children learning? Karson is learning that he can take things he wants when he wants them and that he can take what he wants without asking. Cindy is learning that it is OK—even to be expected—that people can take things from her when they want whether she gives them permission to or not. Fast-forward this learning curve by ten years and introduce the development of a sexual self. Why would this entitlement to take what he wants when he wants it not extend to sexual behaviors and beliefs? Karson at 15 may believe it is OK for him to take sexually from a more vulnerable person, and Cindy is likely to believe that people can take sexually from her without permission.

At Nurture House, we keep the buttons you see below. The "No" button says no in varying degrees of strength when pressed. We also have a "Yes" and a "Maybe" button and are often helping survivors reinforce their verbal boundaries through playing with these buttons. We also have two smaller circular recording devices. A child may speak into the device whatever boundary they want to give voice to and then hear it over and over again simply by pressing the button. This activity delights children who have felt like their voice was previously unheard.

## Containment of Perpetrator Symbols

Some boundaries are established through limit-setting. Other boundaries are created by the child in an effort to exert some control over a monstrous presence. In a previous text, *Play Therapy with Traumatized Children* (Goodyear-Brown, 2010), I gave some case examples of the ways in which children begin to contain perpetrator symbols in order to feel safe to approach them. Below is an expansion of this idea. In addition to containing just perpetrator symbols, clients will sometimes need to contain snapshots or glimpses of the trauma story itself. In this way, the containment is not just around the danger associated with a perpetrator, but the more somatically overwhelming felt sense of the encoded experience or the flood of scary cognitions related to the trauma. I often see this kind of containment needed by children who create play projections related to witnessing domestic violence. Below is a "scene" created on the chair of my old office space by a child who was growing up with an abusive father. This six-year-old boy had chosen the two-headed dragon for the dad figure; a pale, drawn woman for the mom; and a small, naked baby who stuck very close to the woman for his self-object. Once he had chosen these miniatures, he explored the room further until he found the box of cotton balls. He painstakingly unraveled one in order to create "fire" breathing out of the dragon's mouth and explained that sometimes the mom and baby get burned by the dragon. He was quick to explain that the other head does not breathe fire, so sometimes the mom and baby do not get burned. I reflected his words and wondered out loud how the woman and child could know which head was going to be in charge at any given time, adding that it sounded confusing and maybe dangerous, depending on which head was in charge. He nodded vigorously, then explored the room further and came back with the police caution tape. He quarantined off the chair by placing the tape, with the words "enter at your own risk," across the front of the chair. Once he had contained the danger in a visible way, he let out a deep breath and turned away to peruse the other toys (see Figures 4.17 and 4.18). His created boundary contributed to a felt sense of safety that allowed him to drop the hypervigilance and literally turn his back on the danger in order to begin asking the developmental question, what next?

*Figure 4.17* Dragon Burning Mom and Child

*Figure 4.18* The Scene Contained

Jenny, a seven-year-old girl who was bullied by an older boy at school came into her first session with a quiet, cautious demeanor. After ten minutes or so of exploration, she chose an oversized beetle, put it in the sand, and chose a disproportionately small female figure, a tiny girl with downcast eyes wearing a hoody that covers part of her face (I wondered internally if this was Jenny's self-object). We processed the tray as she had created it, and as I wondered out loud how the girl with the hoody felt being so close to the beetle, she chose several pieces of barbed wire fencing and began making a square enclosure to surround the beetle. After a few moments of deep concentration, she sighed and said, "It's not right. I need it tighter." I gestured to our shelving area that holds various fences. Her eyes lit up when she found the length of white picket fence that is bendable. She shaped it carefully until it was just the right size to enclose the beetle entirely and closely. Once the beetle was enclosed, she spontaneously turned her attention to adding in aspects of her life that bring her joy. Once she had provided some containment for the bully, she could build the richness of her world outside of these bullying encounters.

Another containment symbol that has been powerfully used at Nurture House is the wooden house pictured below (see Figure 4.19). There are four numbered doors, each with its own lock and matching key. All the keys are attached to the top of the house so that the power to lock and unlock the doors is always available to clients. We recently had a family in which two preschool-age children witnessed the rape of their babysitter. The two

*Figure 4.19* Perpetrator in the Doorway

children and their babysitter had gone into the wooded area of a park in order to ride their tricycles, a man followed them, and as the babysitter was attacked, she called out for the children to run. It was not until later, as we processed the trauma, that we understood the children's goal as they ran was to find sticks and stones to throw at the "bad man." In the aftermath of the trauma, the children insisted on sleeping with their parents and could not go to sleep unless the door of the bedroom was locked from the inside. They would also double check all the locks of the house before bed. Our locking house became an important part of their posttraumatic play. The children came together for a couple of sessions, and during the safety-building phase, they took turns locking various creatures inside the house. They would decide together how bad a creature was, if it deserved to be locked up, and then decided when each creature came out. Sometimes the creatures were locked in because they were bad, and other times they were locked in for safety from the bad guys. Either way, all four doors were always left locked as the children exited the playroom.

The parents were highly anxious about telling the story of what happened in a way that would bring coherence without scaring the children further. After mom, dad, and I crafted the story, mom and dad were able to tell it to the children while the children were snuggled in between them. Each received bilateral stimulation during one telling of the story and got to listen in a posture of repose for the second telling. After this session, both parents reported feeling significantly less anxious, and the children began to leave sessions with the doors of the house unlocked.

Some of the neglected and abused children who come to Nurture House have described parents with at least two—and often three or four—distinct "moods," if not personalities. These children never know which parent they are going to get when they come home from school or when they wake up in the morning. Children will sometimes draw pictures or tell stories about their parents in which they rename them based on these moods. Sometimes the moods are altered through the abuse of drugs or alcohol. I have had teenagers rename their parents "Dick on Drugs" and "The Alcoholic Ass." When I began practicing play therapy, I had two-headed dragons as well as single-headed dragons/dinosaurs and thought I was offering enough variety. It was a 12-year-old girl named Samantha who taught me that I needed even more nuance. She had been in multiple foster homes after being removed from her biological mom's home because her mom was addicted to drugs and had left her unattended at home on multiple occasions when she was too small to fend for herself. She identified the "cookie mom" as the one who would drink one to three beers, become super happy, and play upbeat Broadway tunes while baking cookies with her and her baby brother. She identified the "angry mom" as the one who had snorted some white powder and had a couple of shots, who might slam the pantry door if she could not find the snack she was after, or who might call her boyfriend and yell at him for several minutes about whatever was upsetting her or, worse, yell at Samantha and her little

brother for not cleaning the kitchen or their rooms properly. Then she identified "blurry mom," who would show up by 9:00 p.m. on school nights and
by noon on Saturdays and Sundays, who would lie in her bed or on the couch,
mumbling nonsense, asking Samantha to bring her some ibuprofen or to turn
down the noise. Samantha knew that her job at these times was to move into
caretaking mode. When I asked Samantha to choose a symbol to represent her
mother, she chose the three-headed creature you see below (see Figure 4.20).

Although she was a very articulate young woman, when I asked her what
she needed to do with the three-headed dragon to feel safe, she chose two sets
of miniature handcuffs and put three of them around the necks of all three
heads, leaving a fourth cuff. She became flustered as she noticed the fourth
cuff dangling. I reflected, "You put a handcuff on each of the creature's heads.
It's hard to know what to do with the fourth one . . ." She decided to bury
the fourth cuff in the ground, as if there were a magical grounding place that
would keep her mother anchored in such a way that she could not wreak
havoc on those around her. Once we both understood that she had anchored
the three-headed beast, I asked if she could choose a miniature to be herself.
She chose the disproportionately small tuxedoed man that you see beside the
dragon in the picture to represent herself. This tiny, buttoned-up male figure
stood rigidly in the shadow of the three-headed creature. I wondered out loud
with her about what it felt like for the guy in the tuxedo to be so small. She
said, "He can't do much." When I began to wonder with Samantha about this
well-dressed figure, she said that it was her job to make everything look OK so
that they would not get taken away from their mom. I said, "I wonder what
it feels like to be in black tie." She said, "Argghh, the tie is so tight, it's like

*Figure 4.20* Three Moods of Mom

it's cutting off his airway. He can't breathe." She began to breathe in a more constricted way in the session as she acknowledged the sensations of the miniature. I said, "I'm starting to feel really tight in my chest as I imagine what it's like for this guy. I need to breathe myself." So I began breathing in for a count of three and out again. I did this quietly as I observed the sandtray. After a couple of moments, her breathing began to resonate with mine. We breathed together quietly, and after 30 seconds or so, she said, "He hates being in this suit!" I said, "I wonder if there is some way to make it safer or more comfortable for him?" Samantha looked, breathed, looked, breathed, and after another 30–45 seconds said, "I know, let's turn him into . . ." and she chose my small warthog that looks like a porcupine: ". . . he's a porcupine." I reflected her choice and then wondered out loud, "I wonder how it helps him to be a porcupine." Samantha immediately replied, "Well, she'll get hurt trying to grab ahold of him." I said, "Oh! I get it. He isn't made to save her. The three-headed creature has got to figure out a way to do that for herself . . . or not." After this session, Samantha's play (keep in mind that she had already been in a couple of foster homes at this point) would start out trying to figure out a way for the tuxedoed man to save the creature. Halfway through treatment, the tuxedoed man was permanently replaced with a warthog. The warthog brought its own set of troubles, however, as new caregiver-friendly creatures in her tray would see the warthog as dangerous. They confused it with a porcupine and acted to steer clear of her poisonous quills. As Samantha and I grew in our relationship, she allowed me to join her team and become someone who could trumpet her true vulnerability and need for connection that was often "miscued" through her learned defenses.

## Conclusion

The number of ways in which containment may be needed and can be provided are legion for traumatized children. The good news is that play therapists are in a unique position to offer these many forms. Play therapists can attune ourselves to the bigness/smallness needs of children within the physical space and redesign these as needed. We can provide physical containers for chunking trauma content. We can provide the symbolic play tools for children to construct containment scenes for perpetrator symbols or protective scenes for self-objects. Most importantly, we can grow our own capacities to function as living containers for those in our care.

## References

Curtis, S., Gesler, W., Priebe, S., & Francis, S. (2009). New spaces of inpatient care for people with mental illness: A complex 'rebirth' of the clinic. *Health & Place*, 15(1), 340–348.

Curtis, S., Gesler, W., Wood, V., Spencer, I., Mason, J., Close, H., & Reilly, J. (2013). Compassionate containment? Balancing technical safety and therapy in the design of psychiatric wards. *Social Science & Medicine*, 97, 201–209.

Gesler, W. (2003). *Healing places.* Lanham, MD: Rowman and Littlefield.

Gesler, W., Bell, M., Curtis, S., Hubbard, P., & Francis, S. (2004). Therapy by design: Evaluating the UK hospital building program. *Health & Place, 10*(2), 117–128.

Gil, E. (2014). *Play in family therapy.* New York, NY: Guilford Press.

Gil, E., & Shaw, J. A. (2013). *Working with children with sexual behavior problems.* New York, NY: Guilford Publications.

Goodyear-Brown, P. (2010). *Play therapy with traumatized children: A prescriptive approach.* Hoboken, NJ: Wiley Publishing.

Goodyear-Brown, P. (2011b). *The Worry Wars: An Anxiety Workbook for Kids and Their Helpful Adults.* Published by the author.

Goodyear-Brown, P. (2013). *Tackling touchy subjects.* Published by author. Nashville, TN. Retrieved from www.nurturehouse.org

Goodyear-Brown, P. & Frew, A. (2015). Short term play therapy for children with sexual behavior problems. In H. G. Kaduson & C. Schaefer (Eds.), *Short term play therapy for children* (3rd ed.). Washington, DC: American Psychological Association.

Goodyear-Brown, P. & Gott, E. (2019). Tech, Trauma Work, and the Power of Titration. In J. Stone (Ed.), Integrating technology into modern therapies: A clinicians guide to developments and interventions. New York, NY: Routledge.

# 5 Metaphor and Medium

Trauma is stored predominantly in the right hemisphere. It is stored iconically and somatically, and for these reasons is often more easily accessed and worked with through expressive mediums and through metaphor than through the stark linguistic narrative. Indeed, when I have a supervisee describe a stuck place in a case with a traumatized child, one of the first things I am likely to explore is what metaphors and what mediums have been used by the child so far. We can often jump-start a child's processing again if we offer a metaphor that has already been important to them but can be further explored through use of a different medium.

Our understanding that "mind" is an embodied and relational process that is constantly regulating the flow of energy and information within the body and within our relationships (Siegel, 2010, 2015) guides all of our interactions with clients at Nurture House. The embodiment of mind is aided by therapeutic tools that invite all of our different somatic ways of knowing. Kinesthetic involvement, which has long been understood as important for children, is part of the growth of a child's mind. If one form of kinesthetic involvement, say fingerpainting, provides a pathway to expanded knowing, then varying our forms of tactile grounding and varying the mediums in which we work offer additional ways of knowing. Play therapists share in the relational process with traumatized children, and as we curiously invite our clients to look at the trauma from a new angle of expression, we can potentially help them integrate different aspects of the trauma. As our field is understanding the limitations of talk therapies more deeply, we are also understanding that there is a powerful right brain to right brain relational resonance that begins to happen as a child creates an image, a sculpture, a sandtray and we witness it with them. The shared experience of what has been created and communicated increases the shared mindsight between therapist and child. When families engage in shared sandtray or art creations, new shared mindsight is offered to the system. Their family lexicon, which up to this point may have been mainly conversational, expands to embrace other forms of communication, impacting both the givers and receivers in new, potentially transformative ways.

## Containment Through Control of Expressive Arts Media

There is a long history of appreciation within the expressive arts therapies for the power of the medium used for creation (Malchiodi, 2013; Malchiodi & Crenshaw, 2015). The medium one chooses for creating can give rich insight into the creator's need for control over their environment and allows them to work with boundaries and containment in increasingly complex ways. One form of quantification, media dimension variables, was set forth by Kagin in 1969. She identified three categories of general assessment related to the materials used: structure, task complexity, and media properties. Each of these three categories gives rise to ends of continuums: unstructured vs. structured, high complexity vs. low complexity, and fluid vs. resistive. How much fluidity or internal resistance a material has can tell us much about the person using it, as can the ways in which they use the material. The eight-year-old child who dumps large amounts of finger paint on paper and then slides it around, all the way to the edges of the page and onto the surrounding table, may be communicating something very different about their own boundaries (or those enforced for them) than the child of the same age who chooses to use acrylic paints with a paint brush. She acknowledged that fluid materials require containment so as not to spill haphazardly all over the place. In fact, one way of beginning to categorize the mediums was to work backward from the container for the medium to the medium itself. One way of looking at media dimensions is to ask what kind of container is needed and how much needs to be contained (Graves-Alcorn & Green, 2014).

The expressive therapies continuum (ETC) (Kagin & Lusebrink, 1978; Graves-Alcorn & Kagin, 2017) delineates four different modes of expression that can be helpful to those of us who integrate the expressive arts into our therapeutic work. The ETC provides a helpful framework that can be adopted by expressive arts therapists and used to better understand where, developmentally, a client's creative energies may be currently directed. Its bottom-up assignations mirror stages in the developmental process that all humans move through and are heavily grounded in the developmental theorizing of both Jean Piaget and Margaret Lowenfeld. Kinesthetic/sensory is the descriptor given to the foundational level of expression, perceptual/affective is the descriptor used for the middle of the continuum, and cognitive/symbolic is the descriptor given to the top layer of the continuum. The fourth level is the creative level and represents an amalgam of all of the other forms of expression. Additionally, variables around whether or not the client wants to have direct contact with the materials or have them mediated are worthy of our attention, as is the distance a client chooses to have from the media.

The clients' need for control over their materials is in essence a need for containment. There can be a powerful neurochemical payoff for a child after wrestling an expressive material into the shape they want it to have: competence as a neurochemical rush. I have had children spend 20 minutes in absolute silence packing a ball of kinetic sand so tightly that when they drop it from a height onto the hard surface of the tray, it does not break apart. Which materials are seen as the least controllable and most controllable have undergone some development over the evolution of our field, and while we can all agree that some materials are more fluid or more resistive, we have also learned to have a healthy respect for children who present with sensory seeking or sensory defensive tendencies. A breakdown of materials that I particularly like was offered by Sue Bratton in a chapter on integrating expressive arts work into supervision of child therapists. She builds on a continuum of expressive arts materials originally offered by Landgarten (1987). Bratton's list, however, includes other play materials that might be found in a fully equipped playroom. She conceptualizes materials from least controllable (things like wet clay/wet sand) to most controllable (things like colored pencils and lead pencils). Between these two extremes, from least to most controllable, are oil pastels and watercolors, dry sand and miniatures, stand-alone figures, puppets (and the storytelling and dramas that accompany these), collage-type activities, model magic and modeling clay, and crayons and markers.

Landgarten was a foundational member of the art therapy community and helped shape thinking around the ways in which art can be used to both understand family dynamics and enter the dynamics to begin to change them. At Nurture House, we offer a whole range of art materials, from fluid to resistive. Since all of the choices listed above are available to children, we are careful to notice which materials children are drawn to during the assessment phase. Their choice of materials and the control they exert over these often inform the therapist's understanding of where they are developmentally in their need for control over their environments.

We can often gain valuable insight into their sensory profiles as well. The child who avoids the sand completely may have some sensory defensive leanings that would be valuable for parents to understand, whereas the child who comes in and immediately buries his feet in the sand is telling us something about his sensory seeking tendencies. This child's trauma work may be mitigated best by deep sensory input. I am remembering Johnny, an 11-year-old boy who lost his 16-year-old brother to suicide. He was entirely shut down and unable to process the way in which his brother died or his subsequent grief. Understanding from his early encounters with the play materials that he was sensory seeking and also felt competency surges through increased challenge, I arranged for our session to be outside near the tire swing. I introduced a giant exercise ball into our work, and he almost immediately began to bounce on it, then attempted to balance on it while getting into the tire swing without having his feet touch the ground. As he bounced vigorously on the ball, he was able to access storying capacities around his brother's death that were not able to be accessed any other way.

I have found that traumatized children often gravitate to the kinetic sandtray in our sandtray room instead of the Jurassic sand or the sparkly white sand. I believe this has to do with how much more controllable the kinetic sand is than the others. It can be packed, molded, brought more easily into submission to the child's will. It can cause a great deal of frustration to an already dysregulated child to try to create a pyramid or a mountain in the flowing sand mediums because they do not stay put. I have had children try to knock over the sandtray before when a shape they were trying to create would not hold in the flowing sand. Justin created the image you see below. He had just been in the second car accident of his life. His mother wisely recognized that the recent accident exacerbated posttraumatic symptoms that had been with him since a much earlier and much more damaging car accident. His mom was also chronically ill and had an understanding that her limited or intermittent availability to her son had made his resolution of the trauma symptoms brought about by the first accident very challenging.

Justin and I had established a rapport, and I had in mind the TraumaPlay™ goal of enhancing his adaptive coping. At the start of a session, we entered the sandtray room. Justin looked at the tray of kinetic sand and said, "This stuff feels so cool!" He started picking up handfuls of it and letting it drip through his fingers. After a few minutes of experiencing the medium, he said, "This is kind of like me . . . it falls apart . . . and I fell apart after the accident." I said, "Sounds like you didn't know what to do after the accident," and he said, "Yeah." I asked him if he could shape the sand into the shape he would like to be when he thinks about the car accident. He said, "I want to be strong no matter what comes, to know I can handle it." He began to pack the sand, and then found an empty rectangular container and packed the sand in it so tightly that when he got it out, it held together. Justin said with delight, "Hey, it's still standing." He then proceeded to make several other large packed columns. He tried putting one on top of another, but the top one was too heavy, and the bottom one began to slide around and break apart. He sighed and said, "Too much weight in one place." I said, "You are figuring out how to build it so it will stand up under stress." He experimented with different shapes and sizes of packed sand and various distributions of weight before he landed on the construction you see below (see Figure 5.1).

He put a flag in the top and called it "Justin's Strong Tower." Then he said, "Let's test it." He took handfuls of sand and let them rain down over the structure. He did this over and over again until he had almost covered it up (see Figure 5.2). We both agreed that his Strong Tower could withstand a whole lot of storms. In a future session, we returned to the picture we had taken during this session and identified adaptive coping strategies for each of the building blocks of his Strong Tower.

One of my new favorite expressive therapy offerings is the Zen water board, which can be purchased in multiple sizes. At Nurture House, we have the large one and several smaller ones (see Figure 5.3). The underboard is covered with a paper that is easily made wet. The "canvas" sits on top of a trough of water with a paint brush. The paint brush can be dipped into the water, and bold, darkening marks can be made on the board. The image is at first very bold and stark, and then, as the water evaporates, the image begins to disappear. I have written previously about the power of the dry-erase board for children with

*Figure 5.1* Solid Coping Foundation

traumagenic stories (Goodyear-Brown, 2010). There is a real power in being able to create an image that is shared in the moment between two people but can be erased at will so that there is no permanent record of it. Children tend to draw images on the dry-erase board when they need more control over the telling. The Zen board is another favorite among children who want to offer the therapist a glimpse of their experience. Then the therapist and child may have a shared experience of watching the depiction on the paper disappear.

Tammy was an 11-year-old girl who had spent her first ten years in an institution in a foreign country before being adopted by a family in Tennessee. She was drawn to the water board but quickly developed a love/hate relationship with it. When she understood that you could make the light paper turn black

*Figure 5.2* The Strong Tower Holds Up Through Heavy Rains

simply by adding strokes of water, she focused all of her concentration on trying to make the whole board black. This required a great deal of work and water, but eventually she succeeded. As soon as she finished the last stroke, however, and sighed with the satisfaction of having darkened all the edges, the first strokes she had made began to dry. The drying creates shades of gray before returning to the baseline color. She spent an entire session working

*Figure 5.3* Creating Glimpses and Watching Them Disappear

to keep the whole surface as darkened with water as she could. At the end, she sighed again and said, "It will always eventually go away." I reflected her frustration and agreed with her assessment. The next session, she went directly over to the water board and drew a picture, in bold strokes, of a small person in a bed and a man looming over the bed, and she drew a penis on the man that was disproportionately large. Then she said, "This was in the orphanage." I reflected her words and thanked her for sharing it with me. She pulled up a counter stool for herself and one for me, and we sat together, watching the image fade. Then she looked at me and said, "This goes away too." This child did most of her trauma narrative work in this way. It seemed to bring her great satisfaction to be able to show me a glimpse of her story and have us hold the space together as we watched it disappear. It seemed to comfort her greatly that the images disappeared on their own, needing no additional effort or intervention from either of us. Since this seemed to be her preferred method for working through her life story, I introduced a few sets of bilateral stimulation just after she had completed each water board. Then I asked her to notice what happened in her body as she watched the image disappear. She described a kind of relief (very articulate 11-year-old) and a weight lifting from her chest. We further installed this sense of lightening with BLS as well.

Children may also approach an aspect of their traumatic experience and work with it in one medium, such as visual art, and then approach it again through a

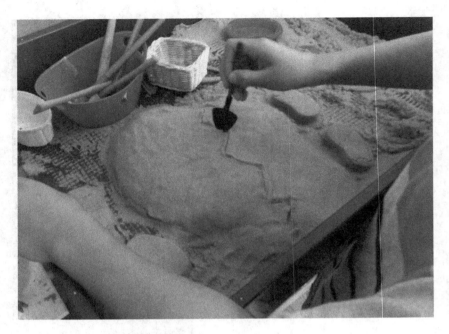

*Figure 5.4* A Broken Heart in the Kinetic Sand

different medium. Penny, a 12-year-old, verbalized early in treatment that she was broken-hearted over some of the abusive things that were said to and about her during her traumatic experience. The left brain verbalization of having a broken heart was one expression of her experience, but I wondered if deeper knowing and, potentially, healing might come from working in some right brain mediums. In the next session, we returned to the phrase "broken-hearted," and I asked if she could show me what this was like for her. She carefully packed the kinetic sand into the shape of a heart and then took the shovel and spent time creating a crack all the way through the center of the heart (see Figure 5.4).

During another session, I offered her canvas and paints, changing up the medium. The painting below depicts a heart with multiple breaks and the words "Would anyone care to mend it?" This opened up an expanded discussion of the hurts she had experienced (see Figure 5.5).

As we continued meeting together, I introduced a thin, wooden, heart-shaped puzzle. The creator had cut it into around eight separate strips of wood, and the pieces could really only be placed beside each other, not inter-locked to create a perfect fit. At first, Penny hid the pieces around the room and insisted that I find them. She gave me almost no help and teased me pretty mercilessly when I failed to find the pieces. Over time, she began to give me hints, and it became easier to find the pieces of the broken heart. At one point, we switched roles, and Penny asked me to hide the pieces and give hints to help her find them. Toward the end of treatment, Penny asked if we could

*Figure 5.5* A Broken Heart on Canvas

glue the heart together so that it would not break again. I felt like this tool had become important enough to her that I was willing to do this. Penny took the glued heart home after our final session as a reminder of our work together.

## Kinesthetic Telling in the Lobby

Randy, a ten-year-old boy who was sexually abused by a biological father who is currently in jail, was waiting in the lobby with his adoptive mother when I came to greet them. Randy had several finger puppets in front of him, and I commented on this, saying, "You found the finger puppets!" He had a cloth monkey finger puppet on his finger when I came in but quickly took it off— almost like he was in trouble—as soon as he saw me. He said, "It's weird how they have the hole in them." He made a yucky face. I responded, "You're not sure about those holes." He put his finger back inside the monkey and made another face. I said, "You look uncomfortable with where the finger goes." He nodded, and his mom said, "Yes, he's mentioned it two or three other times when we've been here." I sat down and, picking up a finger puppet myself, said, "You know, it makes sense to me that you would see these holes differently. Lots of kids who have been sexually abused might think of these holes differently—more like private parts." Randy looked up and seemed visibly relieved that I understood why it seemed yucky to him. The matter-of-fact verbal connection between his play and source of discomfort gave permission for him to share his discomfort and shaped an expectation that it could be contained and given structure. His mom also looked relieved and said, "I didn't know what to say when he would mention it, so it's good for me to see you do it. It's OK to just talk about it?" "Yep, you are one of his primary history keepers now, and as you hold the story and his big feelings, you bring safety and containment to both." Randy requested to be in the sandtray room today, and after I closed the door to the room, I said, "I'm so glad you let us know it was bothering you. Some of the kids that I see have had fingers, penises, or other objects inserted into their bottoms, and I really get that having that happen to you could change the way you see putting a finger into the puppet." "Yeah, Dad put his finger in me." I reflected the language, and the client abruptly moved to the sandtray shelves. We always notice abrupt shifts in attention or changes in the play and may perceive the shift as the client's communication that he has pushed up against the edge of his window of tolerance for looking at or talking about or playing out that scary thing that happened. In this case, Randy seemed to want to give it a different expression. He created his most elaborate sandtray to date. Following is a slightly edited version of the story he told as he created the tray: *Once upon a time there was a castle guarded by snakes, spiders, frogs, monkeys, gorillas, owls, worms, and one bird. There were many big cats (lions, tigers, cheetahs, leopards) who wanted to take over the castle. One giant monkey was assigned to be the leader and stood guard out in front of the castle, watching for the tigers. Inside the castle, the two owls were always afraid of the tigers coming, and so they never went outside. There was a big, gigantic bear hiding, and it was on the tiger's team. The two owls really wanted to go bowling with all the animals* (see Figure 5.6).

Suddenly, the cheetah moved. The owl team attacked the tigers, and they won (see Figure 5.7). *The two owls realized that when you have a team looking ahead and around for you, you don't have to be as afraid,* and the worm was eaten by the gigantic tiger, and the tiger got bigger than anything you've ever seen. *The tiger felt squirmy. He could feel the worm moving on the inside of him.* (I wondered here if some of the somatic expression of feeling like the worm was inside of him was a glimpse of his experience of the sexual abuse.)

*Guards appeared at the top* of the castle (see Figure 5.8), and instead of the gigantic snakes and monkeys and spiders and frogs attacking, *the knights did the hard work of protection. They also saved all the animals from dying. When all the animals were at war, the guards shouted out, "Stop fighting! I will fight them for you! You can just chill and watch!"*

The story was powerful, and after Randy told it, we were both silent for a while, taking it in. I asked if he would like to take a picture of the sandtray, and Randy expressed great pleasure in this idea. I watched him absorb the idea that he could create a photo document of the story. He seemed to enjoy the different perspective he got of his sand world as he looked at it through the camera. He seemed to embrace the distancing effect that the camera provided. Randy got to decide how close up or far away elements of his story would be. He

*Figure 5.6* The Owls Can't Come Out

became the director, showing me what angles he wanted documented, when to zoom in on just the owls, and when to give a longer view of the entirety of the battle. In fact, the photography became another method of titration for Randy, "sitting with" the story but with the added layer of distance provided by looking at the closeness or distance on the screen. He could literally, using two fingers, increase or decrease how far the image was removed from his vision.

*Figure 5.7* The Mighty Battle

Another nonverbal nuance of the narrative involved the miniscule changes you see in the images below. When we got to the part of the story where the owls are scared to come out of the castle, Randy looked more intently, with a photographer's eye, at the placement of the owls inside the castle. It may help us to remember that Randy and his brother were removed together from their sexually abusive father. Internally, I wondered how close Randy was to the edge of his window of tolerance for focusing on this content. However, since the work was all happening in metaphor, the threat level was significantly reduced. I wondered if we might extend the moment of focus by shifting mediums. The world in the sand was one medium, rich with metaphor. The storytelling was a second medium for exploration. Would he appreciate a third medium? I decided to offer. He took a first picture in which the owls are deeper inside the castle and tucked close to the wall. He was not satisfied with the arrangement. He paused, moved the owls closer to the entrance, and took a second picture. It felt like he needed to depict the owls' miniscule risk in coming closer and closer, by degrees, to the entrance of the castle to risk seeing the world differently. He would move them ever so slightly closer to the door and see how it felt. I printed off the pictures he had taken so that we could begin a hardcopy book at his next session. He was unhappy with the quality of one picture. His empty Doritos bag was sitting by the sandtray, and he wanted it cropped out of the picture, and while I am reasonably adept at editing, I could not figure out how to do this. Randy suddenly said, "I know!

*Figure 5.8* Two Guards Come to Help. The First Humans in the Animal World

We can get Brad to do it." Brad is Randy's newly adoptive father, and enlisting his help was a huge shift: it acknowledged both a need for help and his growing internalization of his new dad as a "bigger, stronger, wiser, kind" person who is becoming *his* helper.

Creating the hardcopy involved focusing, sometimes with multiple repetitions, on smaller, chunked pieces of text. This became yet another exposure in the titration process. He cared deeply about which parts of the text accompanied which picture. After we took the pictures, I wondered out loud if we should show the tray to his mom. Randy immediately went to get her and brought her back to the room. I invited her to sit and read the story out loud while Randy ran his hands through the sand. His mom heard the underlying themes of struggle, of isolation, of the owls' helplessness, and their solution to let the two guards come help. She, too, sat in silence, acknowledging the power of the narrative.

Randy's sandtray story was created just after having his concerns about the holes in the finger puppets acknowledged and held. I made the connection between putting fingers in the puppet holes and children who have been sexually abused. When we respect the child's risk-taking as they test for containment, we allow the child to take further risks. As they learn that they can take further risks with us, we give room for implicit experiences to be brought into explicit memory in tolerable doses. In Randy's case, we made the doses tolerable by adding layers of narrative as we explored together. The first "telling" of the story was his non-nonverbal process of choosing and carefully arranging miniatures in the sandtray. The second "telling" of the story was in a verbal but created story that kept the symbols in their character contexts. The third telling of the story, for him, was through the lens of the camera, returning from left brain verbal narrative to right brain pictoral/symbolic telling, but this time with layers of removal that allowed him to begin titrating and tweaking the story, allowing him to move back and forth between deeper and more shallow waters of exploration. The fourth telling involved integrating all the previous parts into book form.

## An Overdose of Flies

Jillian is a four-year-old girl who is being raised by a whole posse of people, including dedicated grandparents and parents who both want to be in her life and who have been wrestling with addiction issues for years. Each of her parents has had abrupt departures—sometimes without meaningful goodbyes—to treatment centers, where the treatment regimens have lasted anywhere from 30 days to several months. Jillian has had addiction explained to her, in developmentally sensitive terms, using several different kinds of explanations. One such explanation talked about how her mom and dad have a switch that does not flip properly in their brains. Most people, after they eat one serving of ice cream, will feel full and know that one is enough. Some people might have a second serving and then feel overly full and stop eating. Her mom and dad's brains do not tell them to stop eating. In fact, if they eat one serving of ice cream, they right away want another, and they will keep eating ice cream even after it gives them a stomachache. Their "stop" button does not work well. Does this explanation get through to Jillian? It is difficult to know, but in my experience, I would say that no explanation from an adult is as impactful for the child as the metaphors they create themselves. Very young children benefit significantly from a child-led play therapy approach, particularly when trying to work through family dynamics that are confusing and for which words may be woefully unsatisfactory as explanation, so Jillian and I engaged in lots of child-centered play. She was drawn to the sandtray and to the symbols that

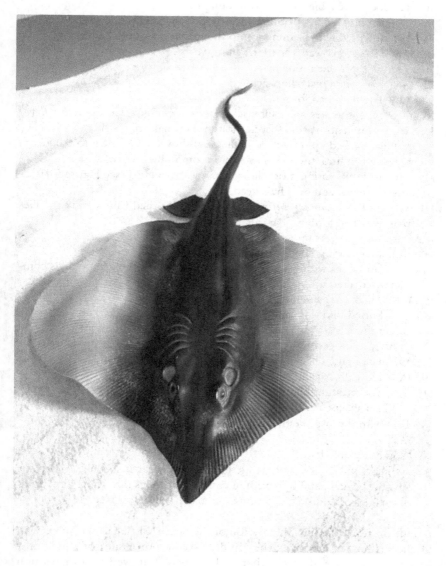

*Figure 5.9* Manta Ray

could speak volumes when words were insufficient. During a season in which one of her parents was away in treatment, she went over to the shelf with the sea creatures and chose a large manta ray (see Figure 5.9).

ME: You chose that one.
JILLIAN: Look, Miss Paris. He has a big mouth [*after flipping the figure over and seeing its underside*].

ME: He does have a big mouth. You found it.

Jillian noticed a set of bugs nearby. She picked up one of the flies and experimented with pushing it into the manta ray's open mouth.

JILLIAN: He's swallowed a fly! He's hungry.

ME: Yep, he's got one inside now.

JILLIAN: It's gonna make him sick.

ME: Oh! The fly is bad for him.

JILLIAN: Yeah, the flies make him sick [*nodding and picking up another fly*].

ME: He just ate another one! Even though they make him sick.

Jillian picked up several more flies and shoved each one into the manta ray's mouth until there was no more room in his mouth.

ME: He just keeps eating more flies [*with a worried tone*] (see Figure 5.10).

JILLIAN: Yeah . . . and now he's dead.

ME: Oh, no [*some concern infused in my tone*]. He died from eating the flies that were bad for him.

JILLIAN: Now he gets buried.

ME: You are covering him up with sand now.

Jillian works with great focus to make sure that the manta ray is completely covered up.

JILLIAN: There [*stepping back from the sandtray*]!

ME: You buried him. He can't be seen now. Only you and I know that he's under there.

After a few seconds of sitting with the reality of not being able to see the manta ray, Jillian abruptly pulls him from the sand.

JILLIAN: He's alive again!

ME: Oh! You brought him back from the dead.

Jillian shakes all of the flies out of the manta ray's mouth, enlisting my help with the last one that is stuck way down inside.

JILLIAN: All out!

ME: You got them all out!

Jillian and I were quiet together, feeling what it felt like to see the manta ray again and to have him empty of the flies that killed him. Then Jillian sighed and started putting flies back in his mouth again.

This overdose of flies, the passion of the death, burial, and resurrection of the manta ray, becomes the child's psychic processing of this family pattern. The pause after the first cycle of play, followed by Jillian's sigh and decision to begin the cycle again, may signal her resignation to the inevitability of this pattern repeating in her family. While she has hope that her parents will come back to her (whether this is in relation to their physical presence or emotional presence), she may be engaging in self-protective preparation for the pattern repeating even as the play helps her make more coherent sense of the pattern as she has already been experiencing it.

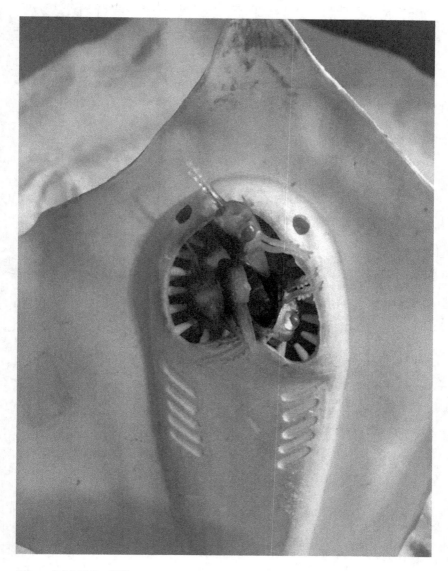

*Figure 5.10* Full of Flies

## Conclusion

Metaphor provides an astonishingly rich wealth of expression. Children never cease to amaze me with the ways in which they will act out the horrors they have experienced in a language removed from reality. Many times the work is done solely in metaphor, symptoms resolve, and children integrate the trauma content. Other times, offering extended ways of working with the metaphors by offering different mediums of expression—paint, sand, photography, puppetry, collage, drama, movement, writing—can bring deepening levels of integration. Because it is the gentlest approach to trauma content, metaphor should always remain prominently displayed on the Play Therapist's Palette and be an offering whenever children or families appear stuck.

# References

Goodyear-Brown, P. (2010). *Play therapy with traumatized children*. Hoboken, NJ: John Wiley & Sons.

Graves-Alcorn, S. L., & Green, E. (2014). The expressive arts therapy continuum: History and theory. In E. Green & A. A. Drewes (Eds.), *Integrating expressive arts and play therapy with children and adolescents* (pp. 1–16). New York, NY: John Wiley & Sons.

Graves-Alcorn, S. L., & Kagin, C. (2017). *Implementing the expressive therapies continuum: A guide for clinical practice*. New York, NY: Routledge.

Kagin, S. L. & Lusebrink, V. B. (1978). The expressive therapies continuum. *Art Psychotherapy*, 5, 171–180.

Landgarten, H.B. (1987). *Family art psychotherapy: A clinical guide and casebook*. New York, NY: Brunner/Mazel.

Malchiodi, C. A. (Ed.). (2013). *Expressive therapies*. New York, NY: Guilford Publications.

Malchiodi, C. A., & Crenshaw, D. A. (Eds.). (2015). *Creative arts and play therapy for attachment problems*. New York, NY: Guilford Publications.

Siegel, D. J. (2010). *Mindsight: The new science of personal transformation*. New York, NY: Bantam.

Siegel, D. J. (2015). *The developing mind: How relationships and the brain interact to shape who we are*. New York, NY: Guilford Publications.

# 6 Playing the Affective Accordion

## Titration in Aspects of Emotional Literacy

### Symbols for Big Feelings

Many of the children that we see in therapy have very little language associated with whatever traumatic events have occurred. Many of our clients with developmental trauma disorder have delays in their expressive or receptive language competencies, and visual representations, especially three-dimensional visual images, of their emotions often prove helpful in providing a different form of integration. Moreover, young children "live" more in their right brains than in their left brains regardless of whether they have experienced trauma, so symbolic expression can often be more accurate than verbal articulation in capturing the essence of an experienced emotion. With some children, we begin emotional literacy work by simply identifying a feeling and then identifying a symbol, a miniature in the playroom, that evokes this feeling or represents this feeling for the client.

I am often combining TraumaPlay™ and EMDR and was using the Thoughts Kit for Kids, created by Ana Gomez, to help George identify negative cognitions. However, he was not able to identify or articulate his thoughts, but he did spontaneously generate these three feelings on small Post-it Notes. I read each word out loud while I attached each feeling word to one of the blank cards that come with the game, and then asked him if he could choose a symbol from my miniatures to represent each feeling. George chose a disproportionately small naked baby to represent the feeling of vulnerability. He chose the nun symbol (who can have sparks fly out of her mouth when she is cranked) to represent angry (see Figure 6.1). Finally, he chose the symbol I call the Invisible Man to personify the feeling of powerlessness he had experienced during his sexual trauma. The pairing of icons with words helps move the processing from left brain to right brain and back again.

### Felt Feelings

Many children come into treatment with a constricted emotional vocabulary. One of the ways we help children expand their vocabulary is to expand their somatic understanding of emotion. How do their bodies let them know that

*Figure 6.1* Fire-Breathing Nun Symbolizing Anger

they are having an emotion? This can be even harder to explore without a tactile anchor, a concrete way to explore the abstraction, so icons are sometimes more useful. A host of icons is placed before the child. These concrete representations of somatic experiences can be as varied as your clinical imagination and can be pulled from multiple sources: you can find images on line, you can cut them out of magazines, or you can draw them yourself. Icons might include fireworks, flames, a tornado, a pair of handcuffs, or even a black hole. The child is offered a visually nuanced set of images that can reflect experience of different emotions, or even the same emotion. In a sampling of three children, all of whom can verbalize that they often feel anxious, one may experience his anxiety as a tornado in his head, another may experience it as fireworks exploding in his heart, and a third may demonstrate his experience of anxiety as a big black ball of heaviness in his stomach. So, the exploration is two-fold: how would you characterize your feeling iconically, and where do you carry that feeling in your body? The set of icons you see below was created by a previous intern at Nurture House, Bethany Berryessa, after she attended a training with Lori Myers, LCSW, RPT-S, and was introduced to felt feelings, Lori's adaptation of an intervention she learned at a TF-CBT preschool learning collaborative.

Sam is an eight-year-old boy who came in to therapy with Bethany because of angry behavior and suicidal ideation. When he was three years old, Sam was physically assaulted by his preschool teacher. Following the assault, Sam was interviewed by the police and removed from the school. Therapy was not recommended for the client at the time of the assault. Five years later, when Sam started psychoeducation around trauma using resources such as the book *A Terrible*

*Figure 6.2* Sam's Somatic Experience of Anger

*Thing Happened*, he became dysregulated and verbally shut down. Bethany offered the felt feelings board, and he was able to show the therapist how his body felt when he was angry. As Sam created his felt feelings board, he described how each of the images correlated with specific physical sensations and, eventually, suicidal ideation (see Figure 6.2). Because of the intervention, he was able to self-regulate and show what he was feeling before being asked to use words.

In the following session, Sam asked to create a felt feelings board to show the therapist how he felt at school when he would try to make cuts in his palms with scissors. Sam asked for a pair of child's scissors in the playroom to add to his felt feelings board. He used an image of a box in his board, which later became an important image for EMDR reprocessing of the event (see Figure 6.3). Because the client felt safe showing his experience (as opposed to having to talk about it), the therapist was able to provide alternative coping skills to replace self-injurious behaviors, regularly assess for safety, and increase physiological and emotional awareness.

## Facial Expressions

Emotional literacy is about not only being able to name an emotion that you yourself are feeling but also being able to accurately identify what those around you may be feeling. This becomes really important between parents and children, as the mirror neurons we referenced in the chapter on the neurobiology

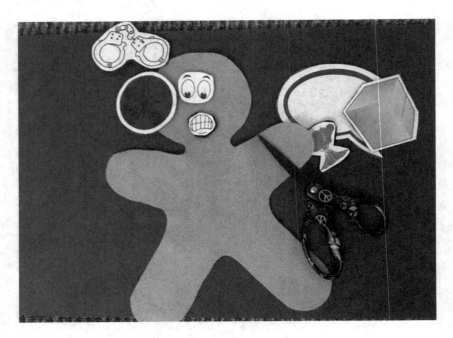

*Figure 6.3* Sam's Elaboration of Self-Harm

of play and trauma can influence the experience of one person's emotions through the expression of another. We know that when a mother smiles at her baby and the baby smiles back, this process is not simply a reflexive one involving a mimicking of facial muscle movements, but rather an exchange of neurochemical experience in which the dopamine released in the mom's brain as she enjoys her baby is simultaneously released in the baby's brain as the baby experiences the mom's smile almost as its own.

When a parent and child are securely attached, they are able to read each other's facial expressions even from a distance. When the attachment relationship is insecure, children may overperceive negative emotions or underperceive positive emotions in the faces of their parents. This is an especially noticeable problem of perception between foster and adopted children and their caregivers. For these reasons, we at Nurture House are continuing to explore ways to work with facial expressions of emotion. There are several interventions that we employ in this process. The newest of these revolves around a dry-erase face that we have installed in the kitchen of Nurture House. Since its installation, it has yielded profound moments of discovery for adults and children alike, both in terms of how they perceive the facial expressions of others and the ways in which their own facial expressions are understood by others.

The first such aha moment occurred even as we were adhering this new tool to the wall. My office manager, Linnea, whom we all agree is the most regulated person at Nurture House, was helping me hang it. I asked if she would use the

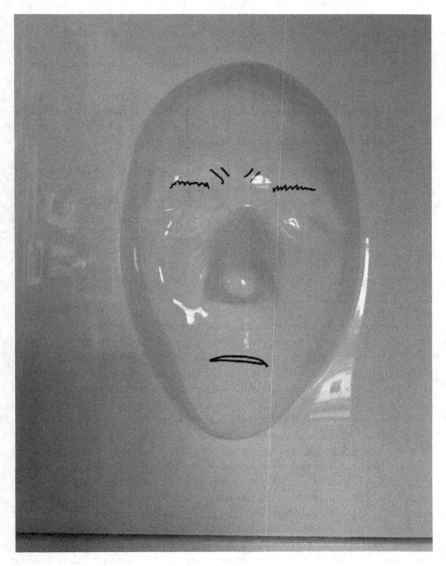

*Figure 6.4* Linnea's Expression of Anger

dry-erase markers to create her own expression of anger on the face. She chose the black marker from the rainbow of colors offered and drew eyebrows and a slightly frowny mouth, then drew two thin, short, slanted lines to represent her disapproving brow (see Figure 6.4). Here is how the conversation went:

ME: That's your angry face?
LINNEA: Yep, I guess so.

ME: This is what makes you so good with people—you don't get super angry.

LINNEA: Well, or I don't show it.

ME: Hmmm . . . that may be good for others but not so good for you.

LINNEA: Well, then it may not really be that good for others [*referencing here the way that unexpressed anger can come out sideways in our most intimate relationships*].

It was not until later that I realized we both had some insight into her physical expression of anger and how this might affect her relationships. Several different prompts can be given with the dry-erase face:

1.  Draw your face when you are . . . (fill in the emotion: angry, scared).
2.  Draw your mom's (or dad's) face when she looks . . . (fill in the emotion).

Once either one of the above prompts has been drawn, two additional prompts can yield powerful information:

3.  Erase the scariest part of the face.
4.  Redraw the emotion the way you would like it to be shown.

At Nurture House, we are often teaching caregivers the SOOTHE strategies, a set of co-regulation strategies meant to provide comfort and regulation when a child is no longer in his choosing mind (Goodyear-Brown, 2010). The "S" in the SOOTHE acronym stands for "soft tone of voice and face." Referencing our mirror neurons again, if we can acknowledge how powerful they may be in generating a shared experience of a positive emotion (such as delight in one another), we must also acknowledge that this same neuronal substrate may provide shared experiences of frustration, anger, fear, and disgust. We keep handheld mirrors at Nurture House to help parents and kids play with this idea together. A mom, for example, holds the mirror out in front of her, and the child gets to draw features on the mirror that augment what the child perceives as mom's communication of anger when she is angry and vice versa. Sometimes this exercise is done in a parenting session where mom and dad take turns showing the anger expressions of the other. It is important that an exercise like this be saved until there is a great deal of safety within the system, as it is meant to be a more concrete way for each party to learn what the other perceives and is in no way meant to be a tool for shaming or teasing. Fear, concentration, anger, and confusion can all look pretty similar. The dry-erase face can be used to compare and contrast parts of the expression of each that are similar and/or unique to each expression.

Soon after we hung the face on the wall, one of my sexual trauma survivors, a 15-year-old girl, was in the kitchen with me. Her mom had mentioned Cassandra's tendency to blunt her affect, to seemingly shut down emotional communication whenever her mom brought any correction. I shared her mom's concern, and she said, "Yeah, I probably look as blank as that face!"

She pointed to the dry-erase face. I said, "Do you think you could show me what you are really feeling underneath the blankness at those times when mom corrects you?" Below is the face she created (see Figure 6.5).

I said, "What would you call this expression?" She replied, "Unbridled fear!"

What she was showing her mom was a mask for intense feelings that actually needed soothing during those moments of correction. We brought her mom into the session and helped her understand her child's face differently.

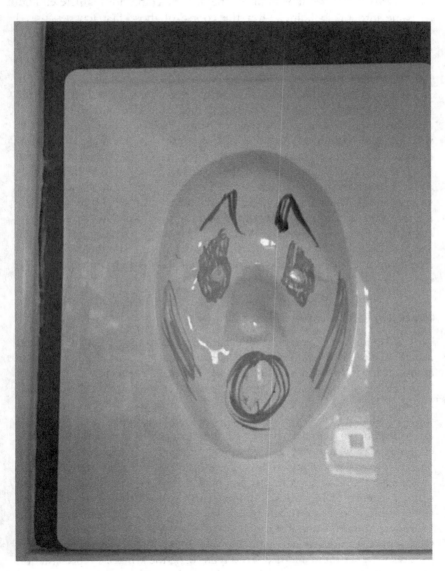

*Figure 6.5* Unbridled Fear

## Symbols as Embodied Feelings

As children begin to feel safe and can tolerate looking at more difficult feelings without believing these feelings make them bad, we often begin to see spontaneous visual depiction or even spontaneous articulation. We keep sticky note pads of emoji starters for times like these—only the eyes are already drawn within a round face. What you see below is the work of Josh, a client who had his first several sessions in his car because he could not regulate enough to come into the building. Over the course of those first few sessions, I would bring several options of play materials out to the car, as Josh alternated between refusing to come inside and wanting to come inside but being too aggressive to safely enter the building. Even the simplest games were sometimes met with intense frustration the moment Josh experienced a feeling of failure for any reason. For example, if he had been tossing a ball and dropped it, he would hurl it across the car. During those times, parents and therapists offered names for some of the feelings being displayed while continuing to welcome him back in his distress. On his seventh visit, when he had become more regulated and more assured that his caregivers could be bigger, stronger, wiser, and kind in the face of his worst behavior, he was able to enter the art room. I offered him an open studio and free rein to use the art materials as he wished, and he chose the emoji sticky pad. Josh spent the majority of the session creating faces to represent the sad self, the angry self, the happy self, the scared self, the cool self, and the loving self. He asked for a piece of paper on which to display them all, and once he had stuck them all on the page, I asked if he could title the page, like you would a book. He thought for a minute and then said, "The real Josh" (see Figure 6.6).

## Emotional Expression Hand Puppets

We have established that many children live more in their right hemispheres than in their left. Therefore, while the left brain may hold the words of an emotion, the right brain may hold the expression of the feeling. Long ago, I bought a set of two hand puppets with many Velcro parts (eyes, ears, noses, mouths, hands, feet). As I was working with a little girl, it became clear that she was not able to verbalize her feelings beyond "bad" and "sad," so I introduced the puppets and their parts. She carefully perused them and chose some eyes that were wide open, with sweeping eyelashes, and a toothy smile to be "happy." In a subsequent session, I introduced the idea of an emotion called "worried," and after exploring the puppet parts very carefully, she chose a set of eyes encased in glasses, a mouth with two top teeth and two bottom teeth exposed—clearly an open mouth—nonmatching legs/feet, and a watch. I wondered out loud about the inclusion of the watch, and she said, "Because there is never enough time to get everything done." When I shared the puppet pictures with her mom and dad (see Figure 6.7), the symbolic representation of her worry helped them shift their understanding of her defiant behaviors in

*Figure 6.6* The Real Josh

the morning. She woke up most days feeling anxious about whether she would be able to get everything done before leaving for school. With her parents' understanding of her pervasive worry came a paradigm shift that empowered them to co-regulate her instead of discipline her in the mornings. This small shift changed the whole tenor of their home.

## Kinetic Involvement Mitigates the Approach to Feelings

As we discussed in the chapter on need meeting, some children are going to mitigate their approach to emotional exploration through the use of their physical bodies. Particularly for latency-age boys, activities that allow them to feel strong in their bodies and that provide a high challenge make for an easier approach to emotional literacy titration. We keep two laminated sets of feelings stuck to the wall in each room of Nurture House, and within each space is a way to target particular feelings on the laminated pages with various projectile toys. In the kitchen, we keep a wooden catapult, a wooden crossbow, and mini marshmallows. We dip the mini marshmallows into pools of glitter glue on paper plates and launch them from the sink. Whichever feeling the marshmallow hits is the one we explore. On a practical note, it is important to coat the marshmallow with something that will leave residue on the laminated feeling

*Figure 6.7* Worried Hand Puppet

face chart; otherwise, the speed at which the marshmallow sometimes flies across the room makes it impossible to accurately judge which feeling was hit.

## Naming Big Feelings Together

Katherine, an eight-year-old-girl, and her mother are in session together. Katherine's mom, a doctor, had developed a drug addiction and went to rehab for a period of time. I had spent time with both Katherine and her mother before her mom went to rehab, helped them through the hard goodbye, made sure attachment anchors were in place, and helped them through a meaningful hello period after mom came home. As is sometimes the case, the role addiction had played in the family was its own kind of anchor and, when removed, led to a destabilization of the family. As it became clear that the family dynamics were shifting, Katherine faced the very real potential of divorce between her parents. Her mom had expressed this fear to me also, but neither mom nor daughter had shared their fears directly with each other. This mom had been working on her own ability to be a safe container for her daughter, and after some discussion, we agreed that she was able to hold her daughter's big feelings about all the changes in the family. Years ago, I created an intervention called The Mood Manicure (Goodyear-Brown, 2002). In my early work, this intervention was often completed between the therapist and the client but is often now offered to a parent/child dyad and facilitated by me. In this case, I felt that the combination of high nurture, connection, and physical touch that Katherine had missed while her mom was away would augment the sharing of big feelings between Katherine and her mom. At the beginning of the session, I ushered Katherine and her mother into the kitchen/art studio and offered them a set of multicolored nail polishes. I invited them to explore the various colors of nail polish and assign each a feeling word. These deliberations took some time, and Katherine's mom gave choice and voice to her daughter throughout the process as they choose feeling words for each color. Katherine chose yellow to be happy, and her mom agreed. Her mom chose blue to be sad, and Katherine went along. They deliberated for a while over which color would be lonely and eventually decided on purple, and they designated orange their "worried" color. After the feeling/color associations were made, mom and daughter painted each other's fingernails, asking first what color the other wanted on each nail. Both parties understood that for each nail painted with a certain color, they would verbalize one situation in which they had felt the feeling represented by the color. Mom took Katherine's hand in her own, pulled out her pointer finger, and said, "Which color would you like on this finger, sweetie?" Katherine pointed to the orange. The action of pairing colors and feelings together and then working with the names of the colors (as opposed to working directly with the feelings themselves first) offers an immediate titration, allowing clients to begin approaching their emotional experiences and sharing them with important attachment figures through a first filter of color, without having to name the big feeling words directly yet. Then it was Katherine's turn to pull out her mom's pointer finger and ask which color she would like painted on that finger. Her mom also chose to have her pointer finger painted orange. I looked at the color/feeling grid and said, "I see that you both chose the color orange. Let's see, orange is worried. So, Katherine, what's one thing you worry about?" Katherine said, while looking at her fingernail, "Well, you and Daddy have been arguing a lot." Mom reflected her words and validated her feeling. I then asked Katherine's mom for one thing she worried about, and she stated, "I worry that the kids are scared when dad and I argue." Katherine shook her head and pointed to the blue, and her mom said, "You feel lonely? When we argue?" Katherine nodded slowly as she held her hand out to have her next finger painted with blue. After her mom painted her fingernail, Katherine said, "I feel lonely because I think you guys will get a divorce and I'll be alone more." Katherine began to cry. Mom reflected her big feeling, acknowledged the possibility that she and her dad might be getting a divorce, validated her daughter's feeling of loneliness, and sat with her in the sadness. Mom and client held the uncertainty together, cried together, and affirmed that they would still be a family and mom and dad would still both love her no matter what. Mom and Katherine left in a much more connected state than they arrived, having faced the big, unnamed possibility of divorce together (see Figure 6.8).

Naming it together brings connection. By the end of the session, both mom and daughter's hands held worry, loneliness, and sadness. Through titrated doses of exposure to difficult emotions (mitigated by physical touch and comforting kinesthetic activity, i.e., painting their nails), their hands became a microcosm for holding each other's big feelings in titrated doses. At the end of the session, both parties seemed visibly relieved that the other had acknowledged the possibility of divorce, that both had permission to feel sad and lonely when thinking about this possibility, and that both felt less lonely than when they arrived because of their shared holding of the story of these feelings.

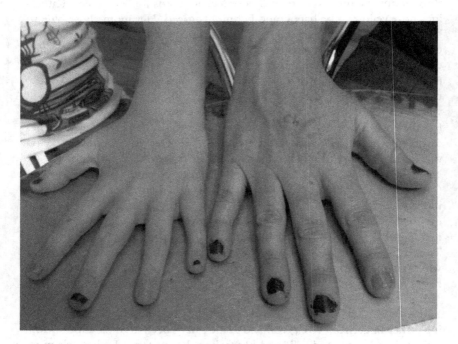

*Figure 6.8* Enhancing Emotional Connections

## Approaching and Naming the Emotions Projected by Others

In the chapter on the neurobiology of trauma, we looked at mirror neurons and the role they can play in affective attunement between parents and children. We learn how to express emotion and how to regulate emotional expression from our caregivers. Even when a foster or adoptive parent has been nurturing and stable, providing modeling of modulated emotional expression, these new interactions may not outweigh the earliest influences of the biological parents' emotional expression as it was perceived by the child. Children who have had several caregivers who have influenced their emotional development can benefit from having a projective way to quantify the emotions of those who influenced them at different stages. In the course of trauma work with families, we are always trying to bring integration to aspects of the life narrative—aspects that include thoughts, feelings, and sensory impressions. Helping children increase their reflective capacity for identifying and naming the primary emotions they "feel" from and with others can further their understanding of self and others.

Dan Siegel's phrase "name it to tame it" (Siegel & Bryson, 2011) is a short encapsulation of the idea that when we can give words (executive functioning/neocortex involvement) to the limbic experience, we can mitigate the need for the body to express the emotions as intensely in a kinesthetic, full body way. The naming of the emotion diffuses its intensity and brings integrations between lower and higher brain regions, opening pathways to the neocortex for thoughtful responses to the emotion being named. When a child has a parent with high emotional literacy who can give words to the child's experience, the attachment between the parent and child is enhanced, the child is anchored, and their shared mindsight evolves. When a parent is themselves disorganized internally, unable to name—or potentially even reflect on—their own emotional responses, the experience of difficult emotions is one parent and child each struggles through alone. The image I will sometimes offer to clients is that of a mother and child wading into the shallows at a beach as the child has his first experience with the ocean. The child is being held by the mother as they wade in together. A small wave comes, and a small, pleasurable excitement is produced as mom and child together get buoyed up and down by the wave. She is holding on tight, and the child is secure and able to enjoy as she enjoys even though it is a little new and therefore maybe a little scary. Let us contrast that with a mother and child who wade into the water and are hit by a wave that is much larger than the mom anticipated. She is knocked off her feet. It is impossible for her to hold on to the child, and he goes tumbling. When a parent has been overwhelmed by a wave of emotion, without any readiness for it or ability to understand it, the child has the experience of being pummeled helplessly by the same wave of emotion but without the anchoring lifeline, the grounding presence, of the parent.

Steven is a ten-year-old boy. His adoptive parents brought him to Nurture House one week after bringing him into their home. He and his siblings lived for the first seven years of his life with a mother who was intellectually impaired and nonprotective and a father who was sexually abusive to all the children in the home, violent with the mom, and physically abusive to the children. Steven was eventually placed in foster care with a safe family who was willing to take all the siblings. While this home was safe, loving, and fun, it would have been impossible for the parents to meet the needs of all the children in the home for the long term. During this foster placement, Steven continued to have intermittent visits with his mom, and sometimes during those visits he would have unsupervised phone conversations with his dad. Understandably, his felt safety, because of the constraints of the system, was never truly allowed to grow. Finally, Steven was adopted by extended family and began work with me. You will read more of his story in the chapter entitled "Holding the Hard Story: Narrative Nuance", but I will share here his work around exploring the emotional lives of his caregivers over time and how his schemas about whether grown-ups are safe and can be trusted began to shift through the reflective activity.

Steven enjoyed art. He especially enjoyed the dot paints and asked to use them every session. While we were working to enhance his emotional literacy, we normalized that all adults have all kinds of feelings. He responded with, "Not my mom. My first mom. She was never happy." I said, "You never saw her feeling happy? Can you show me what feelings you did see in her? I know you like the dot paints. How about you pair a feeling word with each dot color? I'll draw an outline of mom and you can add as many dots for each feeling inside of your first mom as you like." He chose the blue dot paint for sadness and began energetically pounding the paint marker all along the inside edges of his biological mom's body outline. It became clear, after a couple of minutes, that his intention was to color in the whole of his mother's body with sadness. I reflected this, saying, "You are filling her whole body with sadness!" He nodded and then began filling the space outside the figure with blue dots. "You have put blue dots all around her too!" Randy seemed please that I was tracking with his unspoken communications.

He asked to do one for his foster mom and a third for his adoptive mom, so I got to work drawing two more mom outlines. He then put one dot of blue on each of his other two mom figures. He picked up the red paint and completed a similar process for anger, putting much anger in and around his "bio mom" and only one dot on his other two caregivers. The client repeated this process with feeling/color pairings of anxiety, disappointment, and overwhelm. When he was done with all the more difficult emotions, he decided that the last one would be happiness (green), and he ended with a reverse process where he put one dot of happiness on the picture of his bio mom and covered the figures of his foster and adoptive mothers with it.

After he had completed the activity, he sat back, and we sat with our heads together looking silently at all three figures.

Steven spontaneously said, "Hers is colorful [*pointing to the first mom drawing*]. It looks fun, but it's not."

ME: Those are a lot of big feelings . . . pretty confusing for a little boy.

STEVEN: Yeah, I didn't even know 'em.

ME: Well, babies don't have words for their feelings, they just feel.

STEVEN: I didn't have words then, but I have them now!

> Steven was able to bring his current developmental achievements to bear while reflecting on his early experience, and it seemed to bring a sense of competence. He also remarked on how similar the other two mommy pictures looked.

STEVEN: These two look the same [*pointing to the pictures of his foster mom and adoptive mom*].

ME: These two moms do look almost the same.

STEVEN: Maybe most people are like this.

I reflected his maybe and understood that Steven was struggling with his early mental schemas and his more recent corrective emotional experiences. He was trying to wrap his head around his early experience of a disorganized emotional life, his resulting belief that people cannot be trusted to take care of him, his more recent experience with more organized adult caregivers, and what the patterns of care offered by multiple safe adults might mean for how he relates to the world. As he continues to work this out, he will move more and more to felt safety in his current environment. Part of the power of play and expressive arts is that the symbolic representations of content—in this case quantifications of the internal emotional experiences of those in authority over him over time—can be explored for patterns. Then those externalized patterns can be reflected on, tested out, and potentially integrated into new mental schemas that include the potential that the disorganized caregiver was the exception and that most adults will be helpful.

## Helping Children Grow Reflective Capacity Around Emotion

Many of the children I see, especially those who have been adopted from hard places or have had emotionally disorganized parents, have very little ability to identify difficult emotions. My first goal for these clients is to help them increase their window of tolerance for sitting with an emotion. Sitting with an emotion is another capacity that can be expanded through titrated doses. Play and expressive therapies offer mediums that mitigate the approach to difficult emotions and aid in the titration. Almost all creative art materials (from sand and clay to paints and pipe cleaners) can help provide distance while a child works to quantify, reflect, describe, etc., the experience of a particular emotion. Many clients will have explosive behaviors or intense crying jags but then are unable to tell us what they were feeling. Sometimes the resulting shame of having wreaked havoc by hitting, kicking, throwing things, etc., or by making enraged accusations or threats to their parents is so intense emotionally that it blocks their ability to reflect on the emotions involved at all.

Distancing techniques that allow them to manipulate quantities and qualities of emotion can be particularly helpful when shame is involved. James was an 11-year-old boy who was adopted at four after having been in the home of drug addicted parents for a couple of years. He was then removed and placed with older caregivers who had trouble keeping up with a high-energy toddler. He was eventually adopted into a new family with a couple of older biological brothers. When I first met James, he was hiding under a blanket in the car. Having just been released from a psychiatric hospital, he was carrying deep shame and a core belief that there was something desperately wrong with him. The first goals of therapy were simply helping him feel connected, safe, and like a kid who could be enjoyed again. We then worked through enhancing attachments with his caregivers while helping them make the necessary shifts to become bigger, stronger, wiser, and kind for him more of the time. When he felt safe, liked, and connected again (a process that took six months), I introduced some emotional literacy games. Critical tasks in emotional literacy work are setting the bar individually for children from hard places and sensitively remaining within their windows of tolerance for acknowledging and reflecting on big feelings. We first just played a matching game with feeling face cards. Like a classic game of Memory, the cards were all face down, and mom, client, and I took turns turning over two cards at a time until we found a match. There was no attempt to tie actual content from his experience to feeling words; this first "dose of exposure" to difficult emotions was simply finding two faces that matched. He did not even have to say the name of the emotion the first time around, nor did the adults name the feelings the first time around. James has high visual-spatial intelligence, and part of my thought process was to build on his innate strength, giving him competency experiences as he found matches. Competence is experienced neurobiologically as a surge of dopamine (the joy chemical) in the brain, and these competency surges mitigated his exposure to emotions such as *disappointed, frightened,* and *embarrassed*. He was instantly rewarded with being able to remember where he had seen a particular feeling face before. James exhibited a lot of positive affect each time he got a match, and as neither his mom nor I have the same visual-spatial intelligence he has, we were easy to beat. He became more playful along the way, engaging in some light teasing of his mom and me—a form of social engagement I had not seen before.

James asked for this game again the next time we met, but he had ideas about how to make the game more challenging, and he won again. As the novelty of this set of cards wore off, I introduced a second set of feeling cards. These Todd Parr cards display an emotion like "left out" on one side and "connected" on the other. James initially wanted all the cards to have their positive side up but was able to tolerate exposure to several more difficult emotions after we compromised on how many face cards of each kind we would use. To help mitigate his approach to the emotions, I offered him and his mom colored cardstock and a revolving set of scissors, each with a different cutting pattern. As we co-created the activity, we named it Zig-Zag feelings, with a nod toward the strangely shaped scissors and another nod toward the ways in which our feelings can, and often do, move from one extreme to another as we are learning to regulate them. Each of us chose a feeling card and got to work creating a cardstock symbol to represent it. James's mom chose the Todd Parr feeling card *carefree* and created a yellow sun shape with slightly jagged edges to symbolize it. James chose *bored* first and cut out another white cloud to match the rain clouds on the card. Clients are often helped by having an experience of success that may be modeled off an existing representation, like the cloud on the card. Even using emotion cards informs the titration: is the card just a face? is it a specific scenario that might engender that feeling? are you offering only "core emotion" cards or more complicated feeling words, and if so, why? At Nurture House, we keep several sets of feeling cards, some with very simply drawn faces, others with scenarios that might engender that emotion, and some beautiful mixed emotion cards that are artistically drawn and highly evocative. The titration of exposure to feelings vocabulary—naming feelings, recognizing them in others or in ourselves, communicating them to others, and managing them internally—requires titration of the dose of exposure. James appeared to feel a sense of competence when both his mom and I accepted and understood his symbol, and this surge of pleasure, albeit small, scaffolded him to work with a more difficult feeling during the next round. He chose *bored* first, and after experiencing success chose *left out*. The jagged layers of black diamond shapes that he created to expressed left-outness are pictured below (see Figure 6.9). He took great care to create a much smaller and almost identical jagged

*Figure 6.9* Left Out

diamond shape that sat in the center of the larger shape. Very few words were shared as mom, client, and I all absorbed the dark, sinking feeling associated with James's symbol.

## Growing Reflective Capacity in the System

Sydney, adopted domestically at age four after having been in a couple of placements, was referred at age ten when on the brink of being sent to residential treatment. When I first started seeing the family, he was having rages that required physical containment by his parents, runaway behavior, destruction of property, and an inability to tolerate delayed gratification to meet the needs of another family member. He had very little understanding of his life story, a severely constricted ability to reflect on his emotional life, and a lack of trust in his safe bosses. Sydney and both of his parents worked very hard in therapy to make sense of how his early life affected his current behaviors. His parents did work on their own sets of "shoulds"; practiced kind, compassionate responding; met him in compromises; and filled his tank. He also dug in and began finding small ways to help around the house, began to ask for permission or supervision more frequently, and became more regulated.

We began with double sessions twice a week until we were out of crisis mode and then moved to a double session once a week. As the family continued to connect and regulate more and more independently, we titrated sessions down to a single session once a week. Recently, however, the parents described several troubling moments in which the client escalated and the situation was then unable to be de-escalated. This ramp-up in dysregulation happened at the same time every year, and as we began making sense of his history, it became clear that this time of year represented the traumaversary of his abrupt removal from his birth home by police officers. Months into treatment, Sydney and I had developed enough trust that we had been going on walks together. During his previous visit, we had agreed it would be cool to try to catch a fish in the Harpeth River, so between sessions I had gone to the dollar store and gotten a fishing net on a long, skinny pole. At the beginning of the next session, I met Sydney excitedly in the lobby to show him the new butterfly net I had just bought. We agreed we would spend half the session with his mom and then use the net at the river's edge. Mom and Sydney engaged in a bucket filling ritual they had developed for nurturing one another, and then mom asked to talk about a hard thing that had happened at school. Sydney's first response (as it is for many of our children who come from hard places) was to lie. However, mom had physical proof with her and showed it to him. I could palpably perceive his upset as his amygdala became armed. The flight, fight, or freeze response that gets triggered when a perceived threat is near has gained the healthy respect of his parents, and I watched him engage in all three sets of behaviors in succession. First, he froze, hardly moving, barely breathing, staring at the table. Then he ran out of the room, demanding to leave Nurture House and storming out of the building. I serve some families in which if a child left the room, I would need to immediately go after them, but in this case I hoped the secure base we had established would bring him back, and, indeed, he did return to the room quickly. He could not seem to decide what to do with his body. He paced back and forth, opened and closed the door, and repeated himself almost robotically, saying, "Take me home. Take me home." At one point, he screamed over and over again for me to stop looking at him, although I did immediately shift my eye gaze away from him. It was not helpful for either his mom or I to speak, as he was fully in his reptilian brain at this point, and his mom and I breathed through it. At one point, he went and picked up the new fishing net and broke it over his leg. He eventually calmed enough to leave the building, but because of other clients waiting for me, I was not able to help him come all the way back to baseline and engage in any redos or reflection. His mom explained that while many aspects of family life had improved and Sydney had made significant gains in treatment, it was still "impossible" to talk about hard things outside of Nurture House, and accepting responsibility for any of his negative actions was still extremely difficult.

The session had been hard for everyone and had ended in rupture with no repair. This is the reality of our work at times. The next session after an unrepaired rupture is critical and requires the therapist to answer several questions, including, are we at the point in therapy where he can face harder things? Clinicians ask themselves this question, which is one of titration, all the time. And the answer is always case-dependent and largely colored by an additional set of questions: "how strong is the therapeutic relationship?", "is there enough trust and ability to co-regulate for the dyad to move into more treacherous territory?", "what is going on in the child's life outside of the therapy space?", "can the clinician provide safety even if a client escalates?", and "could going through the storm together and coming out the other side together strengthen both the therapeutic relationship and the client's sense of being able to face hard things?" In this case, I thought it was worth a try. After much thought and some peer supervision, I decided to trust and hope that he could meet me in the harder place. When therapy began, it was standard for both parents to accompany him in order to provide enough safe boss coverage. At the beginning of therapy, both parents had always accompanied Sydney, but eventually enough trust was rebuilt between Sydney and his mom and dad that just one of them could bring him without fear of an out-of-control episode happening in the car on the way to or from treatment. Mom and I agreed to move back to double sessions for a short period of time to allow enough time for deeper processing, especially in the event that he had another escalation in session—I wanted to be able to work it through to repair.

When I went into the lobby to meet them the next week, Sydney was sitting calmly, but slightly nervously, on one bench in the lobby, opposite his parents who were sitting together. Another form of titration has to do with how we use our proxemics with children who become easily escalated. I smiled at him and said hello but immediately shifted my gaze to his parents and engaged them (disarming his fear response by not focusing strongly on him at first) while I sidled up to him. This gave him a chance to understand that I was the same person I had always been and was not actively angry with him over breaking my toy. Still looking at his mom and dad and laughing with them about something, I gestured for him to scoot over, which he did, and I sat beside him silently letting him know I would continue to stick with him. After another moment or two, I completed one round of communication with him, tousled his hair (a routine we established a long time ago), and offered him the art room or the sand room. He chose the sandtray room, which signaled that he was in need of the regulation and anchoring the sand provides him. I explained that since our last session was hard, we would be sure to have some fun time today even if we also did some hard work. We went into the sandtray room, counting on the sand to be both regulating and to provide a visual focal point for joint attention that would titrate the family's approach to hard content by decreasing the need for extended eye contact and offering symbols to buffer the work.

Sydney immediately turned his back on us and dug his hands into the kinetic sand. I addressed all three and said, "Last week really helped me understand some things." While I spoke, I picked up a midsized orange metal bucket and put it in the middle of the center sandtray. "I have been seeing so much growth in Sydney, I hadn't been thinking about how hard it is for Sydney to make repairs when he thinks he has injured a relationship. I think it is super scary for him to admit, 'Yeah, I messed that one up.' But it's a real problem because part of what has to grow in us in order to be in healthy relationships is an ability to admit when we made a mistake so people can love us through it. Last week, when Sydney felt caught in a lie, he might have said, 'Yeah, I lied, I'm sorry,' but I think that Sydney still has this deep, deep question about his own goodness or badness. I see kids who believe—really believe—that admitting a mistake is like saying they are bad kids. So right now, when he makes a mistake, I see him like this." (I put a small boy inside the bucket with his head far below the top edge.) "It's like he's in a deep pit with clay-packed walls. It feels impossible to get out of." As I continued to put sand up against the sides of the bucket, Sydney's dad began to talk about sand traps on golf courses (a sport he enjoys with Sydney) and how hard they can be to get out of. Up to this point, Sydney had kept his back turned to us and his hands in the kinetic sand, but once the little boy was in the bucket, he turned around and moved to run his hands through the softer sand, facing us but keeping his hands occupied. I also explained, "Right now, the pattern is that Sydney gets an uncomfortable feeling, and that uncomfortable feeling is so hard to face, he shifts the blame to someone else. It becomes someone else's fault for *making him* feel this feeling, and he gets very angry at whoever caused it. Only, I think he feels even yuckier after he has blown up." All this was said to the parents, with Sydney getting to listen without buy-in or comment.

Then I addressed Sydney directly and said, "I don't believe that you are a kid who really wants to hurt people or destroy their stuff, but if you want to stop, you are going to have to learn to handle uncomfortable feelings." Referencing the parents again, I said, "His window of tolerance for handling uncomfortable feelings and remaining regulated is pretty small right now, but it can be grown with practice, just like weight lifting. While we can't lift much at first, we keep practicing and get stronger over time."

I then introduced Sydney to the mixed emotion cards, a whole stack of beautifully illustrated and evocative cards that carry one feeling word per card, ranging from core emotions to complex feeling words. I said, "Pick three cards that describe the feelings you felt when mom confronted you on your mistake last week and you ended up breaking the fishing net that I had bought for our walk. You don't have to say the feelings out loud, but choose three cards and we will go from there."

Mom and dad and I engaged in some small talk while he chose his cards. I, of course, was simply praying he would continue participating, that the relationships in the room would be both supportive enough for him and meaningful enough to him at this point in therapy to mitigate the hard work. I am not sure what we should call this—it may be a clinical skill—but I believe it has some sort of overarching spiritual component to it, the communication with our whole hearts and minds as therapist to the family that they can do this, that we have them, that we can hold the space for them. It is more than assigning positive intentionality, and there is no pen-and-paper scale that will quantify this quality of interaction, but I feel it over and over again with children from hard places and with the parents who hope so much for them. He chose the three cards without speaking and handed them to me. I set them each upright in the sandtray and then read them aloud: Trapped. Depressed. Hopeless.

After reading the cards out loud, everyone was quiet for a moment, just looking at them in the sand. All of us were struck by the strength and specificity of emotion and level of introspection that Sydney had demonstrated. It felt to me like there was respect by everyone in the room for the risk Sydney took to identify and share those big feelings. Then I said to everyone, "How about each of you choose a symbol or set of symbols that represents each feeling for you and place your symbols near the feeling card?" Thus began some very focused time in which each family member moved inward and perused the shelves. The only negotiations happened around how to place their own experiences of "trapped" close enough to that card while allowing the other two to also have their experiences of feeling trapped nearby. The nonverbal navigation of space paralleled what I believe is the continual navigation of emotional space within the family. As Sydney's mom and dad opened themselves up to visiting their own prior feelings of being depressed, hopeless, or trapped, they connected with their own experiences and naturally became more compassionate toward Sydney's experience.

Sydney picked up a giant snake and coiled it around the *trapped* card (see Figure 6.10). Later, he would tell us, "You know, when a snake is completely wrapped around you and squeezing you, you can't get out."

Sydney's dad chose a castle and said that since he works inside all day, he likes to be outside on his time off and so can feel "trapped" in the house. After reflecting on the snake for a bit, Sydney went and chose a small house with an open front, and after hearing others talk about their symbols, he added a piece of fencing in front. His mom chose a tiny house made from a pumpkin. It was

*Figure 6.10* Sydney's Snake Coiling

interesting that both parents associated the feeling of being trapped with their home. Sydney had put a fence around the *hopeless* card, and he put a skeleton lying sideways within the fence. His dad chose an alligator and a figure that was disabled and talked about the feelings of hopelessness that might come from that. In the area of the tray designated for *depressed*, Sydney had tossed in the John Smith heroic character from *Pocahontas*. I am always trying to be attentive to how clients choose their figures and how they place them in the sandtray. In this case, Sydney looked at me and said, "I'm putting him face down on purpose." Later, he was able to talk about how the guy could not see his way out.

Sydney's mom chose a tiny brass alligator that I had picked up from a street merchant in Nepal as her expression of the feeling card *depressed*. When I asked her about the alligator, his mom said, "Well, it is very small and it is very hard." Sydney also built a section of sand separate from any of the three cards that included four stone walls and a little golfing-type figure of a boy. He later explained that it was a very deep sand trap, validating his dad's interpretation of him as stuck in the sand. All family members had been able to vulnerably choose figures. The symbols added much more richness of expression for each family member and deeper insight by other family members. Taking the temperature in the room, it seemed that mom, dad, and Sydney were more connected at this point than when the session started, and the system seemed to be regulating through the symbolic expression.

## Reopening the Compassion Well

I offered an extension. My prompt was, "Let yourself be drawn to one or more of the symbols in the sandtray." To the parents, I said, "Focus your attention on it and bring your own kindness, compassion, or desire to help to the symbol." To Sydney, I said, "Look in the tray until you find a figure that needs help and figure out a way to bring some help." I was already experiencing the work as very rich, but watching all three of them find a way to hold another family member's distressing feeling symbolically took the exercise to another level. Below is a picture of Sydney's symbol of hopelessness—the skeleton fallen over (see Figure 6.11). After studying the hopeless figure for a while, Sydney's mom chose a fairy and placed her in such a way that she was outside the fence (potentially outside the circle of hopelessness) but holding the skeleton's head (see Figure 6.12).

She said, "I think to be dead is to feel like you don't exist, and I think if the skeleton is held, if the fairy can see him, he will know that he is seen and he is not alone." Wow. Mom's compassion well was reopened as she placed the fairy, and Sydney's hope for possible connection was reinforced by the visual creation of a new connection where there had been only hopelessness before.

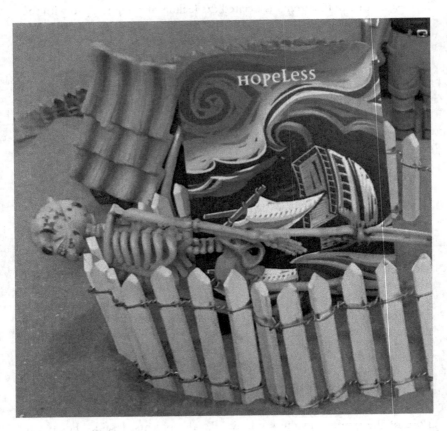

*Figure 6.11* Hopeless

## Conclusion

I called this chapter Playing the Affective Accordion because in my mind's eye I am always looking for the child's growing edge for all aspects of emotional literacy: naming big feelings, tolerating big feelings in a regulated way, talking about big feelings, expressing big feelings, and helping family members communicate them to one another. An accordion allows for great expansion and can communicate more as it is expanded. It has been my experience that children and families greatly expand their window of tolerance for being able to hold their individual big feelings, to express them to one another, and to become larger containers for received feelings when play and expressive therapies become the mediums through which the emotions are handled. I am excited about the possibility of new emotional understandings and new levels of emotional resonance that can be achieved within families if the titration of emotional content is delivered through play and expressive therapies.

*Figure 6.12* Hopelessness Held

## References

Goodyear-Brown, P. (2002). *Digging for buried treasure: 52 prop-based play therapy interventions for treating the problems of childhood.* Nashville, TN: P. Goodyear-Brown.

Goodyear-Brown, P. (2010). *Play therapy with traumatized children.* Hoboken, NJ: John Wiley & Sons.

Siegel, D. J., & Bryson, T. P. (2011). *The whole-brain child: 12 revolutionary strategies to nurture your child's developing mind.* New York, NY: Bantam Books.

# 7  The Nature of Play

The phrase "nature-deficit disorder" was coined by Richard Louv. He wrote a book called *Last Child in the Woods* (2008) that trumpeted concerns around the decrease in children's time spent in nature and the correlation of this statistic to an increase in behavioral problems. One of my life maxims is *those who risk nothing gain nothing*. When children play outdoors, they are training their bodies and brains to take developmentally appropriate risks. Anticipation and prediction skills are honed through various forms of play. Perspective taking, turn taking, negotiation and cooperation with others, integration of sensory systems within their own bodies, accomplished in part by running full speed through an open field or riding a bike along a neighborhood sidewalk—each of these is a natural form of bilateral hemispheric integration that used to happen as a matter of course as children played outdoors. Outdoor play also offers many experiences of competency, and effective navigation of the natural environment leads to increased self-esteem. So why are we not playing outside?

With the age of electronic media, many children are spending upwards of seven hours a day on screens, forgoing the realities—the sights, smells, sounds—of our natural environment for a virtual reality devoid of these sensory experiences. We live in a culture of fear, and as parents hear horror stories about children being taken or hurt, they limit their children's outdoor freedoms more and more. Gone are the days of children grabbing their bikes and going to play with the expectation that they will return by dinnertime. Children who do play outside often do so in the milieu of organized sports. These are good for children—offering proscribed limits and boundaries early in their development—but do not allow for free-form play. Children are amazing, and when left to their own devices in the natural world will create forts, make mud pies, and fight mighty wars, all with dirt, sticks, and trees. The benefits of imaginative and pretend play in the outdoors should not be minimized.

The Disney/Pixar movie *Wall-E* came out several years ago. Most people thought it was a cute, uplifting movie about a computer's quest to find true love. I found it a depressing commentary on where our culture may be headed. In the movie, the remnants of earth's population have relocated to a space vessel in the hopes of one day colonizing another planet. But they have forgotten how to be human, how to interact with the real world. All the citizens float around the airship on levitating chairs, with their attention captured by a screen in front of them. It feels like this is not too far from what we could become if we do not have a radical shift back to valuing the natural world.

This alarming devaluing of outside play in our culture is not limited to families. Whole school systems are choosing to eliminate recess in earlier grades as a response to pressures to achieve more academically. Recess is particularly important for the children we see at Nurture House, who are often having to push through all sorts of somatic and emotional regulation issues to get to their executive functioning skills and be able to perform academically. Especially for children, the time they spend in their bodies during recess provides a form of stress management. We know that, at the least, exposure to sunlight provides regulating vitamin D, and the interactions with peers on the playground allow for many forms of valuable learning that cannot be measured academically.

Nature has been associated with therapeutic healing of the mind, body, and spirit from early Buddhist traditions to Roman baths to medieval monastery gardens to cure cottages. People had an inherent leaning toward the valuing and integration of the natural setting into an overall sense of health, well-being, and a healthy respect for the very real nourishment and equally real destruction that can be wrought on humanity through nature. As our world became more industrialized and people lived in closer proximity to one another in big cities, often with very little thought giving to sanitation needs, the positive and negative effects of the environment became more loudly articulated by such thinkers as Olmsted, who ascribed curative value to pleasurable nature scenes (Pollock-Ellwand, 2010). It is believed that the most powerful separation of nature from healing began with the emergence of germ theory between 1850 and 1920. Germ theory, an understanding that diseases are caused by microorganisms that operate within and multiply or are killed by the introduction of other materials—such as vaccines—into the physical body, minimized the need for a focus on the external environment as a source of healing or disease.

The benefits of nature on our psychological well-being are well documented. The restorative influence of nature on our physiology, and consequently our emotional state, also helps us sustain attention (Ulrich et al., 1991). Exposure to nature can provide a restorative break from direct attention tasks, allowing us to extend attention and perform better when we return to the task (Tennesson & Cimprich, 1995). This appreciation of nature as providing a restorative environment (Hartig, Evans, Jamner, Davies, & Garling, 2003; Herzog, Chen, & Primeau, 2002; Kaplan, 1995) supports the use of nature as a tool of titration, mitigating a child's approach to trauma content. A wise therapist is attuned to when a child is at the edge of his window of tolerance for processing trauma content and is able to offer a shift to the outdoors, offering nature as a co-regulator.

A study conducted through the National Trust in the United Kingdom found that children spend half the time outdoors that their parents' generation did. Shockingly, 10% of study respondents had not been in a natural environment, such as a park, beach, or forest, for over a year. As children and parents play outside together, children learn to take risks, they strengthen their bodies, they are delighted in by their parents, and both parties are refreshed. It is a time of day when the family can be unplugged from devices and open and close circles of communication with one another. An act as simple as pushing a child on a swing set provides rhythmic repetitions of contact as parent and child are physically connected and then the child moves away again. When a child says, "Push me, Mommy, higher, higher!" the touch of the parent is an important part of helping the child to literally reach higher and higher heights. When I have dyadic sessions outside, I will often position the parent in front of the swing. As the child returns, I may extend the moment of connection by having the parent hold the child's feet, lift them higher and higher while smiling into their eyes, extending the suspense, and then let go. The child usually squeals in delight while saying, "Again, again!" If a child needs to feel fully in control of the swinging, we learn that too. Sometimes we sing rhythmic songs or turn taking songs in which the parent sings and the child responds in song, practicing opening and closing circles of communication while the child is somatically regulating in the swing. All this good stuff from a little outside time between parent and child, and yet it is getting harder and harder for parents to make time for this in their daily lives. A recent US study of almost 9000 children found that 50% of preschoolers went without even one parent-supervised outdoor playtime (Tandon, Zhou, & Christakis, 2012).

The profound and beautiful balance through which nature and humanity co-exist is established by the exchange of breath: the trees breathe out oxygen and breathe in carbon dioxide while we breathe out carbon dioxide and breathe in oxygen. If the breathing in and breathing out—both of nature and humanity—are maintained in an attuned fashion, what is achieved? Regulation. As I am writing these words, I am sitting on the front porch, hearing the loud evening sounds of cicadas, watching the pink tinge of sunlight as it slowly fades on the horizon, and feeling that all is well. This feeling is an overflow of my somatic interaction with the natural world around me.

## The Nurture House Nature Play Area

We have designed our backyard as several concentric circles arranged in three separate tiers. We have only a few carefully selected pieces of play equipment in the backyard. The list is as follows:

1.  A tire swing that hangs from a giant oak tree.
2.  A fairy hut that hangs from a tree and can be enclosed on every side.
3.  A JumpaRoo that consists of an inflated tube encircling metal poles that children can hold on to while they jump.
4.  A slackline, covering a distance of 25 feet, that attaches at either end to trees.

5. A spinnable swing seat large enough for two people.
6. A seesaw of some sort. Over time, we have had a traditional seesaw and a seesaw with blowup rubber balls as seats so the child's landing on either side is buffered and bouncy. Most recently, we have added a rocking hammock. This is great for core grounding and balance work.
7. An ENO (a nylon hammock that can be hung between two trees).
8. Bird feeders, as feeding the birds can be a jumping off point for what kind of caretaking each creature needs to live.

## Using Outdoor Equipment to Build Attachment Bonds

Most of the traumatized children we see have trouble trusting their caregivers in the beginning, especially in foster and adoptive situations. Building trust requires vulnerability and need meeting. Foster children who come into adoptive families after the age of three or four can do many of their basic need meeting tasks themselves. They can feed themselves, take care of their own toileting needs, and dress themselves. While these may seem like good things, the independence can actually get in the way of a child recapitulating an attachment relationship with his new parents. It seems counterintuitive to some, but our ability to be autonomous grown-ups begins in being wholly dependent on the other.

It can be difficult to fashion experiences that entice a child into risking trusting a new parent—a safe boss—as these kids can do so much for themselves. At Nurture House, therapists create scenarios with specific outdoor play equipment that invite a powerful payoff for the child after they allow a safe boss to help. The slackline is my favorite activity for this work. Some slacklines come with a second line that is strung above the child's head so they can balance by hanging on to the upper line. At Nurture House, we offer only the bottom line in an effort to create an environment that requires teaming, that requires the child to allow external support in order to be successful. No matter how balanced you are or how athletic, it is nearly impossible to walk the whole course of the slackline without any help from people on the ground. I have parents who tell me at intake that they are saddened by their child's inability to trust them, to allow them to help. Soon after beginning therapy, we explore the backyard together. Sometimes the titration of trust means that the child and I make this exploration alone first, as it is easier for the child to allow me to help than the mother or father who represents the greatest danger of vulnerability based on their previous trust injuries. The desire to conquer the slackline mitigates the discomfort that comes from relying on help. I usually extend my hand and say they can hold on to it to mount the slackline. Some do and some do not. They quickly realize that in order to get balanced, they need the hand. They often let go immediately, wobble, and step back off to the ground. Eventually, many children allow me to support them all the way across the 25 feet. Once the client is able to trust on this level, I invite the parent to join the game and eventually move myself out of the support role, and the parent takes over, providing the anchoring support for the child's competency experience. It is a real win for us all when the child begins to fall, reaches out and grabs his adoptive mom's hand, and I get to say, "She's got you! She was right there to help when you needed her."

One does not have to have a slackline installed to have the benefits of balance and trust work. A sidewalk edge may suffice. A stream with stepping stones can provide a rich environment for this work, as can several stones placed in a pattern on any form of ground cover. My family was recently on a hike in the woods. We enjoy crossing streams together, and having happened upon one, we spent time figuring out where to place stones in order to be able to cross it. Each of us crossed independently, but getting back was trickier business. A couple of the rocks had shifted, and our youngest, Nicholas (age 8), who had previously been wanting to cross independently, said, "I think I need some support for this part." Perhaps because of my heightened focus on the intersection of interpersonal trust and the natural environment, this phrase, which I am sure Nic has said before, resonated differently this time (see Figure 7.1). One of the goals of trauma work with

*Figure 7.1* Trust and Risk in Action

children, as I see it, is to help children grow a healthy balance between independence and interdependence, to be able to enjoy the accomplishments they create independently while simultaneously being able to ask for help when they need it without the expressed need creating shame or doubt. The stream, the slackline, the tire swing (that is strung slightly higher than a child can get into on his or her own)—each represents, embedded in the physical environment, Vygotsky's zone of proximal development and allows for children to do more with the support of a helpful adult than they can accomplish if they remain solidly entrenched in a posture of control that does not allow for help.

A few years ago, we took the whole family on a mission trip to Mexico. Most days were pretty busy, but one beautiful afternoon, we were granted free time to venture out into the city. After taking the kids to the lively Mercado and getting fresh cut mangos with lime (one of my forever favorite foods), we wandered into a local park. What struck me immediately was the difference in size between the playset on this playground and the playsets in America. It was almost twice as tall and made almost completely from painted metal. I climbed with the children all the way to the top (you had to climb upward through a series of tires stacked to create a vertical tube, with just enough space between the tires to afford you a foothold as you squeezed your body through each tube and up to the next level). We were rewarded at the top as we stepped onto a platform, looked over the edge, and realized we were somewhere between 12 and 15 feet off the ground (see Figure 7.2). It was exhilarating and scary, mainly because this seemed twice as high as the American playsets we were used to. I would imagine that American playgrounds have all sorts of safety rules for maximum height. Perhaps this is due to some very careful science that supports a certain height as less likely to cause injury (if you fall from it). It is interesting to me that this playset was in such a public center of a large town in Mexico. Clearly, lots of children came to play on it, and lots of parents, who have the same basic need to protect their child from imminent danger, let them play on it. It did make me curious about how our interactions with the outdoor play equipment offered in different cultures may influence our neurological presets for danger, setting the initial bar for acceptable risk individually based on our early experiences of outside play.

On this same playground, I followed the children through a tunnel made of metal spirals to a platform of monkey bars. I work out regularly doing kickboxing and boot camp, but I still find it very difficult to cross a set of monkey bars. The upper body strength required appears to be too much for me. My son, however, decided to try. The set of three images below shows you his initial swing out onto the path, his concentration in moving from bar to bar with his legs swinging wildly below him, and his eventual arrival at the bar on the other side. The third picture was taken at the instant he reached the final bar, before his neocortex had even realized he was safe. One can still see both the

*Figure 7.2* They Don't Make 'Em Like They Used To!!!

intensity of focus and the vulnerability of risk that were at play as he crossed the monkey bars (see Figures 7.3, 7.4, and 7.5). What you cannot see is the moment after, when his thinking brain caught up and he grinned from ear to ear and said, "I did it!"

*Figure 7.3* Risk

*Figure 7.4* Still Risking

*Figure 7.5* Safely Across

## Desensitization and Sensory Integration in Nature Work

There are many elements of the natural world that can be challenging and/or rewarding for the traumatized child. Some children may be more sensitive to changes in temperature, to the sound or feel of a breeze, or to changes from direct sunlight to a shady area. Whether the exposure involves touching a squiggly, slimy earthworm or packing mud into "clay bombs," play in nature requires children to take risks to experience and enjoy the natural environment. Mud provides an intense sensory experience and is one form of messy play that children with perfectionism and anxiety may shy away from. I was on a walk recently with a child with significant sensitivities to heat and cold. We had been walking on the shadowed side of the street, and he had been regulating well, but as we turned the corner, we came into direct sunlight. It was a profound moment for me to watch that child stop, turn to face the sun, close his eyes, stretch out his arms and say, "There it is!" with such intensity of pleasure. He was drinking in the warmth of the sun and put me in mind of an iguana soaking in the warmth from a heating rock. I turned also and drank in the sun. He took a couple of steps backward—as if he were going to continue our walk backward—with his eyes closed and his face uplifted to the sun. We were on a street where cars could come around the corner, so I offered to be on the lookout and walked very closely beside him for an entire block as he trusted me to protect him and warn him of any impending danger or any need to open his eyes while he absorbed the warmth of the sun. By taking this walk together, I was able to share in this client's experience of being warmed by the sun. Recent sensory ethnographic studies have attempted to examine further the experience of experiencing nature with others by *making sense of our senses* together (Allen-Collinson & Leledaki, 2015). Through therapy, this child gained regulation skills and was able to enjoy the warmth of the sun with me as I helped to keep him safe by being his lookout.

## Creating Art From Natural Elements

This summer, we had a bush around the corner from Nurture House that grew berries. Everyone agreed it would not be wise to eat the berries, but we had tons of fun picking them, smashing them up into "paint" on the rocks, and then painting pictures on rocks, paper, and even leaves. We have another plant nearby that grows "elephant ear" leaves that are as big as a sheet of paper and are smooth on one side and velvety on the other. We have a tree that grows long string bean–looking growths. Children sometimes use these for outdoor sword fights, or they can turn them into paint brushes or writing utensils. Sometimes we create family trees by first having the whole family go into the yard, choose the leaf that fits them best from the variety of trees there, and bring one back to the kitchen. I draw or paint the tree trunk and the branches on butcher paper. I offer brown finger paint and have each family member coat each other's thumbs and fill the tree trunk with bark, which is made from pressing their individual thumbprints into the trunk area, creating a tree completely unique to that family. Each family member decides where their leaf fits on the tree, slides it under the paper, and makes an impression with whatever color of crayon or chalk they choose, and in this way we create their family tree.

## Imaginative or Pretend Play in Nature

One of the great joys of my childhood was making mud pies. I still delight in my youngest when he brings me "chocolate balls" made from mud. His hands and face are covered with dirt. He has experienced his environment and has made a glorious mess! The symbolic nature of young children's play is a critically important part of their development. Nature lends itself to pretend play in ways that facilitate social and cognitive development. Other children pretend there is a pit of fire or a vat of toxic pig snot in the yard and we have to figure out ways around it.

Sometimes the great outdoors become a kitchen. A six-year-old boy finds a long, smooth stone in the yard and says, "This is the griddle." "Oh, it's a griddle," I say. Then he instructs me to go and choose flat, smooth stones to be the pancakes. After a moment, he joins me and explains that we also need to find things to be the syrup, the banana slices, the chocolate chips, etc. He and I enjoy a feast of warm pancakes with delicious toppings—all in nature (see Figure 7.6).

## Metaphors in Nature

Metaphors abound in the natural world. Talking about the things a plant needs to grow—sunlight, water, and soil—often leads directly into a deepening awareness of what we, as humans, need to grow. People sometimes choose to plant a tree in memory of a loved one who has died. The ideas that new life can come from death, that healing can come from hurt, that positive change can come from (and in fact may require) deconstruction first are all rich metaphors to be observed in nature.

## Using Plants to Talk About How to Care for Living Things . . .

Ryan is a client whom I have had the great honor to work with over the course of several years. I met him when he was just on the cusp of turning four. He was one of the children who participated in Camp Nurture. He had an autoimmune disease that had kept him inside for almost the entirety of his life. To my chagrin, I at first thought his parents were exaggerating, and while we diffused essential oils to limit his exposure to germs in public areas, it was not until we went on our first nature walk that I understood. He kept stopping every 30 seconds to notice something else. At one point, he picked up a

*Figure 7.6* Rock Pancake With Chocolate Chips and Bananas: Delicious!

small twig, turned to me, and said, "This is the first time that I have touched bark." Several years ago, I offered a camp for adoptive families who were raising children from hard places who struggled with intense dysregulation and had difficulties connecting and therefore experienced difficulty in every area of family and community life.

Ryan is school-aged now. He is diagnosed with autism, has sensory defensive behaviors, and needs special accommodations to deal with a classroom environment. He has particularly pained responses to loud sounds and to the way certain textures feel on his skin. This child has such severe sensory defensiveness that he often has to wear noise-canceling headphones in loud places or places with lots of ambient noise. On the first day of Camp Nurture, I had to accompany Ryan to the bathroom. A colleague came with us (we have a two-adult rule if we need to help a child with a bathroom routine), and she and I were chatting while he went to the bathroom. When he was ready to flush, though, he hesitated and looked upset. I asked if he was worried about the loud noise the toilet would make when it flushed, and he nodded. I asked, "Could I put my hands over your ears to help you?" Ryan smiled and nodded but still hesitated. I turned to my colleague and said, "Miss Jodi, would you help us make sure Ryan's body feels safe with the flushing? Will you put your hands over my hands on his ears?" Miss Jodi was, of course, willing, and Ryan decided that two sets of hands should be enough buffer between himself and the flush. His look of pride when he flushed was great, and after we washed hands, the high fives all around were good and strong.

Ryan fixates on certain kinds of bugs at different times. In order to meet him in his greatest place of joy, we often spend part of our session outdoors looking for bugs. We had a magical moment a few months ago where he asked to find an earthworm. It was very hot outside that day, but wanting to give yes to his request, we went into the backyard and started turning over stones, one by one, until we found an earthworm wriggling around. This sealed the deal in terms of my trustworthiness in his eyes (no kudos to me, as it was a spontaneous happening, and we have had many overturned stones with no signs of earthworms since), but it also became a shared piece of history unique to our relationship. Ryan was the first child who had ever requested to go earthworm hunting. My surprise and slight bewilderment about how to hunt them, combined with our shared success in finding one, added another moment of delight and shared positive history to our relationship. The neural pathways that will tie me (as therapist) to safety and fun, even when the scary stuff is introduced, are being enhanced every time we have a moment like this. Moreover, though he is fascinated with earthworms, the sensory experience of holding one is challenging to his window of tolerance for sensory input. While he might want to just drop the worm, startled by its slimy wiggling in his hand, he knows it is fragile and pushes himself to hold it safely even with the uncomfortable sensory input.

Just recently, he asked to go find earthworms again. We tried, but this time there were none to be found. However, we were able to use the moment to become more mindful. I said, "It's disappointing when we can't find an earthworm. I wonder if there are other bugs we could notice if we really focused on what else there is to see around us right here and right now." Indeed, as soon as we let the perseveration on earthworms go and opened up to a different kind of focus on the natural world surrounding us, we discovered a giant spiderweb in the backyard with a giant spider (disturbingly big for me) still working on the web. We sat, at a distance, to watch its work. The spider seemed to know it was being watched and ascended to the top of the web (which was attached to a lamppost) to hide. We wondered about why it might feel like it needed to hide and how we might help it feel safe enough to come back down.

A couple of times, Ryan remarked, "It's OK, Mr. Spider, we won't hurt you." I wondered out loud, "Do you think Mr. Spider knows that we are safe?" Ryan watched the web a little longer and said, "I guess not. He's not coming out."

ME: *I* know we are not going to hurt him, and *you* know we are not going to hurt him, but his instinct is to hide if anything unusual comes along. Do you think we are unusual?

RYAN: Yes [*giggling*]!

ME: Wonder what he needs to feel safe?

Up to this point, Ryan had twice gently jiggled the web, pretending to be food caught in the web, hoping the spider would come down to investigate.

RYAN: He probably doesn't want us jiggling his web.

ME: Huh. You think the vibrations are too much?

RYAN: Yeah . . . maybe he needs the wind to stop too.

Here, Ryan sees the sensory stimuli that the spider might be perceiving through his own sensory worldview, giving us rich information about his own experiences and also showing an ability to engage in some perspective-taking, putting himself in the spider's place. We ended up agreeing that we cannot make something come out of hiding before it is ready. We can attempt to send it powerful messages that we will not hurt it, but until it experiences what it needs to feel safe, it will stay hidden. It was a powerful parallel to the journey that this adopted child has been making in coming out of hiding, in sharing himself more and more fully with his adoptive parents.

Sometimes nature just keeps on giving, and after we shifted our focus away from the web, Ryan bolted to the JumpaRoo (a really neat bouncy toy that has colored metal poles converging on a center point). Children hold on to the poles while bouncing on the fully inflated inner tube that wraps around the poles. He noticed some fire ants crawling on the tube, and while I was crouching to look at these, I noticed an interesting beetle-type bug on the yellow pole. It stood out because it is mainly black with an orange stripe around its center. Ryan looked up and carefully focused on it for a few seconds, then his gaze dropped down to the fire ants. Suddenly, he remarked, "There's a baby one!" I crouched down again, and, sure enough, there was a miniature version of the black and orange bug far below on the inflatable tube. He said, "They're mommy and baby!" I said, "Huh, the baby is kind of far away from the mommy. They must know how to get back together again." We then spent time wrestling through the client's separation anxiety symptoms using the mommy and baby beetles on the playset. We were discussing whether we should try to move the baby up onto the pole to be with its mom. After some discussion, Ryan said, "I think we should leave him there. His mommy knows where he is, and she'll come find him." Powerful stuff, and completely separate from anything I could offer organically inside the building. An interaction with nature, with the natural environment itself, provides powerful metaphors for work.

The natural world continues to be the most powerful connector for Ryan. He asks to go on nature walks or bug hunts regularly now, and during our latest walk, we began moving onto a nature path that is not open to regular car traffic, making it a particularly nice place to take children. We had just walked past the gate to this path when Ryan said, "Stop! A snake!"

I glanced to my right, and, indeed, there was a skinny green snake, bright on the black asphalt path. It was about 16 inches long, and I would have probably absorbed it as a long blade of the grasses that grow near the path had Ryan not exclaimed. He immediately leaned closer into me, and I put my hand on his shoulder.

ME: It is a snake. What strong eyes you have. Let's back up a little until we know which way he's going. [*We stepped back three steps.*]

RYAN: My mom would have passed out!

ME: You sound kind of concerned about whether or not she could stick together with you and a snake. Your mom is pretty good at sticking with you in everything.

RYAN: Yeah, she is. I have always wanted to see a snake in real life, and I never have . . . before now.

ME: I have never seen a snake crossing this path, and I have never seen a green snake on black pavement, so this is a first for both of us.

This moment of shared, spontaneous experience will lend itself to a sense of shared and unique history between us, deepen the therapeutic relationship and our shared mindsight, and potentially become part of our therapeutic narrative over time. We stood quietly for a full minute, watching the snake engage in a small but persistent back and forth wiggle with just its head and upper body. It moved back and forth so quickly it looked like an onlooker at a tennis match—back and forth, back and forth.

RYAN: He's moving his head so much!

ME: Yes, he is. Wonder what he's doing.

RYAN: He's checking.

ME: It does seem like he is checking left and right, asking a question with his body.

RYAN: Is it safe? Is it safe? Safe to cross the road [*pretending to be the snake's voice*]?

ME: Yep, it looks like he's saying, "Should I take the risk? Is it safe?"

We watched quietly for several more minutes as the snake continued craning his neck and upper body back and forth, back and forth. Ryan and I talked about how scary it might be for the bright green snake to feel so exposed—clearly visible—against the black asphalt. Eventually, he crossed the road, and once he had slithered into the grass on the other side, it was impossible to see him. It was such a sudden disappearance, we might have each disbelieved what we had seen had we not shared the experience together.

ME: Wow. He went way out of his comfort zone. I don't understand why a snake who is so green and small would risk being on the black pavement. All kinds of predators could see him.

RYAN: He must have really wanted to get over there [*after a pause*].

ME: I think so too. There must be something over there that made it worth the risk.

RYAN: Well, now he is hidden again, so it's OK.

ME: Yep, he gets to decide when he takes a risk—and when he rests.

RYAN: And when he camouflages himself so that no one notices him.

ME: That too.

RYAN: I bet he's trying to get back to his family.

ME: He wanted to be with them pretty badly.

We walked on together for a few minutes and then began identifying recent risks that he has taken: to talk to a friend at school, to accept a change in his schedule, etc. We also talked about whether he ever tries to blend in. He was able to identify times in his classroom when he does not understand what is being asked, but he sits quietly, blending in, instead of raising his hand and asking for clarification. Ryan was able to engage in this level of self-reflection in part because we were in motion (walking together), in part because it was anchored by a metaphor found in nature, and in part because he had received a pleasurable release of neurochemicals as we experienced the novelty of finding a snake in the road.

## Therapeutic Walks

A nature walk may be the most powerful way we have of helping our clients expand their capacity for mindfulness while increasing their overall sense of being at peace. Teenagers, particularly, often have a significantly decreased resistance to connection with their therapist if it is happening in the context of a walk. There are several aspects of walking that buffer and benefit relationship-building with teens. We know that the natural environment—and particularly exercise in the natural environment—increases our sense of well-being (Penedo & Dahn, 2005). The teen does not have to make deep and extended eye contact with the therapist while walking. Attunement can be explored as the therapist tries to match the teen's pace, and challenge can begin as the therapist or child chooses to change the pace. Many times I have been on a walk with a client who appears hardened—even cynical—in the clinical setting but is suddenly disarmed as they notice a beautiful flower, a cool butterfly, or a gorgeous sunset. Tiny bursts of joy (or, more scientifically, tiny releases of dopamine and/or oxytocin in the brain) are engendered by the novel experiences of natural beauty, and the teen can enjoy the moment in joint attention with the therapist. We have an open air market where plants and produce are sold. It is a ten-minute, nicely landscaped walk on a pretty brick sidewalk. Many of my clients request to go there seasonally. We believe strongly in the therapeutic value of nature and kinesthetic involvement in therapeutic work but also understand that stepping outside the building, even to swing in the front porch swing, limits our ability to guard or guarantee the client's confidentiality. Therefore, we have embedded specific permissions in our paperwork to address this issue. Parents decide whether they will allow their child to play in the front yard or backyard or go on a walk with their clinician in the neighborhood. Some parents give permission easily and right away, some parents express that they would like time to get to know the therapist before deciding, and others decide their situation is sensitive enough that they would just prefer to keep all therapeutic intervention in the building. We would recommend that any clinician who will be taking their clients outside the clinical building have these types of special consents in place. In all cases, the consents are just a starting place. The therapist is attuned to the needs of the child and respects the natural environment as both a mitigator and a co-therapist with children who are approaching hard content. Some children come to Nurture House in such a dysregulated state that moving outside the building could be unsafe for them or for us.

I love taking children on walks around Nurture House for so many reasons. The act of walking outdoors together seems simple, and that simplicity is part of what provides the child a sense of comfort. However, there are several skill sets required for both the therapist and the child and several clinical questions that must be answered prior to taking a child client on a nature walk. The three questions we ask before we take any child on a walk include:

1. Is the child/teen able to trust the therapist to be the safe boss on a walk?
2. Is the child/teen able to trust themselves/regulate through being separated from the parent?
3. Does the child/teen have enough access to his/her executive functioning systems at this point in therapy to be able to choose between options?

Let us unpack each of these. The answer to the first question—Is the child/teen able to trust the therapist to be the safe boss on a walk?—is critical to the decision to venture outside of the controllable indoor environment. I have mentioned the permissions we have parents sign when we take children outside. These permissions have to do with helping parents acknowledge the limits of confidentiality that come with being outside the building. However, there are other kinds of permission that are clinical in nature. In terms of the clinical arc of therapy, our clinicians must assess the degree to which any particular client is able to give the therapist permission to be the safe boss on a walk, especially as it relates to following the therapist's directions involving physical safety.

Many of our clients have trauma, neglect, or maltreatment backgrounds that have interfered with their ability to develop a healthy ability to trust and to come up and under the safe boss authority of other grown-ups. There is a sort of invisible string that exists between parents and children. Parents and children in healthy enough family systems are constantly negotiating the invisible boundaries that define the comfortable distance that can be allowed/encouraged between the parent and the child. You can see this in the park, in the grocery store, and in the mall. Little ones will run ahead of their parent, turning back occasionally to make sure they can still see them. If the distance gets to be too large or too uncomfortable for one member of the dyad, they adjust. In some cases, you will see a parent begin to run until they catch up to the toddler. In other instances, the child will slow down or stop and wait for the parent to get closer. In a healthy family system, the safe boss functions as a secure base from which the child can move out in exploration. When the child is hurt, scared, tired, hungry, etc., their attachment system (their need to be close to the caregiver) overrides their exploratory system, and they return to the parent (who is now functioning as a safe haven). Maltreated or neglected children are, by definition, children who have not experienced a secure base from which to move out in exploration, and they have not experienced the safe haven that meets their needs when they are experiencing them. They do not have an internalized sense of the boundaries of physical proximity with their caregivers. Many of these children move out in exploration and continue moving.

Many foster and adoptive parents have looked at me in bewilderment as they describe the way their adopted child will run away from them in the mall or in a parking lot and just keep moving, never checking in and never seeming to feel a sense of danger at being too far away from their parent. As we begin to lay down new neural wiring for what safety feels like through thousands of repetitions of need meeting both during clinical sessions and in assigned therapeutic homework for parents, traumatized children begin to reference their adults differently. There is a sense of accomplishment for both parents and children when the child develops the capacity to entrust the therapist with being the safe boss on a walk. If a breach of trust occurs on a walk, meaning the child is unable to stay within the boundaries set, the conversation goes something like this: "Your body is letting me know that we need to stick together some more at Nurture House before you will feel safe to let me lead you on a walk again."

The second question—Is the child/teen able to manage the anxiety of being separated from the parent?—is often an important consideration, as we see children who are unable to separate from parents at all when they first come to Nurture House. These dyads often meet the criteria for an ambivalent attachment pattern and need some reworking of the relationship, expansion of positive coping, and possibly some play-based exposure work before the child can confidently manage anxiety through self-regulation and/or allow the therapist to become a surrogate co-regulator of anxiety while on a walk. Sometimes the therapist helps the client imagine an invisible string that gets longer and longer during the walk and gets rolled up as they return to Nurture House. Sometimes the therapist helps the dyad make "love connectors" (see Goodyear-Brown & Andersen, 2018) and uses them to aid in separation. This intervention was really created by my son, Nicholas, when he was six years old. He was watching me pack for one of my speaking trips and remarked, "We need more love connectors!" I stopped packing, got on my knees in front of him and said, "We do need more love connectors. What should we use?" I did not truly understand what he meant by love connectors, but as we searched the house for items that might work, he eventually landed on my hair ties. He put one on his wrist and told me to put one on mine. Then he explained that we both had to keep them on the whole time we were apart from one another. The next morning, as I walked him into his preschool, we were holding hands. Our love connectors were touching at the wrist, and he said, "Mommy! Our love connectors are powering up!!" Each night while I was gone, we would FaceTime. He would check that I had my love connector on, and then we would touch them to each other through the screen and "power them up" for another day. We have used this idea in a variety of ways as he has grown, and it has given him (and, OK, me too) a concrete reminder of our connectedness across time and space. Another way to articulate the love connectors, per the lingo of object relations theory, is to describe them as transitional objects. We will sometimes take bubbles, a peacock feather, or a stress ball with us in case the child needs additional aids to manage anxiety at any point along the way.

For children with separation anxiety, the act of moving outside the building, even if we only make it to the backyard or the front porch, is its own form of exposure work, allowing them to feel the discomfort of distance from the parent and manage it in relationship to a safe other. As the client's neural circuits for anxiety become regulated to the external environment, we begin to explore. Moving even a block down the sidewalk allows for the exploration of a variety of plants that are not on the Nurture House property. Currently, there are gorgeous yellow leaves falling from the trees right outside Nurture House, but a few steps away are beautiful deep red oaks. As long as these doses of novelty are titrated, the excitation brought to neural pathways by novel experiences in the natural world further trains the brain to experience new things as interesting and pleasurable instead of overwhelming and scary.

The third question—Does the child/teen have enough access to his/her executive functioning systems at this point in therapy to be able to choose between options?—must also be answered in the affirmative before a child/therapist dyad can go on a walk. Nurture House is one house on a residential block that has been turned into part of the historic district in downtown Franklin, Tennessee. Soon after stepping off the front porch of Nurture House, one must decide whether to turn to the left or the right. To the left is downtown Franklin, with its many shops and crosswalks and a healthy number of people on any given day. To the right is a cemetery (a powerful icon for therapeutic work) and a quaint brick walkway that winds around to the Harpeth River. There are overlooks, rocks for sitting on and watching the river, and an open market that sells fruits and vegetables grown by the Amish as well as plants and pumpkins seasonally.

Simply by exiting the building, a two-choice prompt is set up and an ability to choose is required. Which way do we go? There is no one right way for the choice to be navigated. Does the therapist choose the direction? Does the child? Do they negotiate a process together? Does the child freeze? Does the child ask the therapist to choose? It often puts me in mind of the moment in the Wizard of Oz when Dorothy gets to the crossroads and the scarecrow informs her that "some people do go both ways." However the choice is made, we glimpse the coping strategies the child employs and what sorts of experiences he or she is drawn to in the natural world. Sometimes the choice is made in advance while we are still with the parent in the building. Perhaps we are celebrating a special occasion, like a birthday or a graduation from therapy. When this is the case, we already know our destination, though we may still be making choices about which way to get there. The therapist can set up as many choices on the walk as seem clinically beneficial in extending the client's window of tolerance for choosing. Much like in the playroom, children with significant anxiety can benefit from learning to make decisions for themselves and then experiencing the pleasurable outcome of the choice. This can increase self-confidence and self-esteem. The metaphoric implications of acknowledging the many different paths that may be chosen regularly lead to conversations about choices in other areas of our client's lives.

The physical benefits of walking are well documented (Duncan et al., 2014; Grant, Machaczek, Pollard, & Allmark, 2017; Lee & Buchner, 2008) and include maintaining a healthy weight, weight loss, strengthening bones and muscles, lowering cholesterol, lowering blood pressure, and protecting against type 2 diabetes. The psychological benefits of walking have also been well documented (Rogerson, Brown, Sandercock, Wooller, & Barton, 2016; Brown, Barton, & Gladwell, 2013; Gladwell et al., 2012; Barton, Griffin, & Pretty, 2012; Revell, 2016; Wood, Angus, Pretty, Sandercock, & Barton, 2013) and include elevated mood, stress reduction, enhanced self-esteem, and restoration of certain aspects of mental health (Donaghy, 2007).

Certainly, nature walks can be made even more therapeutically beneficial by introducing mindfulness practice into the natural environment. When I begin to walk with clients, we will often take the first few minutes to notice five things they can see, four things they can hear, three things they can touch, two things they can smell, and one thing they can taste. Other times the client and I will find a particularly beautiful tree and lie down under it, noticing patterns among the leaves, the way they sway in the wind, etc. Sometimes we will hear the wind and then count how long it takes for the wind we hear to make the leaves on the tree move. Lying down in the open grassy field near

Nurture House is conducive to finding shapes in the clouds, encouraging a form of projective testing in the natural world.

The potential positive effects of the natural environment on attention are an especially intriguing avenue for exploration in this chapter, as we see so many traumatized children who have difficulty sustaining attention to necessary tasks while simultaneously overattending to potentially threatening sounds, sights, smells, etc., in their environment. To understand the explanation for positive changes in attention following exposure to the natural environment, we must begin with William James' (1892) distinction between involuntary attention (a form of attention requiring no self-motivation or willful focus) and voluntary attention (attention that requires us to bring our focus to a particular point or topic). Berto (2005) created an experiment in which participants were mentally fatigued by completing a challenge that required prolonged attention. He then showed participants one of three types of images: restorative environments, nonrestorative environments, and geometric shapes. All three groups were then asked to return to the task requiring sustained attention. The only group to make gains in sustaining attention was those who were exposed to the restorative environment images. This research builds on the earlier work of Kaplan and Kaplan (1989), who outlined an attention restoration theory (ART) in which they posited that aspects of the natural environment draw on involuntary attention, which is seen as supporting restoration from psychological exhaustion. For this discussion, restoration is defined as "the process of recovery from a depleted psychological, physiological or social resource" (Hartig, 2007). Kaplan built on James's earlier ideas about voluntary and involuntary attention, equating voluntary attention with an effect he named *directed attention fatigue* (Kaplan, 2001) and equating involuntary attention with *fascination*, a process in which you cannot help but have your attention captured. I feel this process at work in me as I sit on the bank of the Harpeth River editing this chapter. The intensity of focus leads to mental fatigue, but then, all of a sudden, a beautiful dragonfly with otherworldly blue and green coloring perches on a leaf nearby. After watching the dragonfly for 30 seconds or so, I return to focus on the chapter, feeling internally refreshed. According to Kaplan's hypothesis, this enthrallment allows for rest and recovery of the attentional system. In this way, spending time in the natural environment can mitigate the intense attention that is often required for a client to process trauma. Roe and Aspinall (2011) looked at the restorative properties of the natural environment in two groups of people—those with good and poor mental health—as they experienced two different environments (an urban walk and a rural walk). They found that, in general, both groups experienced the greatest gains in positive mood in the rural environment, with people with poor mental health benefitting most greatly. People with poor mental health also seemed to benefit from walks in urban environments (one hypothesis is that the gains have to do with the social context). So many of the children we see have difficulty with attention to tasks. This may come from an organic inattention issue or may be a symptom of trauma or an anxiety-based disorder.

In either case, natural settings can be helpful in extending attention (Berto, 2005; Tennessen & Cimprich, 1995).

There are several additional benefits of walking with a child or teen client. The first is that walking is a very small ask for most children, as it is part of their competency repertoire already and will result in an experience of mastery. Sometimes, despite our best efforts, the introduction of expressive arts or play materials places demands on the client to perform at a certain level of competence or creativity. The simple mastery of walking mitigates the approach to hard topics. It contributes to a sense of normalcy in the child. If we return to the concept of the triune brain, we are targeting the reptilian brain stem first as we walk and engage in self-regulation through ambulation.

Second, the client is getting kinesthetic input and keeping their somatic system occupied in ways that are rhythmic and grounded, allowing for an easier regulation of the limbic brain and potentially more access to cognitive processes. Third, when a client and therapist are walking together, there is a titration of relationship happening. The therapist is not staring at the child or insisting that the scared teen make eye contact while beginning to explore scary stuff. Walking together, the child can set a pace that is comfortable for them, and the therapist can match this, using the movement as its own form of attunement. It allows for the creation of a different kind of space between the client and myself as we become a unit—an "us"—navigating the larger environment.

Even more specifically, for traumatized children, the natural bilateral stimulation that is derived from the simple act of walking aids in the integration of trauma content (Shapiro & Forrest, 2016; Shapiro & Solomon, 1995). In fact, an appreciation of this method of cross-hemispheric integration has led some treatment agencies to invest in treadmills. When walking can be accomplished outdoors, the therapeutic benefits of both exercise and nature can deliver a "double whammy" to clients (Barton, Hine, & Pretty, 2009).

## Metaphors in Nature

One of the treatment goals for traumatized children and their families is to help them understand the defense mechanisms that have been developed to cope with fear-inducing situations. I recently went on a walk with Becky, a 12-year-old girl who has fetal alcohol syndrome and cognitive delays, along with anxiety issues, attachment issues, and significant social difficulties. She often walks with her head down, looking only at the path just in front of her feet. I introduce various ways to help bring her head up, capture her eye gaze for various moments, and engage her in cycles of opening/closing communication loops. During our walk, I began to feel frustrated that she did not seem to be able to move her eyes from the ground. I was breathing through my own desire to strive to have her engage when Becky stopped abruptly. She had noticed a giant furry caterpillar on the path in front of her. I have never seen a caterpillar of such great size. Moreover, it was black and prickly. It was almost as big as the palm of her hand. It looked dangerous and potentially poisonous. We both thought it was "so cool," and the novelty of this discovery pulled her gaze upward. She made eye contact with me and used her words to ask if we could take it back to Nurture House. What a setup! I was able to give her a big yes, but then we had to figure out how to accomplish the job.

We were both in agreement that the bug could be poisonous and we did not want to touch it for fear of it stinging us. I had not brought a bug catcher, a bag, or anything else to pick it up with, so we engaged together in a problem-solving process where we would find different leaves and sticks to try to support the caterpillar, but none of them were strong enough to support its weight while we moved it to its new home—hear the metaphor? Eventually, I took off my hat and used it as a carrier to get the caterpillar back to Nurture House. Once we arrived, we looked up "big spiky black caterpillar" and found out that this caterpillar becomes the giant leopard moth. We read enough to realize that Harry, the name my client gave it, is not dangerous in any way, and then we took turns feeling its spiky body. This action alone was an exposure exercise (for both of us), as we were unsure just how sharp (i.e., painful) the spikes might be, so we were unsure of how much pressure to use. If we used too much force, we might hurt the caterpillar or our fingers, so we agreed to start with a very gentle touch and then try again if we needed to. Another therapeutic benefit was the refocusing that this caused around our somatic "knowing." When she felt it, Becky squealed. Then she went back and felt it again. I asked her what it felt like to her. Becky thought the caterpillar's hair felt like the bristles of a hair brush. Even that required her to experience the caterpillar somatically and then move from that form of knowing to a cognitive knowing, searching her experience for something to equate it to and then verbalizing that to me so I could share it with her. Her desire to communicate this unique experience to me mitigated her approach to finding the right words, which is her scary stuff. She is self-aware enough to recognize that her words sometimes fail her, and that she cannot articulate things as well as her peers.

We spent time looking for a proper container for Harry and then began transferring Harry from my hat to the container. Harry curled up in a ball and played dead. My client was alarmed at first and said, "We killed it!" I explained that sometimes when creatures get scared, they have a defense mechanism that has them freeze and pretend to be dead. As we reviewed together the ways in which small creatures protect and defend themselves against threats of harm, she said, "Once he's in there (the container), he will start moving again." We put him in the new environment, and sure enough, after a couple of minutes, Harry began to move. As we engaged with this caterpillar, the immobilzation system, identified in Stephen Porges' polyvagal theory was unpacked in a child-friendly way. We saw a state of hypoarousal/collapse as the caterpillar curled up and froze. While a scientific explanation may have been too wordy for Becky, she was given permission for her own immobilization responses as she learned about the caterpillar's way of coping with fear. Becky asked how long it would take for the caterpillar to become a giant leopard moth, and when we looked it up together, we found that while different caterpillars can take different lengths of time to "become," this one takes many months of hibernation. Using the rich metaphors offered to us by our exploration of the natural world, we normalized states of hyperarousal and collapse, and we normalized the truth that growth and change can take different amounts of time for different people. As Becky and I looked at pictures of what the prickly caterpillar would become, we both exclaimed at the beauty of the giant leopard moth, which is white with black spots. We marveled at how something so prickly and hairy could become something so beautiful. We looked at two or three more images and then came across an up-close image of the moth. A finger was holding up the wings to reveal a body that was rainbow-colored.

BECKY: Wow! I thought he was only white and black.
ME: Yeah, me too. You have to look really closely to see all those colors.
BECKY: We can only see them because he's holding up his wing.

It struck me, when I heard Becky say this, that these words describe one of my constant hopes in work with traumatized children and their parents—that I can hold up the child in a different light so they can see themselves differently and the caregiver can see the beauty of the vulnerable parts that are often covered up.

Another metaphor in nature presented itself to me last week as I was working with a boy who had recently been adopted and his newly adoptive father. For the purposes of our narrative, we will refer to them as Jack and his dad. Jack, age ten, had a sexually abusive father who ended up being prosecuted and is currently imprisoned, a nonprotective mother, and an interim foster placement of several years with a safe family before he was placed with his adoptive family. While he had known the "dad" of this family for several years in an extended community network, Jack had been in his new adoptive family for about five months at the time I began seeing him and at this point called his new father figure by his given name, Ned.

One day I went out to meet Jack in the lobby. The weather was perfect, and I offered to take him on a walk. He hesitated and then said, "Ned too?" I was delighted that he wanted his new father's company and said, "Of course! Ned, do you want to join us?" It was heartwarming to see how Jack's preference for Ned's company affected Ned. I find that adoptive parents are often offered such mixed messages by their adopted children that they become very unsure of their importance. When adopted children can give signals of preference, adoptive parents receive the reinforcement that often comes normally in a day in, day out attachment cycle between biological children and their parents. The biological child may, for example, scream and cry and have to be held and rocked for a long time before bed. Certainly, nursing mothers are sacrificing sleep on a regular basis to get up every two to three hours to answer the hungry cries of their babies, but then they go out to run an errand, and when they return, the child smiles widely and reaches out their arms to be picked up, clearly preferring the mother to all others. This cyclical reinforcement helps the child feel safe and secure but also provides repetitive experiences for the parent of being preferred and vitally important. The reinforcement for the parent of their own vital importance sustains them through the long watches of the night.

My guess was that the offer of a walk—which requires a deeper level of trust both on the part of the clinician and the client—was mildly anxiety-provoking for this child and he felt more anchored by having his father figure with him. We began walking toward the river. As we moseyed, I asked Ned to tell me one thing he has seen Jack really grow into since he has known him. Ned talked about how he used to startle easily and had trouble asking for what he needed, but how he has become much more relaxed at home and uses words more often now. We began talking about trust and shared with Ned Jack's decision to continue calling him "Ned" until he "felt" like he was ready to call him Dad. This was a topic that Ned and I had been over together already, and he was well prepared to reflect Jack's need to feel more connected and to build more trust before he would be ready to call him Dad. Ned exerted no pressure, and I also gave permission for the client to take his time, explaining that he will know if and when he is ready to call him Dad, and that Ned does not "need" this from him. Ned will continue to show up and meet needs and delight and connect regardless of what Jack calls him. As we approached the bridge, we stopped and looked at the water.

Jack immediately said, "There's a beer bottle." Both Ned and I had to search the river for several seconds to see the sunken bottle, but Jack had noticed it right away. Ned said, "Jack always notices details." I said, "I bet he does. Children who have grown up with some scary things happening to them or around them learn how to constantly be scanning the environment for any signs of danger. Sometimes this is cool because kids can feel really ready for whatever comes, but it can get really tiring to have to be constantly on guard." Jack listened quietly to this and took one unconscious step closer to where Ned and I were standing. Jack exclaimed, "Look!" and then showed me the flashes of silver that were happening in the water below. Ned watched, and when he saw the same flashes, he was able to explain that the fish who are gray on top, in order to blend in with the water, have a shiny color on their underbellies. When they become excited and flip over, you can see them. Jack said, "I'd like to catch one." I said, "We can go down there if you want." Jack was very excited about this idea, and Ned agreed that it would be fun to see the fish up close. This entailed climbing down lots of big gray rocks to get to the edge of the river.

Let us be clear about one thing. There is no graceful way to climb down a rocky slope. As we struggled down, we held on to each other, we held on to the rocks above us, and occasionally we placed a foot on a rock that wobbled. If you have ever been rock climbing at the edge of a body of water, you have experienced the feeling of placing your foot on a surface that you thought was stable, solid, supportive, and then having it disappoint you, as it was not able to hold your weight, to properly support you—do you hear the metaphor yet? We got all the way down to the water's edge, myself rather ungracefully in my flip-flops, and Jack spent a few minutes trying to catch the fish. We talked about how hard they are to see, with their protective camouflage, and how the fish only show themselves when they feel like there is no threat—hear the other metaphor? We watched a couple of fish spontaneously jump out of the water, delighting all of us, creating a moment of shared enjoyment of a novel experience (another win in creating positive shared history between Jack and his dad), and then it was time to climb back up to the road. As we walked back to Nurture House, I said, "You know what I noticed while we were down there? Sometimes I would place my foot on a rock that I thought was stable, and it would hold me. Other times I placed my foot on a rock that I thought would hold me, and it wobbled. It was unstable. But both rocks looked equally big, equally solid, so it's hard to know which is which." I stopped talking and kept walking. There is something powerful in itself about walking together in the same direction with someone else. Bilateral stimulation is occurring in an ongoing manner within both participants, and spontaneous thought, conversation, reflections, wonderings, and new associations can be made. Moreover, for highly resistant or shut down children, the lack of necessity to make ongoing eye contact or have intensive focus on the relationship can mitigate the client's approach to hard questions. In this case, a few moments after I fell silent, Jack said, "It's hard to tell which ones are stable and which are not." I reflected his statement and then said, "And yet we've got to step on the rocks to get down to the river. Which ones do we step on?" Jack's biological dad was unstable and abusive. Now that he has a "new dad," how will he solve the very real problem of answering the very important question of whether this dad is trustworthy, is solid enough, is stable? As we walked, Jack thought about it and eventually said, "I guess you have to look for how deep into the ground it goes. If it's really deep, it won't move." I looked at Ned, who seemed to hear the unspoken layer of our communication, and said, "I think that is a very wise answer, Jack. You just have to see how deep it goes and then you'll know if it's stable."

Sometimes the children see the metaphors in nature before I do. Recently, I was on a nature walk with an older boy who is wrestling with some complex emotions related to his younger brother. Both siblings (we will call them John and George) were removed from a neglectful home and placed together into foster care. Although they are now in a safe home together, their physical safety is not becoming "felt safety" at the same pace. Both children showed up at their new home dysregulated and, according to their foster parents, "bouncing off the walls." The boys were often aggressive with one another with a ferocity these parents had never experienced with their own children. One of the dynamics at play here was the trauma bonding that occurred in their abusive home. The ways in which these trauma bonds manifested often included aggression and control between the boys. For the purposes of this conversation, I will define a trauma bond as an ultimately maladaptive but probably necessary pattern of relating in relationship to one another in a traumagenic environment.

The older of the two boys, John, is a deep thinker, has a personality that appears to be matching the culture of the foster-to-adopt family very well, and is able to soothe and entertain himself much more frequently than his brother. As we have worked together, John has been able to gain more and more control over himself in his interactions with his brother. However, his brother continues to hurt him and make threatening statements during many interactions. During this walk, John and I began exploring the agency's decision to place them together in a home. We talked about the pros and cons of this decision and about how things might look/feel/be different if they had been placed separately. We walked and talked until we were standing on the Bicentennial Bridge. The water completely caught our attention, as there had been a big storm the night before and the water level was significantly higher than normal. The river was moving more quickly than we had ever seen before. As we stood on the bridge and looked, we realized that the water in the distance was the river rushing as it moved around to the right and the water in front of us was the much slower moving (almost still) tributary. John and I watched the water for a while, and then John spontaneously said, "The raging part back there, that's like George, and this part is like me." Wow. The power of metaphor in nature. I think when we run out of words to explain our experience, we return to the natural world. The more often we are interacting with the natural world, the more expanded the potential metaphoric languaging options we offer to our clients (see Figure 7.7).

## Focusing Our Attention

Sally is an eight-year-old girl I began seeing after her father died of a sudden cardiac episode. Sally has processed through a significant amount of grief, including a belief that it was her fault that her father died—if she had found him sooner, if she had known how to help her dad would still be alive. Her family is restabilizing, learning how to live in the world again, and her mom is learning how to do the job of both parents. Her mom, who had been a stay-at-home mom until her dad died, got a full-time job and worked sometimes from before school until almost dinnertime. Sally's mom talked about her irritation at the bickering that goes on between Sally and her brother and bemoaned the fact that they were always irritated with each other.

On a particularly beautiful fall day, Sally arrived for her session and asked to go on a walk. We walked down to the Harpeth River and stood for a time at the overlook. Sally and I both sat quietly for several minutes with our arms resting on the fence and our chins resting on our arms. Sally said, "It's so deep." I was surprised to hear this, as the river is fairly shallow on a good day and on this day had been without rainfall for a week and a half. I responded by

*Figure 7.7* The Raging and the Still Water

saying, "It looks really deep to you." Sally said, "Yeah. See right out there in the middle, you can see the reflection of the tree in the water." I looked and was able to refocus on the reflection. Sally then said, "I'm trying to figure out if that branch is really in the water or if it's a part of the reflection." Again, I needed a moment to refocus my attention to notice the branch she had identified. I validated her observation, saying, "It is hard to tell whether it's a separate branch or part of the picture." We both stared deeply for a while, shifting our visual focus between the bottom of the river, where the branch outline could be seen, and the top of the water that reflected the tree above in its mirror-like calm. After a few moments, Sally concluded that it was indeed a separate branch. As I lessened the intensity of my focus in and on the water, I realized that there was a whole swarm of gnats buzzing just slightly above the surface. As the light hit them, they became glaringly obvious and engendered a desire to immediately step away from them. However, neither Sally nor I had previously noticed they were there, such was the intensity of our focus on what was below. On the walk back to Nurture House, Sally remarked more than once on how she had not noticed the gnat swarm at all. I wondered out loud if our ability to overlook annoying things might depend on where we put our focus. Sally commented that sometimes her brother was like that annoying swarm of gnats—an irritation, a constant blur of movement. Sally said, "Yeah, and I didn't even notice it when I focused on the water." We then began pairing potential shifts in focus with Sally's management of her brother's irritating behaviors.

Experiencing the changing of seasons in nature offers rich fodder for metaphor work. We have had a delayed spring this year, and the earliest trees were beginning to bloom as I went walking with Christy, a teenage client. Her therapeutic work right now is related to taking risks in social engagement. People are often confused by her, as she looks like she should be socially adept, and yet every experience of meeting someone's eyes (especially if they are not well known to her), exchanging pleasantries, or asking for a certain item in a restaurant or store is perceived as a risk for her. So we go on walks together and decide what store we might enter, what question we might ask, what social engagement we will complete during a session. Recently, during one of these walks, she either opened or closed a circle of communication at least seven different times, a new record for her, and spontaneously said, "I enjoyed this walk, Miss Paris," when we returned to Nurture House. I have felt that she carries some shame in relation to her social self, as she is self-aware enough to understand her own social interaction patterns and recognizes that socially she is much younger than her peers. As we were walking, we passed a house where a beautiful cherry tree was blooming. The sun was out, the wind was gently blowing, and petals were mingling with the air currents—so many petals that we stopped walking to see where they were coming from. We stopped and looked up. A man came out on the front porch. I smiled at him and said, "What a beautiful tree you have! It graces the neighborhood." He said, "Thanks. I don't actually know what kind of tree it is." Christy remarked, "I think it is a cherry tree." When he went back inside, she commented that it had not been too hard to talk to him. Then she noticed the dormant tree next to the blooming one. Below is both the picture of the two trees and our conversation about them (see Figure 7.8).

*Figure 7.8* Late Bloomer and Early Bloomer

CHRISTY: Maybe that one's dead. They should cut it down before it falls over."
ME: "Hmm . . . let's see," and walked closer to see it up close.
She and I noticed at the same time that there were the tiniest little buds begin-
ning to show on the branches.
ME: "I think it is alive. It's just taking longer to bloom." I was struck by the
profound parallel between the way we had misunderstood the tree and the
way my client, in her delayed development, is sometimes misunderstood.
CHRISTY: "Well, I guess we have to wait for it. Maybe it will bloom by next
week."
ME: "Yes, we will need to be patient."
CHRISTY: "Patience sucks."
ME: "It does suck sometimes to have to wait."
Christy sighed and said, "We'll come see it next week," and we left the tree
to grow.

## Making Nature Soup

Social engagement can be explored and expanded in an outdoor context in
lots of ways. On day three of Camp Nurture, every child had grown in relationship with their buddies enough to be able to go outside. All the children
and buddies followed, and it almost felt like a mini group marathon. The
target was a wooden playset with two large trees that provided shade for it.

The big trees were dropping their seed pods, and one child picked up several, then asked his buddy to hold his shirttail like a sack, and the child dumped scores of seed pods into the shirt basket. Another child noticed what was happening and started gathering leaves. I commented on the green of the leaves and the brown of the seed pods. I said, "Wonder if we could find some things in nature that have other colors . . ." The children brought their focus even more connectedly to the present moment and immediately found smaller yellow leaves from a bush and tiny red seeds from another plant. We lined up all of these natural elements next to each other, and then I said, "Each of these is like an ingredient. I am thinking about a story called Stone Soup. In the story, the soup starts out with only a stone, but eventually each villager brings an ingredient to the soup, and it becomes a delicious soup and makes enough for everyone to feel full." One of the campers said, "Yeah, let's make soup!!" and an intern ran back to the building and brought me back a clear plastic bowl with water in it. I sat down on the floor of the fort of this playset, and the first child came and dumped in the seed pods. The second child came in and put in the leaves. Another child went and got little red seeds that were on the ground, and pretty soon every child had added an ingredient. This may seem like a very simple exercise, and it was certainly unplanned, but let us reflect for a moment on what skill sets and what presence of mind were necessary in order for each child to participate.

First, the bowl itself, the container for all the ingredients, was a shared space. Everyone had to negotiate and compromise about what ingredients would go into the shared space and in what quantity. Second, each child had to bring an offering and had to want to share the space (there was, of course, no requirement that everyone participate, but by this time their individual static was quieted enough that everyone was interested in exploring being together—at least for intermittent intervals). Third, each child had to become highly present, grounded in the natural world, ensconced in their natural environment and noticing the details of the world around them. After the seed pods and leaves were taken, other children spent time looking for "an ingredient" with a different color or a different texture. They would then bring the ingredient back, offer it to the group, and negotiate its addition to the soup. The final product was a really magical-looking soup with green, brown, yellow, and red colors, shiny and dull finishes, rough and smooth surfaces—all the differences making the soup more interesting. You can probably hear the richness of metaphors. All the children seemed proud of what they had created and asked that we bring the soup into the group room. We had several more hours of camp that day, and during our last Nurture Group of the day, we reflected on the experience of making the soup. Everyone took turns, if they wished, talking about what they had added and why (although the why was minimal). After all the children (supported by their buddies) had shared about their ingredient, one of our four-year-olds asked, "What did you add, Miss Paris?" As we all sat and pondered this, one of our older campers, one who had been very difficult to reach, spoke up and said, "You are the bowl. You hold us all together." The profundity of what had happened among us, among our connections to each other, and in our understandings of the natural world and our role in it was profound. That camp was three years ago, but that experience and that child's ability to function in the natural world and make contact with the environment around her and then extrapolate to a different developmental plane that involved synthesis, symbolism, analysis, and metaphor were profound to me and reinforced the power of nature in healing and its continuous communication with us, if we are willing to listen.

## Warmth

I am often made physically uncomfortable by air conditioning. I do not enjoy feeling cold. In fact, I find it painful. I often walk outside in between sessions in the summer just to feel the "hug" of the heat and regulate myself. If I have a similarly cold-natured client, I will invite them to join me. We may even lie down on the driveway to absorb the warmth of the concrete, behaving almost like iguanas on their hot rocks. This also serves as a very informal assessment tool for whether a particular child in my care is able to translate information from lower brain regions into self-care actions. Can they read the signals when their body is telling them they have absorbed enough heat and are ready to return to the air conditioning, or have they been conditioned to disconnect mind and body? Children from hard places, especially those who may have had an injury to their developing brains in utero through drugs, alcohol, or excessive stress hormones, will have impairments in their ability to "sense" when they are too hot or too cold. The reptilian brain stems of traumatized children—responsible for body temperature as well as several other regulatory functions of the body—are often compromised during development. These children do not know when they are overheated. This potential for internalized feedback from the natural world is becoming more and more limited as our society moves toward a disconnection with the natural environment.

## Conclusion

Nature is a powerful force for regulation and a powerful ally in the world of metaphor. Since we have expanded our outdoor play options for children at Nurture House, we have seen families approach harder content faster as the child swings and the parent and I push them back and forth. We have seen children who I thought were fairly concrete in their ways of relating to the world end up using a powerful metaphor to describe their sense of their family dynamics or their own sense of self. Walking together in the same direction has long been a way for humans to experience the world and increase shared mindsight together. For all these reasons, the role of nature is given special attention on the Play Therapist's Palette.

## References

Allen-Collinson, J., & Leledaki, A. (2015). Sensing the outdoors: A visual and haptic phenomenology of outdoor exercise embodiment. *Leisure Studies, 34*(4), 457–470.

Barton, J., Griffin, M., & Pretty, P. (2012). Exercise-, nature- and socially interactive-based initiatives improve mood and self-esteem in the clinical population. *Perspectives in Public Health, 132*(2), 89–96.

Barton, J., Hine, R., & Pretty, J. (2009). The health benefits of walking in greenspaces of high natural and heritage value. *Journal of Integrative Environmental Sciences, 6*(4), 261–278.

Berto, R. (2005). Exposure to restorative environments helps restore attentional capacity. *Journal of Environmental Psychology, 25*(3), 249–259.

Brown, D. K., Barton, J. L., & Gladwell, V. F. (2013). Viewing nature scenes positively affects recovery of autonomic function following acute-mental stress. *Environmental Science & Technology, 47*(11), 5562–5569.

Donaghy, M. (2007). Exercise can seriously improve your mental health: Fact or fiction? *Advances in Physiotherapy, 9*(2), 76–89.

Duncan, M., Clarke, N., Birch, S., Tallis, J., Hankey, J., Bryant, E., & Eyre, E. (2014). The effect of green exercise on blood pressure, heart rate and mood state in primary school children. *International Journal of Environmental Research and Public Health, 11*(4), 3678–3688.

Gladwell, V. F., Brown, D. K., Barton, J. L., Tarvainen, M. P., Kuoppa, J., Pretty, J. M., . . . Sandercock, L. (2012). The effects of views of nature on autonomic control. *European Journal of Applied Physiology, 112*(9), 3379–3386.

Goodyear-Brown, P. & Andersen, E. (2018). Play therapy for separation anxiety in children. In A. A. Drewes & C. Schaefer (Eds.), *Play-based interventions for childhood anxieties, fears, and phobias*. New York, NY: Guilford Press.

Grant, G., Machaczek, K., Pollard, N., & Allmark, P. (2017). Walking, sustainability and health: Findings from a study of a Walking for Health group. *Health & Social Care in the Community, 25*(3), 1218–1226.

Hartig, T. (2007). Three steps to understanding restorative environments as health resources. In C. Ward Thompson & P. Travlou (Eds.), *Open space, people space* (pp. 163–180). New York, NY: Taylor and Francis.

Hartig, T., Evans, G. W., Jamner, L. D., Davies, D. S., & Garling, T. (2003). Tracking restoration in natural and urban field settings. *Journal of Environmental Psychology, 23,* 109–123.

Herzog, T. R., Chen, H. C., & Primeau, J. S. (2002). Perception of the restorative potential of natural and other settings. *Journal of Environmental Psychology, 22,* 295–306.

James, W. (1892). *Psychology: The briefer course.* New York: Holt.

Kaplan, R. (2001). The nature of the view from home. Psychological benefits. *Environment and Behavior, 33*(4), 507–542.

Kaplan, R., & Kaplan, S. (1989). *The experience of nature: A psychological perspective.* New York, NY: Cambridge University Press.

Kaplan, S. (1995). The restorative benefits of nature: Towards an integrative frame-work. *Journal of Environmental Psychology, 15,* 169–182.

Lee, I. M., & Buchner, D. M. (2008). The importance of walking to public health. *Medicine and Science in Sports and Exercise, 40*(7), S512–S518.

Louv, R. (2008). *Last child in the woods: Saving our children from nature-deficit disorder.* Chapel Hill, NC: Algonquin Books.

Penedo, F. J., & Dahn, J. R. (2005). Exercise and well-being: A review of mental and physical health benefits associated with physical activity. *Current Opinion in Psychiatry, 18*(2), 189–193.

Pollock-Ellwand, N. (2010). Rickson outhet: Bringing the Olmsted legacy to Canada. A romantic view of nature in the Metropolis and the Hinterland. *Journal of Canadian Studies, 44(1), 137–183*. Retrieved from https://doi.org/10.3138/jcs.44.1.137

Revell, S., & McLeod, J. (2016). Experiences of therapists who integrate walk and talk into their professional practice. *Counselling and Psychotherapy Research, 16*(1), 35–43.

Roe, J., & Aspinall, P. (2011). The restorative benefits of walking in urban and rural settings in adults with good and poor mental health. *Health & Place, 17*(1), 103–113.

Rogerson, M., Brown, D. K., Sandercock, G., Wooller, J., & Barton, J. (2016). A comparison of four typical green exercise environments and prediction of psychological health outcomes. *Perspectives in Public Health, 136*(3), 171–180.

Shapiro, F., & Forrest, M. S. (2016). *EMDR: The breakthrough therapy for overcoming anxiety, stress, and trauma.* New York, NY: Basic Books.

Shapiro, F., & Solomon, R. M. (1995). *Eye movement desensitization and reprocessing.* Hoboken, NJ: John Wiley & Sons.

Tandon, P.S., Zhou, C. & Christakis, D.A. (2012). Frequency of parent-supervised outdoor play of US preschool-aged children. *Archive of Pediatric Adolescent Medicine, 166*(8), 707–712.

Tennessen, C. M., & Cimprich, B. (1995). Views to nature: Effects on attention. *Journal of Environmental Psychology, 15*, 77–85.

Ulrich, R. S., Simons, R. F., Losito, B. D., Fiorito, E., Miles, M. A., & Zelson, M. (1991). Stress recovery during exposure to natural and urban environments. *Journal of Environmental Psychology, 11*, 201–230.

Wood, C., Angus, C., Pretty, J., Sandercock, G., & Barton, J. (2013). A randomised control trial of physical activity in a perceived environment on self-esteem and mood in UK adolescents. *International Journal of Environmental Health Research, 23*(4), 311–320.

# 8 Humor, Novelty, and Shared Delight

## Humor

Humor is another mitigator of the approach to trauma content on the Play Therapist's Palette. Embracing humor and other expressions of delight can help with the dance toward and away from the trauma content while it further enhances the therapeutic relationship. Humor is likely to be appreciated in a sequence that moves from playful verbal interaction to a smile to laughter. Laughter is most frequently a shared experience and is likely to draw people closer together (Provine, 1992; McBrien, 1993; Scott, Lavan, Chen, & McGettigan, 2014). Laughter itself is broken up into protohumor (a slightly milder form of enjoyment) and laughter that expresses true enjoyment. It is often accompanied by the Duchenne smile (Duchenne, 1990), which is a combination of the activation of the zygomatic major muscles of the mouth and the contraction of the eye muscles. While many women strive to avoid wrinkles around their eyes, I see them as the trail of true enjoyment in a person's life. Duchenne laughter, which is the only form that expresses strong emotions and can have emotional health benefits, is closely associated with the Duchenne smile.

The physical health benefits of humor are real (Hasan & Hasan, 2009). From biblical references to laughter as good medicine to recent research finding that people who embrace humor tend to live longer than those who do not (Seligman, 2004), the health benefits of humor have long been supported. Laughter, it seems, has the same health benefits whether it arises from simulated or spontaneous humor (Mora-Ripoll, 2011). On a purely biological level, when someone laughs, their muscles become activated, oxygen is sent to the bloodstream, the heart pumps more vigorously, and they experience a sense of well-being. Charles Darwin made the persuasive argument that a behavior as pervasive, loud, and sometimes physically uncomfortable as laughing must serve some sort of survival function or it would not have remained in our behavioral repertoire over time. He suggests that laughter functions, evolutionarily, as a social signal of happiness and builds group cohesion that is adaptive in communities (Gruner, 1997; Wild, Rodden, Grodd, & Ruch, 2003).

Positive emotion, rather than cognitive understanding of a joke, is most strongly associated with an increase in flexibility and problem-solving abilities. The perception of humor activates the dopamine-based reward centers of the limbic system, enhancing an experience of pleasure. Humor can alleviate stress (Abel, 2002; Cann, Holt, & Calhoun, 1999; Cann & Etzel, 2008; Lefcourt et al., 1995; Lehman, Burke, Martin, Sultan, & Czech, 2001; Newman & Stone, 1996; Overholser, 1992; Prerost, 1988), mitigate depression (Nezu, Nezu, & Blissett, 1988), and positively impact our resilience in the face of adversity (Berg, 1995; Keltner & Bonanno, 1997; Lefcourt, 2001; McGhee, 2010). When humor is shared between two people, it acts as another form of "love connector," becoming a bridge between the pleasurable feelings of each participant and creating a shared history around these. The neurochemical boxing match that aids in trauma recovery was discussed in Chapter 2, but the power of laughter as an ally in this neurochemical grappling deserves special attention. Since cortisol is often released in massive and damaging quantities in our bloodstreams during intensely stressful moments, oxytocin can be a balancing agent in trauma recovery.

The bottom line is that we need as much oxytocin on board as we can get as we heal from trauma. Oxytocin is released during nursing between a mother and a baby, creating a deeper sense of connectedness. However, oxytocin can be accessed in several other ways, one of which is humor. Neuroscientists have found that laughter stimulates the 5-hydroxytryptamine 1A receptor. This receptor, once activated, stimulates the release of oxytocin. Interestingly, the neurochemical process is compared to the experience people have when taking ecstasy (Thompson et al., 2007). Oxytocin itself inhibits the release of cortisol (Spitzer, 2001) and modulates the amygdala's response (Kirsch et al., 2005) to signs of danger in the environment. These inhibitions decrease the traumatized child's felt sense of danger and potentially the need for hypervigilance—at least in the moment of shared humor. In fact, laughter involves auditory cortices (Woodbury-Farina, 2014) and activates our social engagement system, and oxytocin strengthens our attention to social cues and can function as a powerful bridge to an internalized sense of felt safety with the laughing partner.

The use of humor in a relational context can moderate the psychological distress caused by stressful life events (Fritz, Russek, & Dillon, 2017). When traumatized children or families are coming for treatment, they are often full of trepidation about how deeply, how quickly, or how overtly they will be asked to approach the trauma content. The use of humor mitigates the approach, providing a form of titration of positive relational interaction in the approach to hard things. When I hear laughter happening between a clinician and a child, parent, or family in a treatment room of Nurture House, I believe the healing process is being activated neurochemically while shared mindsight about whatever has been found humorous is being grown between the therapist and the client relationally.

Laughter can become the great equalizer in a session, giving humanity to a therapist who may have otherwise been perceived as more of an expert or technician than a human being and helping counter the hierarchical relationship that sometimes occurs in therapy (Nasr, 2013). While humor and laughter do not always occur simultaneously, when we find something humorous, we are likely to laugh or at least produce a precursor for laughter—a smile. Believe it or not, these intuitive ideas that smiling is good, that laughter is good medicine, are only now starting to have scientific explanations. Researchers are beginning to map the neural correlates of laughter and humor, and some are even interested in how this works specifically in children. There are at least 16 different forms of smiles, but the Duchenne display is the smile of enjoyment There has been some fascinating research on the difference between authentic expressions of positive emotions and those that are forced. In fact, these two different forms of expression seem to have separate neural substrates (De Paulo, 1992; Ekman, Friesen, & O'Sullivan, 1988).

Children, especially, are often able to discern the difference between a truly authentic expression of enjoyment made by one of their safe grown-ups and the Pollyanna, falsely cheery smile of the teacher welcoming the child whom she knows is going to require several corrections throughout the school day. The more deeply I explored the research on laughter, the more clearly it became a mystery. Many authors have posited the neuroanatomical structures involved in laughter, but there is as yet little agreement. However, "nearly all authors agree that there must exist in the brain stem a final common pathway for laughter, integrating facial expression, respiration and autonomic reactions" (Wild et al., 2003). After synthesizing the current research, Wild et al. posit that the frontal and temporal regions are involved in how we perceive humor and trigger laughter and musculature changes within the face that are mediated by the ventral brain stem. I and other therapists find humor to be especially effective in mitigating a teenager's approach to difficult content in family therapy sessions (Gladding & Wallace, 2001).

Early writings on humor often talked about the potential perils of its inclusion in a therapeutic context (Kuiper, Grimshaw, Leite, & Kirsh, 2004). However, more recently, people have begun citing a sense of humor as a hallmark of optimal mental health and as a brother or sister to processes like creativity, spirituality, and flexibility (Gladding, 1995; Myers, Sweeney, & Witmer, 2000; Seligman, 2004). There are widely differing perspectives about the potential role of humor in therapy. Some therapists believe there is no place for humor in therapy, and that the potentially deleterious effects of misusing humor or minimizing/making light of a client's actual psychic pain could bring injury to the client. Others, however, see multiple benefits of using humor in therapy. Family values, cultural expectations, and resistance levels must all be explored prior to introducing humor. Experiments have shown that humor can expand creativity and influence cognitive processes toward more flexibility, as humor often brings together two concepts that would normally be incompatible. Some researchers call this incongruity theory (Cardeña, 2003). Some researchers have also posited that because humor involves being able to keep more than one interpretation in mind at the same time while wrestling with incongruity, it is to growth in cognitive flexibility as a hand weight is to building arm muscles. To grapple with humor requires perspective shifting and can allow for a stressful experience to be reframed or evaluated from a less threatening position.

In some laboratory experiments in which groups of participants have been shown gruesome images from accidents, the participants who reported the lowest levels of disturbance around the imagery were those who used humor to cope. In these studies, the participants self-reported lower levels of distress. Objective measures such as heart rate and skin conductance provide evidence of reduction of the physiological effects of stress. We have all been on the receiving end, however, of humor or laughter that is ominous or even cruel (Provine, 2000).

My approach during supervision, when the issue of the therapist's use of humor comes up, is to orient the conversation to a question: whose need is being met here? Therapists may intuitively use humor when things get "too heavy" in therapy. If the "too heavy" is referring to the client's window of tolerance for processing hard things, then kudos to the therapist. He or she is judging, in an attuned fashion, when a dose of oxytocin and the buffer of connection with the therapist in a playful way will mitigate the approach to the scary stuff. If the "too heavy" has more to do with the therapist's window of tolerance, it behooves us to expand our own containment capacities while learning how to regulate ourselves more intentionally to meet the client's need.

The contrary impulses we all wrestle with—between self-focus and other focus, between overreactions and underreactions—in our attempts to get needs met and our constant dance of attunement, rupture, and repair in all of our interactions with other humans leave us no choice but to embrace the ridiculous. Being able to embrace this allows for a distancing as well as a potential laughing at the self that makes room for a change and shift in the way we see ourselves and the ways in which we deal with each other. To this extent, I believe the potential of using humor as a tool in therapy is worthy of further exploration.

Gladding and Drake Wallace (2016) talk about humor as a constructive expression of strong feelings, that humor provides a safety valve for the release of emotions that might otherwise be difficult to hold. With the advent of functional magnetic resonance imaging (fMRI), people are beginning to study what happens neurologically in the presence of something humorous. In one study, participants were engaged in event-related fMRIs while being exposed to episodes of Seinfeld or the Simpsons. The researchers were hoping to separate the neural substrates of humor detection from an appreciation of humor. The most interesting finding, for the purposes of this discussion, was that the amygdala and bilateral regions of the insular cortex were activated during humor appreciation. In other words, in order to appreciate humor, one has to have neural pathways that support the expression of affect (Moran, Wig, Adams, Janata, & Kelley, 2004). The amygdala's role in our reaction to traumatic events has been expanded upon earlier in this volume as well as in *Play Therapy with Traumatized Children*. If the amygdala can be triggered for fight, flight, or freeze responses by any number of environmental triggers, perhaps our trauma responses could be mitigated by building up the positive neural network of amygdala-facilitated humor responses. One of the fascinating questions that is evolving from this research is whether there is a common structure that is activated in both crying and laughing, but perhaps to varying degrees. This is purely my hypothesizing now, but it seems to me that intense emotional responses, like those expressed following grief, often vacillate between tears and laughter; in fact, the only constant is the intensity of both. Moreover, tears are viewed by many of my child clients as a lot less desirable than laughter, so perhaps laughter—and the humor that accompanies it—might be the most accessible way to help child clients begin working through window of tolerance issues related to intense emotion.

Goldin et al. (2006) believe that there is a cathartic effect of humor that helps people feel more present, more vigorous, and more equipped to solve problems and utilize their creativity to achieve life goals. Seven forms of humor that have been categorized and are potentially helpful in therapy are 1) anecdotes, 2) jokes, 3) puns, 4) stock conversational witticisms, 5) irony, 6) hyperbolic statements, and 7) self-enhancing or self-effacing statements. Multiple scales have been created over time to try to get at the heart of humor. The Humor Styles Questionnaire (HSQ) (Martin et al., 2003) looks at four styles of humor, two of which can create iatrogenic effects (aggressive and self-defeating) and two of which are edifying and may positively enhance relationships between people and contribute to a healthy sense of self (affiliative and self-enhancing).

There are also five forms of humor that are generally agreed to be negative. These forms often create uncomfortable feelings and could certainly create iatrogenic effects in a therapeutic setting. They are 1) satire, 2) sarcasm, 3) dark, grim, or depressing humor, 4) teasing, and 5) blue/risqué. These forms of humor often involve sexually inappropriate references and therefore have no place in the therapeutic relationship. While humor can bring people together, anyone who has been on the receiving end of a sarcastic or meanly intended joke knows that humor can also be used to diminish or shame. The decision about whether to integrate humor into therapy will have to do with answering the following questions: does the use of humor promote client well-being? and does the incorporation of humor promote positive change in the client or client system? Part of the decision may depend on how humor is already being used in the family system, and for this reason, measures such as the Relational Humor Inventory (De Koning, & Weiss, 2002) may be useful.

Sarcasm, for example, may be an aggressive form of humor or an affiliative form based on the context. Some of the context has to do with how much more cognitively developed the one providing the sarcasm may be than the one receiving it. For example, many teenagers feel more connected to their parents as they share in some form of sarcasm together—perhaps making light of a situation at school or "ragging on" a pop star and their latest exploits. The teenager may feel that their intellect is being respected when they are sharing a joke with someone who "gets" them. However, the child who does not yet have the cognitive sophistication to understand sarcasm may simply feel confused, minimized, or even shamed by not understanding the joke made by the caregiver or older child. Parent-child interaction therapy, a protocolized model intended to enhance the relationship between parent and child, increases child compliance, decreases parental stress, and relies heavily on equipping parents with a new set of skills to use in communication with their children. To this end, the first didactic training session with parents offers them the PRIDE skills (praise, reflection, imitation, description, and engagement). A handout is also given regarding what kinds of interactions are to be avoided during special playtime. Commands, questions, and criticisms are the three broad areas of communication to be avoided. Interestingly, as "criticisms" are unpacked, parents are asked to avoid using the words "don't," "stop," "not," "quit," and "no" during the five minutes of special playtime, but parents are also asked to avoid using sarcasm. The way I have explained this to parents is that young children understand that there is a joke and feel like the joke is at their expense but do not understand the content, so it just makes them feel bad. For a parent who needs a deeper explanation, it can be helpful to explain that most children do not understand the meaning of ironic statements until they are around six years old and may not perceive irony as funny until they are eight or nine, as it requires what Piaget termed a "theory of mind," an ability to infer the beliefs or intentions of the other.

Humor can be used to enhance group identity and is used in family, group, and individual therapy with various theoretical underpinnings (Galloway & Cropley, 1999; Isen, 2003; Jacobs, 2009; Fry & Salameh, 1987). There are several clinicians and researchers who have put forth models for using humor in therapy (Berlyne, 1972; Fox, 2016; Franzini, 2001). There are some who have used humor as a projective technique, asking the child client to identify their favorite joke, the idea here being that the content of the joke may reveal important intrapsychic conflicts or unresolved issues. Freud talked about humor as one of our most sophisticated and freeing defense mechanisms (Martin & Lefcourt, 1983). In fact, Freud saw humor as evidence of self-actualization, an ability to evoke pleasure out of excruciatingly painful situations (Goldin & Bordan, 1999).

In 1987, Mosak articulated five therapeutic uses of humor. Humor can help establish therapeutic rapport between client and therapist (Mosak, 1987; Richman, 1996). Richman talks about laughter in a session as enhancing our human commonalities and as therefore a powerful tool for building rapport and establishing trust. Some clients come to therapy with either a fear or a distorted (to the point of intimidation) view of the therapist and his or her expertise. When a hierarchical relationship is in danger of being established, humor or shared laughter can be the great equalizer (Chapman & Chapman-Santana, 1995; Goldin & Bordan, 1999). While some have given a nod to the potential use of humor in building rapport, others have created specific interventions that harness the power of the ridiculous for change in family systems (Sultanoff, 2013).

In our own family, there are times where one of us (myself, my husband, or one of my three children) has become dug in to a behavior or a point of view that is truly arbitrary. If we can insert humor at the right moment, we immediately take ourselves and our arbitrary dug in behavior less seriously. Ricks et al. (2014) designed an intervention, Laughing for Acceptance, specifically for high-conflict and highly resistant families. The intervention consists of having each family member create a comic strip related to the conflictual interactions. After all family members have shared, the clinician asks a series of questions geared toward reflection and giving permission to find humor in the patterns of interaction and that offer creative solutioning potentials. Odell asks people to engage in silliness, in part to interrupt the pattern of family interaction. Examples have been given of humor as adaptive coping among trauma survivors (Garrick, 2005; Lipman, 1991).

Children lacking a sense of humor tend to have higher conflict in relationships, lower self-esteem, and higher anxiety and other subsets of depressive symptoms (Erickson & Feldstein, 2007; Fox, Dean, & Lyford, 2013; Yarcheski et al., 2008). A sense of humor is associated with a sense of being able to impact the world, access to positive coping skills, and more positive relationships with others (Erickson & Feldstein, 2007; Fox et al., 2013; Yarcheski, Mahon, Yarcheski, & Hanks, 2008). Understanding a client's use of humor can reveal important information about his approach to emotions more generally.

While there is limited empirical evidence to date supporting the therapeutic benefits of humor in therapy, many potential therapeutic uses seem possible. Insight can be gained, hope can be ignited, and positive affect can mitigate a client's approach to difficult content. In fact, in a controlled study, humor presented in hierarchical scenes was found to be as effective as traditional systematic desensitization (SD). Of particular interest was that the hierarchical scenes were not paired with relaxation training—an aspect of SD work seen as necessary to counterconditioning the fear response (Ventis, Higbee, & Murdock, 2001). As humor is often associated with dopamine release in the brain, this "feel good" neurochemical reaction may render the need for relaxation training unnecessary. As humor is paired with difficult emotional or phobic content, emotional toxicity can be leached away from the scary stuff. Abrami's definition (2009) of humor is the "capacity to appreciate and derive some pleasure from that which is incongruous, ludicrous, absurd, or unexpected" (p. 7). There is an American Association of Therapeutic Humor and a whole journal given to the topic of humor.

## Delight

In my case conceptualization, I always include consistently expressed delight in the client (whether this is a child, teen, or parent) as the most powerful mechanism of change in clinical work. Infants who are securely attached can spend long periods of time absorbing and returning the loving gaze of their parents. Toddlers and preschoolers who have had caregivers delight in them really do see the world as their oyster and other grown-ups as interested in everything about them. Each of my children would run up to participants in my trainings and tell them everything they had for breakfast and everything we were going to do when the training day was over. They simply believed that the other would care. When children have not been able to absorb multiple experiences of delight expressed by a safe boss, they are often playing catch-up, and one of the therapist's roles is to provide delight. We do this in part with a loving eye gaze, the warmth we infuse in our tone of voice, and even the kinds of smiles we provide as we are saying hello in the lobby.

## The Safety of Sound . . .

What we have always known to be comforting, the prosody of the mother delighting in her baby, now has multiple layers of scientific explanation, including Porges' polyvagal theory. Our social engagement system is activated by the vagus nerve embedded in the brain stem and connected to the musculature of the face and head. The muscles of the inner ear are a powerful antenna in the neuroception of safety. Our nervous system evolved to perceive the low-frequency bass sounds that have historically been associated with the vocalizations of an aggressive male. The tones of a male voice startle us, whereas melodic female voices are comforting (Devereaux, 2017). Another fascinating piece of the puzzle is that the same activation of the vagus that controls eyelid movement also tenses the stapedius muscle in the inner ear, which enhances our ability to hear the human voice. When the inner ear is tensed, the ossicular chain is activated, and it dampens the prominence of low-frequency sounds in the environment, helping us hear the human voice. When this age appropriate prosody is accompanied by a Duchenne smile, the child has the message "you are delightful" communicated to them in an amplified way. The child then squeals and smiles in response, and the delight circuit is completed, as the mother also feels delighted in. An important part of communicating delight is making sure your tone, cadence, etc., is age appropriate. When I am traveling, I often get on the phone to say goodnight to each of my three children, who span in age from eight to 16. People have commented that they can tell which child I am talking to based on the tone, volume, and cadence of my voice. Attuning ourselves to the communication of nurture and delight appropriate for each child in our care is part of the job of the trauma therapist. Paying attention to the child's physical body and celebrating who they are by decorating them (Theraplay) is one of my favorite activities (see Figure 8.1).

Children who come to therapy with their amygdala armed and with deeply held beliefs that they are damaged goods or bad kids can be positively shaken by our initial delight in who they are. The delight might begin in the lobby with an observation of their beautiful brown eyes, it may erupt spontaneously as the child beats me repeatedly at getting a ball through a hoop, or it may be expressed as I notice the child's propensity for drawing. Authentic delight

*Figure 8.1* The Love Is All Over Their Faces

requires us to set the child's bar in a place that is attainable. One of my biggest wins is when I can help a parent reposition the bar for a child in their care to a level where they can celebrate the steps or risks a child is taking to grow.

Delight can be communicated through nurturing touch, such as high fives, fist bumps, or hugs. Delight can also be communicated in the shared novel experience. My family and I travel quite a bit, and it never ceases to surprise me that the moments we remember most about our trips, the shared mindsight that is most deeply concretized, are always the places/people/experiences we happened upon, not the ones that were planned into the itinerary. The unexpected, the serendipitous happening, can build shared delight. Shared delight can be powerfully enhanced by inviting shared novel experiences into a family. Many examples of this are given in the chapter on nature, as the natural environment is always changing and offering elements of surprise in the plant life, the animal life, even the sky. Some of my hardest clients have resulted in some of my most novel toys, as they needed an anchoring in novel, often kinesthetic, experiences to mitigate their approach to hard things.

Tools like funny videos and funny snippets of songs can be powerful therapeutic aids. I particularly enjoy the apps that take your face or voice and morph them in some way, making your voice sound like a chipmunk or like you have ingested too much helium, or extending your forehead so you look like an alien, or bulging your eyes. I have a dear friend who has moved away, and in order to stay connected, we will each occasionally create a video in which we are singing each other a song but then morph the voice quality or the facial expression in a way that embraces the ridiculous. We each laugh out loud when we receive these, each feeling remembered by the other and more closely connected through the moment of shared humor. The Photo Booth application on Mac computers has provided many moments of silliness with clients who were otherwise very serious. Dress up itself can be a way to bring spontaneous laughter into a session. In the tradition of Patch Adams, some hospitals have experimented with therapeutic clowning and have found a reduction in both children's perceived pain during procedures and parental anxiety (Koller & Gryski, 2008; Wolyniez et al., 2013; Dionigi, Sangiorgi, & Flangini, 2014). Fake teeth and silly hats are great ways to bring giggles as are apps that morph one's voice or appearance (see Figures 8.2 and 8.3).

Another way to delight in a client system is to celebrate expanded competencies. When a client grows in any area that they were hopeful to grow in, we get to celebrate with them. The child who was not able to sleep in his own bed by himself, the teenager who had wanted to break up with her abusive boyfriend but had not been ready to before, the client who was willing to try working in the sandtray even though she had proclaimed herself to be without creativity— each of these provides a moment of sheer delight as the therapist witnesses and reinforces the courage and the expanded competence of the client. I had a recent family in which we identified the young man's goal as being able to do things that are uncomfortable for him. We created a beautiful poster and gave the family a set of cutout butterflies. Each time the client did anything that was uncomfortable for him, from brushing his teeth to tying his shoes, the parents wrote it on one of the butterflies. At the next session, I asked his parents to recite for me out loud each of his moments of pushing through discomfort to do a hard thing anyway. At the end of the session, he asked for another set of butterflies. I am not, at heart, a behaviorist, but delight can be communicated in many ways—including a celebration of competence.

With children who come to therapy already feeling like they need to perform well to be valued, a tool like the one described above might create an iatrogenic effect. With that child, we simply delight in his being—being with him in a grounded, perfectly contented way in session no matter what the content. To these children, I want to communicate through both my verbal and nonverbal actions that if they never accomplished anything ever again, they would be delighted in just the same simply because they were created.

The final way I believe we delight in traumatized children is by *staying present in their pain*. This idea, when whittled down, involves a constant communication of this message: *I see what you are showing me, you can show me more, and I will be delighted with whatever you show me because it embodies the great trust and holding we have developed between us.*

My most powerful example of this was given to me by a four-year-old child, Thomas, adopted at birth. He had significant medical issues and lots of sensory defensiveness. When I met him, he seemed to always feel like the world was coming at him, not that he was a participant in it.

*Figure 8.2* Fake Teeth

He was a participant in a week-long camp experience in which regulation and learning to use his voice to express needs were the most important goals. Thomas learned more about his internal cues for regulation, and he found some of his first words at camp. He had started the week with very few words,

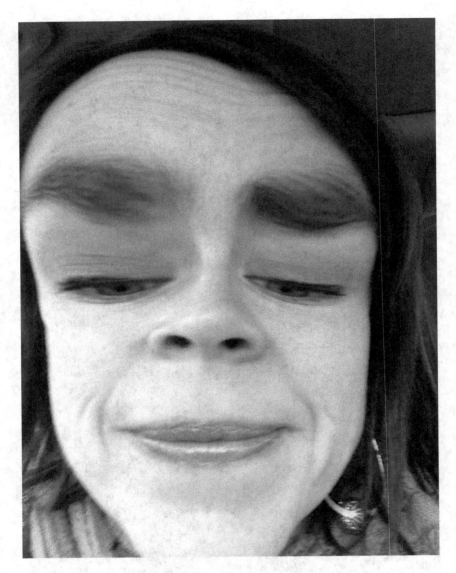

*Figure 8.3* Changing Face and Voice App

grown in his vocabulary and self-regulation over the course of the week, and then decompensated on the final day of camp. When it was time to leave, he was picked up by his father to be carried out to the car. He began to hit his father with a toy he had just received as a transitional object and began to revert to old behaviors. I put my arms around Thomas and his dad together and said, "Use your words, buddy. I know you can. I am right here to listen, and I want to understand." His face crumpled up, the aggression left him, and he began to wail, "I . . . will . . . miss . . . you!" This was the first honest expression of pain at a separation that Thomas had ever offered. All of the caregivers involved teared up, held the big feelings, and validated the pain of saying goodbye to people who had become important to them. There is a form of delight that can be expressed in sharing someone's pain. In this case, it was a privilege to help Thomas find the words to express his pain. His parents delighted also in his ability to communicate his pain so effectively, and even delighted in the pain itself, as it provided evidence that the relationships had been meaningful for this child.

## Conclusion

While there continues to be debate within the larger field of therapy, I believe that most play therapists would acknowledge the powerfully positive role that humor has played in their relationships with traumatized children and families. This would be a good topic for expanded discussion in another volume. Most play therapists have also seen how, when a child pushes up against the edge of his window of tolerance for grappling with difficult content, the introduction of a novel toy or stimuli can help to extend the window for the work. Delight, and the many ways it is communicated, continues to be my preferred and primary change agent in work with traumatized children and parents. Delight in the child, delight in the parent, delight in the pain, and delight in the healing are all held together by the wise play therapist.

## References

Abel, M. H. (2002). Humor, stress, and coping strategies. *Humor, 15*(4), 365–381.

Abrami, L. M. (2009). The healing power of humor in logotherapy. *International Forum for Logotherapy, 32,* 7–12.

Berg, D. V. (1995). Building resilience through humor. *Reclaiming Children and Youth: Journal of Emotional and Behavioral Problems, 4*(3), 26–29.

Berlyne, D. E. (1972). Humor and its kin. In J. H. Goldstein & P. E. Mcghee (Eds.), *The psychology of humor: Theoretical perspectives and empirical issues* (pp. 42–60). New York, NY: Academic Press.

Cann, A., & Etzel, K. C. (2008). Remembering and anticipating stressors: Positive personality mediates the relationship with sense of humor. *Humor, 21*(2), 157–178.

Cann, A., Holt, K., & Calhoun, L. G. (1999). The roles of humor and sense of humor in responses to stressors. *Humor: International Journal of Humor Research, 12*(2), 177–193.

Cardeña, I. (2003). On humour and pathology: The role of paradox and absurdity for ideological survival. *Anthropology & Medicine, 10*(1), 115–142.

Chapman, A. H., & Chapman-Santana, M. (1995). The use of humor in psychotherapy. *Arquivos de Neuro-psiquiatria, 53*(1), 153–156.

De Koning, E., & Weiss, R. L. (2002). The relational humor inventory: Functions of humor in close relationships. *The American Journal of Family Therapy, 30,* 1–18.

DePaulo, B. M. (1992). Nonverbal behavior and self-presentation. *Psychological bulletin, 111*(2), 203.

Devereaux, C. (2017). An interview with Dr. Stephen W. Porges. *American Journal of Dance Therapy, 39*, 27–35. Retrieved from https://doi.org/10.1007/s10465-017-9252-6

Dionigi, A., Sangiorgi, D., & Flangini, R. (2014). Clown intervention to reduce preoperative anxiety in children and parents: A randomized controlled trial. *Journal of Health Psychology, 19*(3), 369–380.

Duchenne, G. (1990). *The mechanism of human facial expression.* New York, NY: Cambridge University Press.

Ekman, P., Friesen, W. V., & O'sullivan, M. (1988). Smiles when lying. *Journal of Personality and Social Psychology, 54*(3), 414.

Erickson, S. J., & Feldstein, S. W. (2007). Adolescent humor and its relationship to coping, defense strategies, psychological distress, and well-being. *Child Psychiatry and Human Development, 37*(3), 255–271. Retrieved from https://doi.org/10.1007/s10578-006-0034-5.

Fox, C. L., Dean, S., & Lyford, K. (2013). Development of a humor styles questionnaire for children. *Humor: International Journal of Humor Research, 26*(2), 295–319. Retrieved from https://doi.org/10.1515/humor-2013-0018

Fox, E. (2016). The use of humor in family therapy: Rationale and applications. *Journal of Family Psychotherapy, 27*(1), 67–78. Retrieved from https://doi.org/10.1080/08975353.2016.1136548

Franzini, L. R. (2001). Humor in therapy: The case for training therapists in its uses and risks. *The Journal of General Psychology, 128*(2), 170–193. Retrieved from https://doi.org/10.1080/00221300109598906

Fritz, H. L., Russek, L. N., & Dillon, M. M. (2017). Humor use moderates the relation of stressful life events with psychological distress. *Personality and Social Psychology Bulletin, 43*(6), 845–859. Retrieved from https://doi.org/10.1177/0146167217699583

Fry Jr., W. F., & Salameh, W. A. (1987). *Handbook of humor and psychotherapy: Advances in the clinical use of humor.* Sarasota, FL: Professional Resource Exchange, Inc.

Galloway, G., & Cropley, A. (1999). Benefits of humor for mental health: Empirical findings and directions for further research. *Humor: International Journal of Humor Research, 14*(2), 163–179.

Garrick, J. (2005). The humor of trauma survivors: Its application in a therapeutic milieu. *Journal of Aggression, Maltreatment & Trauma, 12*(1), 169–182.

Gladding, S. T. (1995). Humor in counseling: Using a natural resource. *Journal of Humanistic Education and Development, 34*, 3–12. Retrieved from https://doi.org/10.1002/j.2164-4683.1995.tb00106.x

Gladding, S. T., & Drake Wallace, M. J. (2016). Promoting beneficial humor in counseling: A way of helping counselors help clients. *Journal of Creativity in Mental Health, 11*(1), 2–11. Retrieved from https://doi.org/10.1080/15401383.2015.1133361

Goldin, E., & Bordan, T. (1999). The use of humor in counseling: The laughing cure. *Journal of Counseling & Development, 77*(4), 405–410. Retrieved from https://doi.org/10.1002/j.15566676.1999.tb02466.x

Goldin, E., Bordan, T., Araoz, D. L., Gladding, S. T., Kaplan, D., Krumboltz, J., & Lazarus, A. (2006). Humor in counseling: Leader perspectives. *Journal of Counseling & Development*, *84*(4), 397–404. Retrieved from https://doi.org/10.1002/j.1556-6678.2006.tb00422.x

Gruner, C. R. (1997). *The game of humor: A comprehensive theory of why we laugh.* New Brunswick, NJ: Transaction Publishers.

Hasan, H., & Hasan, T. F. (2009). Laugh yourself into a healthier person: A cross-cultural analysis of the effects of varying levels of laughter on health. *International Journal of Medical Sciences*, *6*(4), 200–211.

Isen, A. M. (2003). Positive affect as a source of human strength. In L. G. Aspinwall & U. M. Staudinger (Eds.), *A psychology of human strengths: Fundamental questions and future directions for positive psychology* (pp. 179–195). Washington, DC: American Psychological Association.

Jacobs, S. (2009). Humour in gestalt therapy-Curative force and catalyst for change: A case study. *South African Journal of Psychology*, *39*(4), 498–506. Retrieved from https://doi.org/10.1177/008124630903900411

Keltner, D., & Bonanno, G. A. (1997). A study of laughter and dissociation: Distinct correlates of laughter and smiling during bereavement. *Journal of Personality & Social Psychology*, *73*(4), 687–702.

Kirsch, P., Esslinger, C., Chen, Q., Mier, D., Lis, S., Siddhanti, S., . . . & Meyer-Lindenberg, A. (2005). Oxytocin modulates neural circuitry for social cognition and fear in humans. *The Journal of Neuroscience*, *25*(49), 11489–11493. Retrieved from https://doi.org/10.1523/JNEUROSCI.3984-05.2005

Koller, D., & Gryski, C. (2008). The life threatened child and the life enhancing clown: Towards a model of therapeutic clowning. *Evidence Based Complement Alternative Medicine*, *5*, 17–25.

Kuiper, N. A., Grimshaw, M., Leite, C., & Kirsh, G. (2004). Humor is not always the best medicine: Specific components of sense of humor and psychological well-being. *Humor: International Journal of Humor Research*, *17*(1/2), 135–168.

Lefcourt, H. M. (2001). *Humor: The psychology of living buoyantly.* New York, NY: Kluwer Academic.

Lefcourt, H. M., Davidson, K., Shepherd, R., Phillips, M., Prkachin, K., & Mills, D. (1995). Perspective-taking humor: Accounting for stress moderation. *Journal of Social and Clinical Psychology*, *14*(4), 373–391.

Lehman, K. M., Burke, K. L., Martin, R., Sultan, J., & Czech, D. R. (2001). A reformulation of the moderating effects of productive humor. *Humor*, *14*(2), 131–161.

Lipman, S. (1991). *Laughter in hell: The use of humor during the Holocaust.* Lanham, MD: Jason Aronson.

Martin, R. A., & Lefcourt, H. M. (1983). Sense of humor as a moderator of the relation between stressors and moods. *Journal of Personality and Social Psychology*, *45*, 1313–1324.

Martin, R.A., Puhlik-Doris, P., Larsen, G., Gray, J., & Weir, K. (2003). Individual differences in uses of humor and their relation to psychological well-being: Development of the humor styles questionnaire. *Journal of Research in Personality*, *37*(1), 48–75. Retrieved from https://doi.org/10.1016/S0092-6566(02)00534-2

McBrien, R. J. (1993). Laughing together: Humor as encouragement in couples counseling. *Individual Psychology: Journal of Adlerian Theory, Research & Practice*, *49*(3–4), 419–427.

McGhee, P. (2010). *Humor: The lighter path to resilience and health*. Bloomington, IN: Author House.

Moran, J. M., Wig, G. S., Adams, R. B., Janata, P., & Kelley, W. M. (2004). Neural correlates of humor detection and appreciation. *Neuroimage, 21*(3), 1055–1060. Retrieved from https://doi.org/10.1016/j.neuroimage.2003.10.017

Mora-Ripoll, R. (2011). Potential health benefits of simulated laughter: A narrative review of the literature and recommendations for future research. *Complementary Therapies in Medicine, 19*(3), 170–177. Retrieved from https://doi.org/10.1016/j.ctim.2011.05.003

Mosak, H. H. (1987). Guilt, guilt feelings, regret, and repentance. *Individual Psychology, 43*(3), 288.

Myers, J. E., Sweeney, T. J., & Witmer, J. M. (2000). The wheel of wellness counseling for wellness: A holistic model for treatment planning. *Journal of Counseling & Development, 78*(3), 251–266.

Nasr, S. J. (2013). No laughing matter: Laughter is good psychiatric medicine. A case report. *Current Psychiatry, 12*, 20–25.

Newman, M. G., & Stone, A. A. (1996). Does humor moderate the effects of experimentally induced stress? *Annals of Behavioral Medicine, 18*(2), 101–109.

Nezu, A. M., Nezu, C. M., & Blissett, S. E. (1988). Sense of humor as a moderator of the relation between stressful events and psychological distress: A prospective analysis. *Social Psychology, 54*, 520–525.

Overholser, J. C. (1992). Sense of humor when coping with life stress. *Personality and Individual Differences, 13*, 799–804.

Prerost, F. J. (1988). Use of humor and guided imagery in therapy to alleviate stress. *Journal of Mental Health Counseling, 10*, 16–22.

Provine, R. R. (1992). Contagious laughter: Laughter is a sufficient stimulus for laughs and smiles. *Bulletin of the Psychonomic Society, 30*(1), 1–4.

Provine, R. R. (2000). *Laughter: A scientific investigation*. New York, NY: Viking.

Richman, J. (1996). Points of correspondence between humor and psychotherapy. *Psychotherapy: Theory, Research, Practice, Training, 33*(4), 560.

Ricks, L., Hancock, E., Goodrich, T., & Evans, A. (2014). Laughing for acceptance: A counseling intervention for working with families. *The Family Journal, 22*(4), 397–401.

Seligman, M. E. (2004). *Authentic happiness: Using the new positive psychology to realize your potential for lasting fulfillment*. New York, NY: Simon and Schuster.

Scott, S. K., Lavan, N., Chen, S., & McGettigan, C. (2014). The social life of laughter. *Trends in Cognitive Sciences, 18*(12), 618–620. Retrieved from https://doi.org/10.1016/j.tics.2014.09.002

Spitzer, P. (2001). The clown doctors. *Australian Family Physician, 30*, 12–16.

Sultanoff, S. M. (2013). Integrating humor into psychotherapy: Research, theory, and the necessary conditions for the presence of therapeutic humor in helping relationships. *The Humanistic Psychologist, 41*(4), 388–399.

Thompson, M. R., Callaghan, P. D., Hunt, G. E., Cornish, J. L., & McGregor, I. S. (2007). A role for oxytocin and 5-HT 1A receptors in the prosocial effects of 3, 4 methylenedioxymethamphetamine ("ecstasy"). *Neuroscience, 146*(2), 509–514.

Ventis, W. L., Higbee, G., & Murdock, S. A. (2001). Using humor in systematic desensitization to reduce fear. *The Journal of General Psychology, 128*(2), 241–253.

Wild, B., Rodden, F. A., Grodd, W., & Ruch, W. (2003). Neural correlates of laughter and humour. *Brain, 126*(10), 2121–2138. Retrieved from https://doi.org/10.1093/brain/awg226

Wolyniez, I., Rimon, A., Scolnik, D., Gruber, A., Tavor, O., Haviv, E., & Glatstein, M. (2013). The effect of a medical clown on pain during intravenous access in the pediatric emergency department: a randomized prospective pilot study. *Clinical pediatrics, 52*(12), 1168–1172.

Woodbury-Fariña, M. A., & Antongiorgi, J. L. (2014). Humor. *Psychiatric Clinics, 37*(4), 561–578.

Yarcheski, A., Mahon, N. E., Yarcheski, T. J., & Hanks, M. M. (2008). Psychometric evaluation of the interpersonal relationship inventory for early adolescents. *Public Health Nursing, 25*(4), 375–382.

# 9  Holding the Hard Story
## Narrative Nuance

## The Dance Toward and Away From the Trauma Content

One of the foundational concepts of the TraumaPlay™ model is the therapist's stance of respectful titration in our approach to the scary stuff, whatever that might be for a hurting child or family. Children will show us or tell us their story through glimpses and snapshots. We can honor and extend these moments of holding their trauma content through play therapy, expressive arts work, sandtray work, songs, creative writing, verbal storytelling, and the attaching of content to prop-based game play. Children who might only share so much on their own can go further and deeper with the help of a therapist armed with the Play Therapist's Palette. This chapter will provide a multitude of case examples demonstrating how we move towards coherent narratives with clients by weaving in the creative tools that can extend their window of tolerance for holding the hard story.

### Holes in the Story: Using Jenga in Life Narrative Work

A colleague of mine at Nurture House, Shelby Henson, had been pressing into finding new ways to kinesthetically mitigate a client's approach to life narrative work. She began to recognize the process of Jenga play, pulling out blocks, as leaving holes in the structure that compromises the integrity of the building. She used the metaphor of holes in the structure to mitigate an approach to Ray's life story work. Below is an excerpt from a session in which she introduced this play to Ray, a ten-year-old adopted at three. As we worked on it together, we talked about how memories can be confusing. Sometimes we have an actual memory encoded in autobiographical memory. Other times someone tells you the story of what happened, and this telling becomes a memory. Still other times you are not sure about something you think you know and you may have a question about it. We decided to attach colors to each kind of memory: orange for things kids were told, purple for autobiographical memories, and red for questions.

SHELBY: Today, we're going to play Jenga . . . and here's another one of my metaphors [*laughing at herself*]. You know we've been working on creating your life story, right? We will get to add to that today by playing Jenga. From what you know about Jenga, what happens when you start taking out more and more blocks?

RAY: It falls.

SHELBY: Exactly. That's just like people's stories. The less information—or in this case, blocks—that we have, the weaker the story becomes and can lead us to feeling yucky. The empty spaces are like unanswered questions. So today, as we play Jenga, we're going to start writing your story with as many memories that you have to make your story stronger.

RAY: Sure.

SHELBY: For each memory you tell me, I want you to tell me if it's something you remember, it's something someone told you, or if it's something that you aren't sure happened or if you have a question about it.

RAY: Okay.

SHELBY: All right, let's start! You go first.

RAY: I'm going to make this hard! [*Ray chose a hard one to pull out.*] Ahh, got it!

SHELBY: Wow, you are making this hard. Okay, tell me your very first memory.

RAY: Well, I guess it would be that I was born. I don't remember it, but I know it happened. So I guess use the orange marker [*to indicate someone told him*].

SHELBY: Got it. Okay, my turn. [*Shelby pulls one out.*] That one was too easy. Okay, your turn again.

RAY: Next thing would be going into foster care [*pulling out block*]. But I don't remember that either because I think I was like one or something. But I know it happened. That would be orange.

SHELBY: Yeah, it's hard to remember things when you're that young. Do you know why you went into foster care?

RAY: Yeah, my mom couldn't take care of herself. I think she had like some blood clot issues or something.

SHELBY: Got it. Mom couldn't take care of herself [*repeating back as she writes*]. Okay, I'll go [*pulls out block*]. Okay, your turn!

RAY: Next, I guess it would be when my mom now offered to adopt me [*pulling out block*]. I don't know exactly when that happened.

SHELBY: It's okay if you can't remember all the dates and ages. We're going way back in time.

RAY: Yeah, I don't remember it all happening, but I've just been told that. So that one would be orange too.

SHELBY: You've done a lot of thinking back to your past today! I think this is a good place to pause this game and go play our outside game.

Shelby had earlier made a compromise with this client that they would spend part of the session on life narrative work and part involved in more full body kinesthetic play. Shelby kept in mind at all times while working with this client the need to slowly stretch his window of tolerance for looking at or talking about hard things while mitigating the approach with lots of play and pleasurable sensory input.

Shelby facilitated a nuanced titration of narrative rehearsal here. Notice that every "memory" that her client identified was a piece of his history that was told to him by others and held for him by others, some of whom he no longer has contact with. Part of what this means is that eventually helping the caregivers hold the hard pieces of his narrative with him will be necessary because it will solidify their positions as history keepers for him. With this new dimension of history-keeping will come a deepening of the attachment relationship with his adoptive parents and more coherence in his internal narrative as well as more coherence in the caregivers' narrative that they hold for him. This client's first pass at life story work was rendered more "doable" by the client because he already knew how to "do" Jenga blocks, as Shelby sought to mitigate the approach to difficult content with the mastery experiences of kinesthetically pulling blocks from the stack without them falling. In essence, the client's success in pulling out parts of his story, including foundational parts of his story around why his mom chose to let him be raised by other parents, allowed him to pair competence-enhancing play experiences with the hard work of narration and coherence building. Shelby reflected on her work with him during the week following and wondered if he was ready to hold some of the less comfortable pieces of his narrative. She then tweaked the activity to be less predictable and to include a metaphor for memories being less neatly placed.

In the next session, Shelby chose to use something different than the Jenga blocks to continue to create his life story. She used a bag of blocks that were all different shapes, sizes, and colors. She explained to Ray that no one's life looks like a perfectly crafted Jenga tower, with every block being the same shape, size, and color. She said that most people's lives look like something created out of the bag of blocks I brought in. She had client start to build a "Jenga" tower using the different sized, shaped, and colored blocks to continue the life narrative work.

As Ray was building, he used all of the flat and similarly shaped blocks first to build the base of the tower. As the tower grew, he was left with using blocks that were cylinders, ramps, triangles, etc. As he struggled to make them fit and stay, Shelby reflected, "Sometimes there are pieces that just don't seem to fit." She then asked him, "What would you like to do with those pieces?" He said, "Can we just get rid of them?" She told him that while that seemed like the easiest solution, we had to use all the blocks somehow. She compromised and told him he could place them to the side for a while, and then he could make the decision for what to do with them.

After the client completed the tower using the "normal" blocks, they then continued to write down and label the memories that he could remember from his past. After he had written down several for the day, Shelby readdressed the blocks that had been put to the side. She said, "So, these blocks over here are a lot like the stories in our past that we'd rather just get rid of or ignore, but we can't just delete them. We can acknowledge that they're here, and we can also choose what we do with them." She then asked him, "So what can we do with these blocks?" He said, "Well, we can add to them to have them make more sense." She reflected, "Yes, I remember seeing you stack two ramps on each other going opposite ways to make them flat." He agreed. Shelby went on, "So with your story, maybe that means adding more information around confusing things to make them make more sense?" He agreed. Shelby continued, "Yeah, we can do that. Also, we can know they're here and just try to put them on the tower and make them fit the best we can and know it just may not look perfect." Ray nodded, "Well, it sure isn't perfect . . . but it's kind of cool."

## Trauma Narrative in Creative Writing

When trauma processing is occurring, whether it is in sandtray form or in art, clay, or storytelling, there are frequently shifts to a somatic focus for a piece of the work. An example follows: Jenny, a 16-year-old girl who was sexually abused by a significantly older cousin during multiple family gatherings, had shared the verbal narrative of her memories of the abuse on two or three separate occasions. As we grew in trust together, she was able to acknowledge that, for her, referring to what had happened to her as "my abuse" provided a level of removal from the actual events that kept her regulated, but she acknowledged that this did not help her with resolution of her PTSD symptoms. Jenny brought creativity to many other areas of her life, and we began titrating doses of exposure to previously unexplored experiences of the trauma through creative writing. Jenny's family has allowed her writing to be shared here in the hopes that it might be of value to other survivors and the clinicians who work with them. This particular excerpt is her memory of one abusive moment.

*"Blow," he would say.*
*"Harder," he would yell.*
*As a four-year-old, this made no sense. What does he mean? I am blowing. Why was he screaming at me? What was I doing wrong?*

*Many times he would yell at me to do things.*
*"Take off your clothes," he would say.*
*"Let's play a game."*

*What was this feeling I felt? Was it pleasure? Or was I uncomfortable? I was a girl, he was a boy. They are supposed to kiss and hug, but what was this? Why do boys and girls do this?*

*It doesn't matter, he is my cousin. And I look up to him. He is like my best friend. I'll do whatever he wants. Even if his mouth tastes like Takis.*

*Yet, this feels weird. Why do I have to put his penis in my mouth? Why does he have to put his mouth on my vagina? Why does it feel good?*

*We are breathing so hard. Why are we underneath blankets? It's hot in here. Can I take them off for a second? I can't breathe.*

*Wait! Don't pull the blanket back! I need some fresh air!*

*Done. Finally . . .*

*Ooh! What's that bag of color packet things? Lipstick? Cool! Hold on . . .*

*Wait, you want me to wear it? What color? Red? I'm not allowed to—okay.*

*What, Mom? Oh, I'm putting on red lipstick for dress up with Ty Ty. Yes, it's just
    pretend.*

*Ugh. I can't get this on my lips! Jo Jo—can you help me? Thanks.*

*Now to our game. Again.*

*"Blow," he would say.*
*"Harder," he would yell again.*
*As a five-year-old, this still made no sense. What does he mean? I am blowing.
    Why was he screaming at me? What was I doing wrong?*

*Auntie, is that you? Hi! What's wrong? You look scared. Ty Ty and I were just
    playing a game. Why are you yelling at him? You're scaring me! What's
    wrong?! Why are you wrapping me in a towel? I can just put my clothes back
    on. Please tell me what's wrong! Can't you see I'm crying?*

*Auntie? Uncle? . . .*

*. . . . . . . .*

*Momma's home! Are you gonna tell her what happened? Come on, tell her! Tell
    her! Why won't you tell her? . . .*

*Should I?*

*. . . . . . . . . . . . . . . . . . . . . . . . . . . . . . . . . . . . . . . . . . . . . . .*

*As the film reached its end and the memory faded . . . so did a child's faith in the
    protection her family once provided her.*

Sometimes the richness of what is possible through writing is underappreciated. Jenny read this narrative out loud to me once and said it felt good to read it out loud. I offered to provide bilateral stimulation as she read it again, and we did so. As she continued processing, she became at one point hyperfocused on the feeling of Ty Ty's hot, wet breath on her neck. She had a somatic reactivation of a body memory. This is an example of a memory stored in the body—an implicit memory—being activated as she explored a chain of associations that were deepened by her experience of first writing the narrative and then reading the narrative out loud while we were engaged in bilateral processing. It could then be integrated. When clients reexperience somatic intrusions as part of their processing, I offer an alternative nurturing experience for that part of the body. In this case, I offered the client essential oils, and we applied a particular scent to her neck that resonated for her with strength, clarity, and hope. We sent some home with her in a scent jar to continue engaging in self-care. We also began creating a ritual in which she could put some of the strong/hopeful scent on the palm of her hand and place the palm of her hand over her collar bone, giving herself the measure of compassion that she would give to any of her friends.

Even this kindness to self was a process that had to be titrated, first by imagining that she would provide this kind of nonjudgmental comfort and healing to a friend and giving herself small doses of the same compassion she would have offered to others. Jenny had so much self-loathing related to her sexual abuse that she engaged in self-injurious behavior when I first met her. She was eventually able to show me the cuts on her arms and allow me to take physical care of them, providing healing lotion and Band-Aids to the affected areas. Her mom also gave this care at home when Jenny found things just too hard to handle. Eventually, as we processed through her trauma, integrating parts, offering compassion, and fostering self-compassion, she began to take a different kind of care of herself. Jenny stopped wearing long sleeves, and she would come in with violent red writing on her arms and wrists (see Figures 9.1 and 9.2). Instead of cutting herself, she would bring a red ink pen to school and draw lines on her arm. Eventually, the lines of red and the violent words began morphing into a recognition of her own numbness and slowly into hopeful words and images, such as *fearless and brave* and *individual growth* and *not a spirit broken*. She also began to draw smiley faces, hearts, and other symbols she perceived as hopeful.

Healthier digestion of difficult content is the goal and desensitization and/or diffusion of the memories is an important part for the work, as is the creation of a coherent narrative. As Jenny was moving toward an ability to morph her urges to inflict harm upon herself into more positive uses of voice, we began doing some expressive arts–based work through an internal family systems (IFS) lens. After Jenny had been stabilized and was experiencing some success in using her adaptive coping, I explained to her that we all have many parts to ourselves, and that sometimes we can isolate these parts from one another, but as we build bridges and start to give room for all of the parts of us to have a voice and be heard, we can begin to integrate them. I asked her to focus on the moments just before she cuts. I wondered aloud if she could name the part of her that wants to cut. She paused just a moment and then said, "Harley." I then asked Jenny how old Harley was. She paused, visualizing this "Harley" part in her mind, and said, "She's five." I then simply asked if Jenny could focus on the Harley part of her and draw a picture of her (see Figure 9.3).

If you look closely at the picture, you will notice several things. First, you will see the handprint of the 16-year-old Jenny layered, almost like a watermark, on the paper—perhaps representing the struggle for dominance between the five-year-old self and the 16-year-old self. You may also notice the erasure marks and ghostly half present legs that were first drawn and then erased as she became more and more aware that the Harley part was on her knees, hands

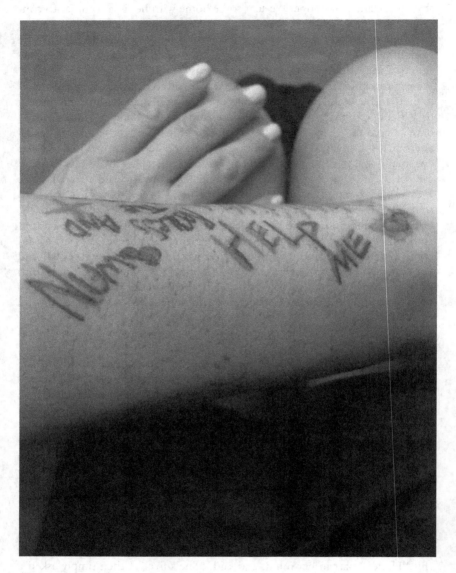

*Figure 9.1* Recognizing Numbness

tied behind her back, with her mouth sewn shut. I asked Jenny if she could try and talk with Harley. What would she say to her? It was remarkable, really, the transformation that occurred as this now 17-year-old saw her cutting behavior as a much younger part of herself. She began to speak to her Harley part and said things like, "It's OK. I know you are scared. We are in this together. I know you didn't have any power then. But we don't have to hurt ourselves.

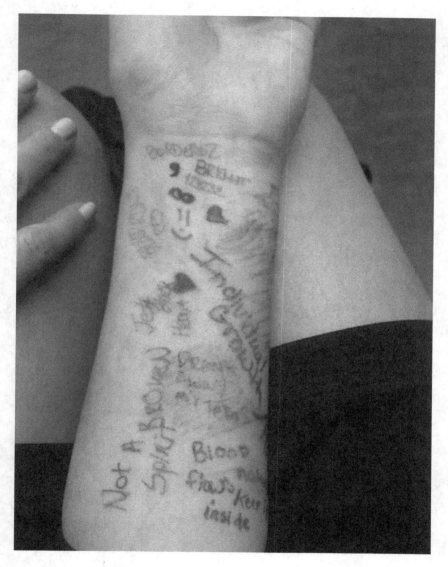

*Figure 9.2* Individual Growth

We will get through this together." And ultimately, "If you could just come to me when you are scared, I will take care of you . . ." This was the beginning of a new season of self-compassion that had not been present before.

As we continued to process together and integrate parts of the self into a functional whole, Jenny began to be kinder to herself. The family ended up having to make an abrupt move out of state before her therapy was complete.

*Figure 9.3* The Five-Year-Old Harley

Normally, when I am completing a termination process with a client, especially a teenage client, I engage them in the Before, During, and After intervention outlined in *Play Therapy with Traumatized Children*. In this case, I tweaked the format to account for our piece of work and to help her articulate (in trans-hemispheric language) her treatment goals for her new therapeutic relationship. I asked the client to draw a picture that represented her experience of herself before therapy (see Figure 9.4).

She drew a self-portrait and then cut out another square of paper, scribbled on it, and placed it over her face. Then she took another square of paper and wrote "Just Another Body" on it (see Figure 9.5). She talked about the dual meaning of this statement for her. She talked about how disconnected she had been from her body and recognized after all our work that this dissociative coping had been really helpful to her during her abuse but was no longer needed. She also talked about how people see her tall, curvy, beautiful body now and appreciate her for just her body, not bothering to even look for her true self. I then asked her to draw a picture to represent how she saw herself now, at the end of our time together. She chose to take the picture she had drawn and make some changes. She began by removing the face covering, and then she crossed out the words "Just Another Body" and wrote her name and the words "A Human Being" (see Figure 9.6).

She talked about how she had come back into her body and felt comfortable in her own skin now. We had also done some work around humans "being" vs. humans "doing," and she felt equipped to engage in mindfulness practices when she started to become anxious. I had taken a picture of the first image, and we put the two side by side. We looked at them together, and I asked,

*Figure 9.4* Jenny Accepts Harley as a Safe Part

*Figure 9.5* Just Another Body

"What do you notice about these two pictures?" She immediately said, "My hands. They're behind my back." I acknowledged having noticed this too and wondered out loud what it meant for her. She thought for a moment and then said, "I guess, although I'm more me now, I don't interact with the world the way I want to. I'm not very confident with others." I asked if she wanted to draw a new picture or change this one in some way to show what she would like to see for herself.

She created a new picture, filling the page with more of herself than she had done in her previous drawing. She has one arm up in front of her, making the peace sign, and her other hand is on her hip/back pocket. Her shirt, in this picture, covers her whole midriff. She then wrote on the paper "take it or leave it" (see Figure 9.7). Jenny was quick to say, "I'm not being mean. It's not like a snotty take it or leave it. It's just me being secure and confident about me and willing to reach out to others." It never ceases to amaze me how much our right brains want to tell us about what we need, what we most deeply desire. Drawing the pictures first and then bringing verbal articulation to the drawings gave her a richer, wiser reflection and allowed for a more fully felt process of goal setting for her next therapy relationship.

## Posttraumatic Play and Storytelling

Trauma narrative can be done in sand, in art, in puppetry, in music, in dance, and sometimes in plain storytelling. This next case example is an example that illustrates the resolution of a discrete traumatic event. The approach interweaves trauma-informed play therapy with storytelling and EMDR. I received a call one day while I was on vacation. A set of twin three-year-olds had been on an outing to the park with their babysitter. They took their scooters and began moving along a bike path attached to the park. A man who had been parked in the parking lot followed them and attacked the babysitter. She screamed to the three-year-olds to run away. They did run away, but we did not understand until I started working with them that they went to search for rocks and stones to throw at the "bad guy." When their mom and dad came for our initial session, they were in crisis. The boys were asking questions that the parents did not know how to answer. They were deeply afraid that they would say the wrong thing and further traumatize the children, so they did not answer the questions at all. The children were deeply afraid, and these previously independent sleepers not only insisted on sleeping with their mom and dad but would also ask repeatedly for their parents to lock the bedroom door at night to keep the bad guy out. The family was asking for immediate relief, so I brought the twins in together for the first couple of sessions. In both sessions, they responded in fairly classic posttraumatic play scenarios, choosing perpetrator figures and multiple jails and jailing the bad guys (see their containment work in Chapter 4). These more purposeful moments of play were peppered within significant stretches of dysregulated play. I wanted to give them more mastery experiences, particularly around the restoration of their sense of power in the environment. At home, they seemed to be grasping for power by assuring themselves that the doors were being locked, so I began hunting for a way for them to play this out.

I found a wooden house-shaped box with a door on each side and four keys, one that fit each of the four locking doors. I simply added this toy to the playroom where we had already been meeting. The more regulated of the two twins gravitated toward it and began locking and unlocking the doors. He

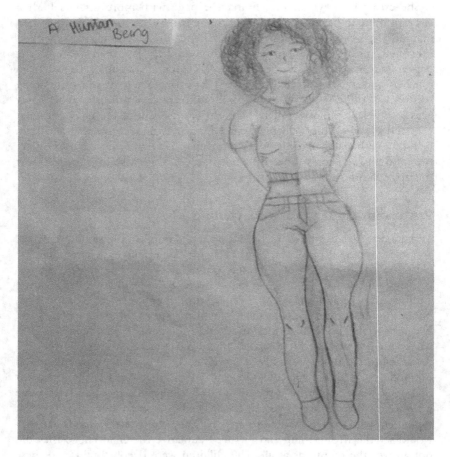

*Figure 9.6* A Human Being Front View

put symbols inside each door. Eventually, he left some unlocked and locked others. When he went home, he was much less focused on making sure the bedroom door was locked.

Meanwhile, I asked the mom to record the children's questions and I asked the parents to come together, and we crafted an appropriate story. Sometimes this process is long and grueling. How to talk about rape in a way that is developmentally appropriate can feel like an impossibility to parents whose whole goal to date has been protecting their children from ugliness. The goal is not to say everything "just right" but to craft a story that is able to be absorbed and held by both the parents and the children. These parents have given me permission to share the story we crafted in the hope it will inform the work of other therapists who are helping to speak the unspeakable.

*Figure 9.7* Take It or Leave It

### The Story of What Happened

One day Laney was babysitting Robert and Ronnie. All three went to the park to feed the ducks. They played on the playground also, and then Laney took them to ride their bikes on the trail. They were riding their bikes, when Laney said, "It's time for lunch." They turned their bikes around to go back to the truck. A man walked past Robert and Ronnie as they were riding their bikes. The man threw Laney to the ground. Laney knew the man was dangerous and was making bad choices. She wanted to keep the boys safe, so she yelled, "Run, boys!" She hoped they would run to safety. But the boys wanted to help, so they looked for rocks and sticks and couldn't find any. They got a little lost, and then they found Laney again. When they came back to Laney, they saw the man on top of her. He had his hand over Laney's mouth because he didn't want her to scream for help. But she did scream anyway and kept screaming for help. The boys also saw that Laney's pants were down and wondered why. The man made a bad choice to take her pants down without her permission and to touch her body without her permission. The man hurt her and then jumped up and ran away. Laney jumped up too, fixed her clothes, and said, "Boys, it's an emergency." They left the bikes, ran to the truck, and drove home quickly. Laney ran into the house screaming. Daddy came running and started helping right away. They were back in their safe place at home. Daddy called the police, and police officers came, an ambulance came, and a fire truck came. Mommy rushed home and helped Laney go to the doctor while Daddy stayed with the boys. The doctor helped Laney's body feel better. Jennifer came back from her trip to help take care of the boys and to help the family feel safer. A couple of days later, they went to see Laney, and the boys saw that she was safe with her mom. Now Laney is safe with her family just like Robert and Ronnie are safe with their family.

The End

The posttraumatic play expressed prior to the EMDR session was important preparation, and yet these children needed an overtly articulated story of what happened that could be held by the whole family. Once this story had been told, most of the posttraumatic symptoms receded immediately.

## Nurturing Narrations

Hands are powerful on many levels. They are the primary instruments by which we touch other people. Handshakes, high fives, and even hugs begin with outstretched hands. But they can also represent our ability to do things—typing, cooking, building, tending. I am remembering a very important session for a dad and his newly adopted daughter. We engage in a process at Nurture House that we call Nurturing Narration. It combines Theraplay principles with trauma narrative work. It was our first session of nurturing dyadic work, and we were checking for hurts. The dad worked in construction, and as we began to pay attention to his hands, we saw the "stories" from his work all over them. The little girl was unused to nurturing touch, and I thought it best to model what she and I would do with her dad first. I asked if I could have his hand, and as I inspected it, I noticed an old scar and a new scrape. I offered a little lotion for the old scar, and while I was taking care of the scar, I asked the dad to tell me the story of how he got it. He had cut it on a piece of heavy machinery when he used to work in a car manufacturing plant. The little girl piped up with "I didn't know you made cars!" I commented, "You didn't know that about daddy. Sometimes when we pay attention to the hurt places, we get to hear new things about each other." Sally thought about this for a moment and then rolled up her pants leg to show us both the scar on her knee. I offered to put some lotion on it, and she scooted toward me, bringing her knee closer.

I applied the lotion while asking, "How did you get this scar?"

She looked at her knee while saying, "My brother was playing with his pocket knife, and he was mad that I was changing the TV to watch my show and not his. He threw the knife at me, and it hit my knee."

I kept a hand on her knee and reflected her words to her. "He was playing with it, and when he got angry, he threw it at you. It cut your knee. I imagine you were bleeding."

"Yeah, I was bleeding." She moved slightly closer again.

"You were bleeding and you needed help."

"Yeah, I ran to Mom's room, but she was asleep and told me to stop crying and get a Band-Aid, but I couldn't find them."

"Then what?"

"Well, I didn't want to wake her again. She'd be mad. So I just used a paper towel."

"So, you helped yourself. Daddy, that day she had to help herself, but now you are here to help her." Then, to mitigate this potentially threatening interweave, I shifted my attention to concrete things: "I have five different kinds of Band-Aids in this jar."

Her dad took the jar from me, opened it up, and dumped out the Band-Aids. He said to his daughter, "Which one do you like?" She scrambled into his lap and carefully perused all the Band-Aids, finally landing on a My Little Pony Band-Aid.

Her dad said, "I'm sorry that you had to handle that hurt all by yourself. From now on, I'm here to help with hurts," and he put the Band-Aid over the scar.

The little girl jumped up and said, "More Band-Aids! More helps!"

I said, "You want to find some other things to put Band-Aids on."

She ran to the circus tent where the puppets were kept and brought back several. She would choose one, say how it was hurt, and then offer that body part—a beak, a paw, an exposed tummy—to her dad and tell him which Band-Aid to put on the doll. Her dad put each Band-Aid on with care. When she was finished, she arranged them all among the comfy pillows of our snuggle nook. She and her dad looked at them together, and she said, "They'll be OK now."

We have found that this approach, Nurturing Narration, invites storytelling by pairing it with nurturing touch (aimed at actively taking care of hurts), and caregiver engagement often elicits parts of the child's narrative that might not have been accessible without these supports. There is a risk a child takes when they give us more of their story. When the story is received and listened to with interest and compassion by a safe adult, the risk is rewarded. The reflection of the child's story by the safe grown-up meets three goals: 1) it lets the child know that the therapist is not overwhelmed, immobilized, or left speechless by the information; 2) it can bring an additional level of organization or coherence, as the child can listen to the story being told by another; and 3) it can help structure and support the child in sharing more of the story. I talk a lot about the various ways we communicate to children "I see what you are showing me, and you can show me more." It is the same with telling.

In other cases, there may be events in the child's life that are confused or misunderstood or have not been narrated for the child. I was recently in session with an adoptive family. Derrick, the nine-year-old boy, had just become regulated enough for us to begin making sense of his history. At the beginning of the session, I was joining with each family member, and the mom expressed relief about an adoption outside of their family that was, after much time, energy, hope, and disappointment, going to be able to happen. The mom said, "So, we finally got a yes! The kiddo can be adopted!" Derrick, who was sitting on the pillows with a cover partially over his head, popped his head out and asked loudly, "Am I going to be adopted?" His mom said, "You already are adopted. You are a part of our family forever." Derrick replied, "What is *adopted*?" As his mom began to explain, Derrick was struck by the phrase "tummy mommy" and blurted out, "But I grew in your tummy!" I offered these words: "It sounds like Derrick is really asking to better understand his story of being adopted. Let me run and get my 'tummy mommy' figure, and you guys can start picking out sandtray miniatures to be mom, dad, and Derrick." I explained that, for Derrick, the story would be easier to grasp and remember if it was told with concrete symbols of the various players. I offered two separate sandtrays and set them up in different parts of the room so that tummy mommy's story, as she was pregnant with Derrick, could be played out in a different location than his parents' story. His parents acted out going to an adoption worker and asking to be parents, waiting expectantly, and then getting news that a baby had been born early in a hospital in another state. They learned that this could be their baby if they could drop everything and come and get him. Derrick blurted out, "We need a car . . . I'll get it." Clearly, he was now very invested in the story. I moved back over to the tray with "tummy mommy" and clarified, "So, Derrick was born in a hospital over here?" His mom nodded. I asked, "Did you meet tummy mommy?" His mom said, "Nope. Oh, I guess that's an important part of the story. Tummy mommy knew she couldn't keep Derrick and said goodbye quickly to him and left the hospital. I think she was sad about saying goodbye and wasn't able to stick around." I asked, "Who took care of Derrick while you were driving to get him?" She replied, "The nurses were with him, and he spent one night with a nice woman named Gina." As his parents narrated, I would stop and ask clarifying questions. As we took the story step by step, Derrick asked questions and ran to get more miniatures. Afterward, his parents shared their astonishment that there were so many places of confusion or misunderstanding in his story. More than likely, he will need several more repetitions of narration before he really has the coherent story inside of him. To this end, I had his mom take pictures of the various scenes as we went along. They will print them out, and we will type up the narrative and make it a family exercise for the family to create the hardcopy version of the story that can be told again and again.

The ways in which narratives are created and retold are vast, and different children may need different ways to access their story after the initial telling. I was helped in my understanding of this by Loto, a nine-year-old boy with Down syndrome who was internationally adopted after living on the streets of his hometown for several years. Loto's developmental differences and his slow acquisition of English as a second language made communicating story challenging. Prior to entering treatment, his parents had created a timeline complete with important dates for him. His mom requested a consult, as Loto had begun engaging in a behavior that was bewildering to her. She described a frequent scenario in which she is driving and Loto begins to shout out from the backseat, "2009! Born!" His mom will say, "Yes, you were born in 2009." Other times he will just shout out a date, but in both cases it had become perseverative. His mom was exhausted with trying to understand what he was asking for each time he made an exclamation and was even more exhausted with repeating herself. I explained that Loto seemed to be seeking assurance regarding the parts of his story he remembers and may be asking for more elaboration or deeper coherence regarding other parts of his story. To this end, we decided to narrate his story in the sandtray first and videotape it so he could rewatch the video narrative at will. We also took pictures and made a hardcopy book and then had the parents record their voices reading the book. When done this way, it can even be fun to include a sound like "bing!" at the end of each page to indicate to the child that it is time to turn the page. This allows for the parent and child to snuggle together and read the story (my preference) and for the child who needs more repetition than the parent can tolerate to be able to hear the story retold in a read-along fashion.

This is where another critically important aspect of narrative work must be understood. The narrative itself and the ways in which the narrative is delivered must honor the needs of the system. You might be saying, "Don't you mean the needs of the child?" Nope, I mean the needs of the system. To unpack this further, we must be asking a series of questions all along the way as we are crafting, concretizing, and conveying the story. Are we answering the question asked? Is the information being shared in a developmentally appropriate manner? How much information can be tolerated by the child right now? How much information can be tolerated by the caregiver right now? In family work, the caregivers become the holders of the story, and if they cannot buy into the narrative that is shared, then they will be likely to go way "off script" in a moment when they feel triggered. Finding the balanced, nuanced story that will work for the system becomes the sustainable goal.

The delicacy of this balancing act is often brought front and center in divorce cases that are highly conflictual. We find ourselves sometimes working with divorcing parents for several sessions, holding the big feelings related to the hard truths of the grown-up story so that they can tolerate the withholding of developmentally inappropriate information from the shared narrative both parents support for the child. Most child therapists would agree that a simple explanation that does not assign blame is best—something like, "Mommy and Daddy are not able to get along together anymore and are going to live in separate houses. We both love you very much, and you will have a forever place with each of us." In other cases, the story might go like this: "Mommy and Daddy have been having some grown-up problems and have decided to divorce." These are the simplest constructions but can often be painfully difficult for parents to implement. The dad who is divorcing mom because she had an affair may want the child's understanding to reflect his own judgment—that mom's "bad choice" caused the divorce. The mom who has spent ten years as a stay-at-home mother and wakes up to a husband who says, "I don't want to be married anymore," may really want her children to understand that she did not choose the divorce. Parents who have a deeply rooted faith-based belief that divorce is wrong may have even more trouble with creating a narrative they can live with. Often, I am entering into these negotiations with this guiding question: "What is the story that will provide as much truth as possible at an age appropriate level in a way that can be consistently repeated and held by both parents and child?"

One of the dimensions of narration that my parents struggle with the most is summed up in the small distinction between using the word "couldn't" and using the word "wouldn't." Sometimes a parent is gone because of mental illness. Perhaps that parent has bipolar disorder and does not take the medicine that is prescribed. If it can be argued that bipolar disorder can be regulated through medication, does it become a "choice" on that parent's part not to take the medication, or is it part of the pathology? Does the story for the child become "Daddy couldn't take his medicine regularly" or "wouldn't take his medicine regularly," and which version is more helpful to the child? Get ten therapists in a room and they will give you ten different answers. In some cases, a child may have taken on the burden of blame, and using language about choice in reference to the parent may be freeing for the child, providing relief that he did not choose to be without his parent. For other children, however, the idea that a parent chose something, whatever it might be (drugs, alcohol, an affair), over the current family system might reinforce an already pervasive sense of rejection. How does this difference in language affect the child? This tiny little difference in language is the ballgame for many people.

Many parents feel strongly that the other parent had choices and chose to abandon the family.

I have worked with several parents with the same story: "I was at work when I got a call from the school saying that my daughter wasn't picked up from school." In all of these cases, this phone call marked an abandonment by the other parent, either due to alcoholism, drug addiction, or mental illness or the other parent having abruptly left without a word and remaining unreachable for days, months, or years. In such cases, there is a myriad of narratives available to the remaining dyad. For one dad and his son, the narrative became that mom had a sickness in her mind that made it difficult to take care of herself or others. At one point in treatment, it became important for the sense of abandonment that each felt to be acknowledged by the other. I drew a large figure for the dad and a smaller one for the son and gave them Band-Aids. I asked if they could show me any hurts on the inside or outside that needed a Band-Aid. They both put Band-Aids in their midsections, slightly below their hearts, to represent the mom/wife leaving. The child also put Band-Aids for concrete hurts on his legs, and the dad put a Band-Aid on his head and talked with me briefly about how much harder it was for him to trust than it used to be. This art activity became the anchor for rich nurture and holding of big feelings between dad and son.

## Who Am I? Enhancing a Child's Internalization of a Parent Who Has Died

The image shown below was created by the child referenced earlier in this book. This child found his father in his study no longer breathing. He called 911 and provided CPR to his dad for a significant period of time before the ambulance and emergency workers arrived. He believed, deeply, that his father had died because he had not been strong enough to give the compressions properly. He was helped to tell his story by the kinesthetic involvement and concrete nature of domino people, lining them up and knocking them over. He used these domino people to work through the blow-by-blow of what happened that night and to arrive at the conclusion that there was nothing he could have or should have done differently. He resigned himself to the fact that his father had, in essence, already died at the time his son found him. While absorbing this truth was initially painful, it ultimately brought him relief and absolution. This resolution of guilt opened him up to celebrating the parts of himself that are like his father. We began by taking a precut hand template and attaching it to paper. We asked his grandmother (who had brought him into the session) to help and had his grandma paint the client's hand. He put his handprint inside his dad's, and then we talked about how many parts of his dad lived on in this child. He talked about how his blue eyes came from his dad. He remembered his dad telling him to brush his teeth and talked about the love of sports he had learned from him. At the end of the session, I asked him to title his creation, and he decided to call it My Daddy in Me (see Figure 9.8).

*Figure 9.8* Encoding Transmission of Positive Attributes

## Anchors for Activation

What do I mean by the phrase "an anchor for activation"? In play therapy, the miniatures, toys, and other play materials are all made available to the child in the context of a safe therapeutic relationship. The trauma-informed play therapist is able to be with the child in play at the same time he or she is always watching for the moments of "quickening." In these moments, a child might become more intense in their attention, sometimes by intensely rejecting or embracing an object. In the moment of intense rejection of an object or symbol in the playroom, the therapist's antennae go up, and they pay closer attention, as there are several possible responses.

One approach would be to simply reflect the action and the child's embedded emotional relationship to the object. The child may be communicating feelings of intense anger, and the child may simply need to have these feelings and their actions reflected to them. For example, if a five-year-old girl picks up the male baby doll and throws it across the room (and has just had a new baby join the family who is taking most of mom and dad's attention), that child may simply need the play therapist to reflect the intensity of her anger with the new baby by reflecting her action and holding the big emotion that comes with it. The therapist might say, "You seemed so angry at that baby when you threw that baby doll across the room."

Another approach would be to view the rejection of the object as a rejection of parts of the self, parts that may need to be approached with curiosity and compassion in order to heal and become fully integrated into the child's sense of self. Through an internal family systems (IFS) lens, there may be emotional parts of the child that have been exiled. Through a trauma narrative lens, there may be an activation of trauma content that the child is hopeful to avoid. In some cases, the traumagenic material may include feelings about themselves at certain ages or stages of their life. Tracking these moments of rejection can be very helpful in understanding where the unresolved, incoherent life narrative moments remain.

Case example: Sally is an 11-year-old girl who was adopted by her family at birth. However, the in utero threats were real and persistent during her gestation. She generates various realities around how she was left, all of which have an undertone of self-loathing and deserved abandonment. She and her mother have been seeing me for help with their attachment relationship. When they first started seeing me, Sally would become almost feral whenever her sense of shame was triggered. Her shame was triggered if she was told no, it was triggered whenever her mother tried to teach her something new, and it was triggered whenever her sibling was the focus of attention, even briefly. She was set up, both neurologically and psychologically, to perseverate and often did so on negative thoughts about herself. She believed she was stupid. She believed she was ugly. She believed that no one liked her. She believed she had been left in an alleyway because she was trash. These were pieces of her narrative. She misperceived many looks given to her by her parents as looks of loathing even when they were attempting a soft, loving eye gaze.

The first long period of work was simply spent helping this little girl develop trust with her adoptive mom. As trust was developed, we began tackling harder things in sessions. One day Sally and her mom accompanied me to the sandtray room. Sally began her work by making a zoo where each animal was

*Figure 9.9* The Baby in Mother's Womb Activates the Client

caged separately, had its own space, and everyone got along. As she perused the sandtray shelves, she noticed the figure I have of a baby inside the translucent torso of its mother (see Figure 9.9). Her loud and immediate exclamation was, "EWW! That's gross!" I commented, "That seems really yucky to you. Mom, can you hand it to me?" Her mom handed me the figure, and I held it cupped in both of my hands, silently gazing at it. Sally looked surprised and carefully watched me for a moment before turning away. She continued choosing animals for her tray. As she remained kinesthetically engaged elsewhere, I spoke to her mom, saying, "You know, mom, I've been thinking a lot lately about how babies can either be made to feel really, really safe and welcome in their mommies' tummies or not safe or wanted. When a mom isn't sure how it will go to have a baby—maybe she is worried about whether or not she can take care of it, whether or not she will be in trouble for being pregnant in the first place [*this related to this child's birth story*]—she can release a lot of stress hormone into her bloodstream, and it can make the developing baby feel unsafe." Sally, who had given no visible sign that she was listening, stopped what she was doing and asked, "There's a chemical for that?" I said, still looking at the baby, "Yep. The fancy name is *cortisol*. Cortisol can make it hard for the growing baby to feel good, even in the womb. One of the first things that babies hear is the heartbeat of their mother. If a mom feels pretty calm and content during her pregnancy, the baby hears this [*I thump my hand rhythmically against my own chest*]." At this, Sally turned to look at me, and her jaw literally dropped. "You mean babies can hear inside?" "Yep, and if they have a mom who is scared sometimes or stressed sometimes, what they hear is [*I thump my hand several times very quickly and then slower in a dysregulated pattern*]. Babies can have a hard time feeling safe when the sounds don't even stay the same." Sally looked directly at her adoptive mother and said, "That explains a lot!" I said, "Looks like this means something for you." "Yeah, I'm scared all the time. I didn't know it started back then." I returned my gaze to the symbol in my hand and said, "Sometimes the baby thinks there is something wrong with her, when really the mom was just so stressed she couldn't make it safe for the baby." I shift my focus to Sally and say to her mother, "Mom, how 'bout you hold the baby while Sally tells me about her sandtray." Her mom held the baby for the rest of the session.

In the above case example, Sally experienced a moment of activation in relation to an object in the room. This provides rich material for work. She vehemently rejected a sandtray miniature of a baby in its mother's womb. Sensing the intensity of rejection as a moment of activation, I chose to extend the moment of interaction enough for us to work with it. Since the question of why she was given up for adoption has not been fully resolved for her and her narrative of how it happened keeps changing, the vulnerable baby exposed through the translucent womb sparked feelings of fear and self-loathing. Whatever the core question that needs to be answered for trauma resolution to begin, we begin the dance toward and away from the question in part through these moments of activation. Here, Sally does not take the time to examine her feelings of discomfort when she attends to the baby in the womb but pours her rejection of the vulnerable parts of herself into the symbol. The good news is that if we track the moments of intense rejection or embrace of symbols in the playroom, we can begin to track the moments of the life narrative that need help to become more coherent and integrated into the child's story.

As soon as Sally would begin to get some clarity about one aspect of self, it would seem to drift away, and a new worry would become the defining force in her life. Earlier in our work together, as the worries had been so interruptive and overlapping, I offered her the Worry Worms activity (Goodyear-Brown, 2010) as a way to categorize and bring some executive functioning voice to the anxieties she carried. I hid rubber worms around the room, and each time she found one, she would verbalize another worry that she carried. The number and kinds of worries filled the whole page and continued on to another page and included the following:

1. Being treated like a baby.
2. Sleepovers.
3. That all my clothes will be ripped.
4. That I won't get enough food.
5. That I'll be embarrassed in front of my friends.
6. That if I keep eating I'll get fat.
7. That my sister will get hurt.
8. That we'll have to go to the doctors.
9. That I might be touched inappropriately.
10. Getting blood drawn.

You will notice themes among her worries. She described these worries almost like a Ferris wheel in her mind—they all revolved in her head and got recycled many times a day, with different worries being most vivid at different times. As her attachment relationship with her adoptive mother became more trusting, she began to articulate these worries to her mom while she was having them. Having a connecting co-regulator in her mom really helped take the intensity out of them. We began to understand that the worries came in waves, but she could decide which ones to ride and know that eventually that worry would shift, become smaller, and recede into the background. This new way of relating to her worries equipped her to move beyond full-time management of her anxiety to deeper questions related to her core sense of self.

After several sessions that focused on building self-compassion and narrative integration through play, Sally seemed ready to approach her sense of self more head on. I asked her if she could create a sandtray to show me how she saw herself. She chose a beautiful princess and began draping her with handcuffs. The room remained very quiet, her mom and I both maintaining a respectful silence while she worked. When she was finished, she stepped back and said, "OK." The image you see below was her central creation (see Figure 9.10).

I asked if she could describe the princess now. Sally said, "Trapped." We all continued looking at the figure. After another silence, Sally said, "Each handcuff is a trap." I said, "Oh, she has a lot of traps." Sally said, "Yeah." Then she pointed to one and said, "This one is the stuff in my head from the computer." She pointed to another and said, "This is the way I always think my friends

*Figure 9.10* Self-Object Covered in Handcuffs and Weapons

are laughing at me." She pointed to a third: "This is me always thinking I'm ugly." The client had moved from a strictly symbolic/metaphoric approach to storytelling to an ability to articulate her own reality in a left brain linguistic way after creating and experiencing the symbolic representation with primary weight given to right brain ways of knowing. The hemispheric integration that we were beginning to see in her work mirrored the integration of parts of the self that were beginning to be allowed to exist and be named. We spent time working with each of the traps.

Sally had been exposed to some pornographic imagery and was internally harassed by pictures that had morphed together in her mind and could not easily be shaken. She described a very specific image and then said, "It's hard to describe." I asked if she could draw a picture of the image that stuck in her mind (see Figure 9.11). She perceived herself as being forever tainted by what she had seen. The image is that of a mouth-like opening with teeth at its center and, below, the words "My mind is forever swollen with the memories of thy sin." The word *swollen* may describe her experience on multiple levels. It may reference the pleasurable sensations she experienced in her own physical body when looking at the images, the images themselves, the sense of being "too full" of this content without a way to digest it, or a sense that she was being gnawed at by the perseverative nature of the visual images, rehearsing them again and again in her mind's eye. She has gone through cycles of blaming the creators of the imagery and then blaming herself for looking at them. We have since worked with this image, helping it lose some of its power over her.

Sally's belief that her adoption meant she was dirty or bad was so pervasive that even moments of attempting to make a positive self-statement were followed immediately by negative ones. She would often gravitate to extremes in language, saying in the same breath, "Everyone is jealous of how pretty I am,"

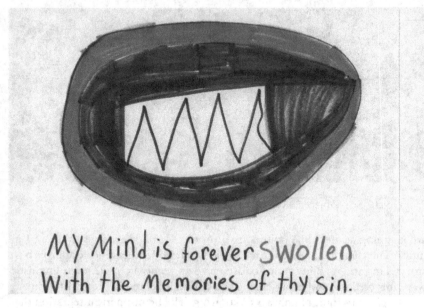

*Figure 9.11* Hard to Forget

followed immediately by "I am so fat." It felt like Sally was unable to reflect on her feelings of brokenness and questions of worth in a meaningful way, but rather ping-ponged back and forth between extremes of grandiosity and deep self-condemnation. Expressive arts materials can be very useful in beginning the work of integration around self. In Sally's case, I simply asked her to create a self-portrait. Her first response was, "I'm a terrible drawer." I offered all the materials and said, "There is no right or wrong way to do this, and you can use any of the materials that you see." Her self-portrait (see Figure 9.12) was fascinating.

She spent time painstakingly trying to make the two sides of her face symmetrical. She actually asked for a measuring tool and tried to measure how far each of the eyes was from the midline of the face. And yet, the finished version is an extremely out-of-balance face. One cheek is carrying a bruise, possibly reflective of her belief that she is damaged goods and her extrapolated belief that everyone can see the damage when they look at her. Her mouth is placed so off center that it almost looks like it is in her cheek and reflects feelings of deep uncertainty. When I asked her to describe what the expression in the drawing was communicating, she said, "What else is gonna happen to me?" If you look closely, you will also see that there is a person reflected in her eyes. When I noticed the figure drawn in the eyes, her response was, "That's just a person." This may indeed be true for her and may portray her feeling that her sense of self is defined by the "other" who is seeing her and is shaped by what they are communicating or how she perceives their communications. She is

*Figure 9.12* Unintegrated Sense of Self

easily buffeted by the feelings, actions, and facial expressions of others, as she often overperceives disgust and underperceives admiration. Since Sally's sense of self is constantly tied to how she believes others are experiencing her and she is carrying the core question "What else is gonna happen to me?" she is constantly in a hypervigilant place, leaving little room for the development of trust or an autonomous self.

One day, her mom shared with me about a birthday party Sally had attended. During the party, she told the other girls that she "had been left in a box in an alley" when she was little. All the other girls cried, but the narrative she had told was not the truth, and this concerned her mom greatly. I said to Sally, "It sounds like you are asking the question 'Please help me understand my early life better.'" Sally looked relieved, and we decided it was time to dig deeper into her questions about her birth parents. We agreed that mom, dad, Sally, and therapist would all sit down together and create a book that answered her questions in ways that were developmentally appropriate for her. We began slowly, asking Sally simply to create a cover for her life book. Below is the first cover she created (see Figure 9.13).

Shattered. This was a powerful expression of her anticipatory fear that her birth story would prove that she was broken beyond repair. For me, it resonated most strongly with the shifting shards of self that we had been identifying in the playroom.

Before we began the life book, I asked her if she could draw her birth mother the way she has pictured her. Sally got right to work and drew the "svelte, skinny woman in the skin tight pants" (Sally's description) you see below (see Figure 9.14). She had been carrying a story inside her in which her birth mother was a 16-year-old "slut"—again, Sally's word, not mine. Notice that this figure is also without facial features. After saying the word *slut*, she then carefully wrote on the page "Had sex with her high school sweetheart."

Sally also believed that her birth mother did everything she could to attract male attention, including wearing skintight pants and shirts that showed her midriff and making sure her long hair was always styled. The simmering rage she felt for the betrayal by her birth mother is hinted at in her characterization of her mom as "the freaking woman who didn't give a _____ for me." When asked to fill in the word out loud, Sally mouthed "shit."

I began her life story work by asking her to generate a list of questions. She had lots of questions, ranging from "Where was I born?" to "Why did she leave me?" She was very surprised to understand that her mother had been in her 30s when she was born and had been in a relationship with one man, and that they had broken up before he knew her birth mother was pregnant. I sometimes have to be reminded of how children absorb (and don't) different parts of their narrative. Sally had been rehearsing one story in her head and heart for so long regarding where she came from that it was very difficult to overwrite this distorted narrative with aspects of the real events as they unfolded. During the first session in which we began answering some of these questions, she made the only cognitive shift around her birth mother's age. Having been told her birth father did not know she existed did not change her internal narrative.

*Figure 9.13* First Cover of Sally's Life Book

Next session, she spontaneously said again, "I hate him. He left us." She had rehearsed the story of how her birth father had left both her and her mother so many times that simply correcting the distortion verbally was not enough. We played it out in the sandtray—the sequence of events from birth mom and dad being together to them breaking up to mom's learning she was pregnant to

*Figure 9.14* Sally's Internal Story About Her Biological Mom

Sally's being given to the people who have been mom and dad ever since. We were four sessions into rehearsing the story differently before Sally was able to interrupt her own old story with "No, wait, he didn't know about me."

As we did the laborious work of helping her piece together the story of her beginnings and how sought after she had been by her current parents, we saw her internal coherence and her ability to trust her mom and dad increase. She is still an emotionally labile child and can become anxious very quickly, but now she seeks out her mom for support and can allow her mom to hold parts of her story with her that she used to try to hold herself. In a recent sandtray, I asked Sally to create her family in the tray. Many times children begin a tray with one configuration and then move the figures as they tell the story. Below

*Figure 9.15* Family of Origin

*Figure 9.16* Internalized Nuclear Family

are two pictures taken as Sally was working in the sandtray (see Figures 9.15 and 9.16).

Notice that her birth mother figure has remained the same since her initial connection to this figure. She said, "This was my birth mom and my birth dad. I made him a helicopter because he left. Well, he was gone before I was born. I mean, he didn't know about me. The baby is me after I was born." Notice that she still has the chronology confused but is able to self-correct as she works. She then went and got a second, smaller sandtray and placed it inside the first. She created her current family in that tray, and she moved her father figure over to the other side of the tray. She is the girl with the cowboy hat, her father is a heroic figure, her younger sister is happily roller skating, and her mother is her cheerleader and placed so closely behind her you can only see her hands and pom poms from the angle at which I took the picture. This shift was so striking to me. There were very clear boundaries now between her nuclear family and the rest of the world. Her birth mom and birth dad are clearly apart from one another and are still present but are on the fringes of the world. It seems that the life story work—which allowed her to ask questions and have her parents answer them—allowed her to experience them as her history keepers in a new way. Their new roles as history keepers opened Sally more and more to accepting them as her safe bosses and to an overarching new sense of them as her nuclear family. We are further down the road now in her healing

*Figure 9.17* A Day at the Beach, Emerging Boundaries

process. Recently, I invited her to create a world in the sand. She created a tray that was built around one of her favorite memories with her family—a trip to the beach. She put two chairs in the shade where she and her mother could sit. She included "monuments" for the family to see during their trip and a safety fence to keep the water at bay (see Figure 9.17). We begin to see new boundaries emerging and more organization and coherence in her creations.

## Future Circles

I have become enamored of circular sandtrays in the last few years. They offer a kind of containment that is reminiscent of the mandala and offers the hope of completion by making a journey all the way around the circle. Because the circle offers no beginning and no end, it encourages the perception of trauma recovery as revolutions or iterations of experience, spiraling toward deeper levels of integration and/or further distance from the traumatic event. The sandtray pictured below was created by a mother and her son Jimmy. The mother had been driving with the son in the backseat when they were in a car accident that resulted in injuries. The family was unsure about how to process the trauma together and brought the son to Nurture House. I worked through the TraumaPlay™ model with them, and when we arrived at the trauma narrative piece, I checked in with the mom for any important details that Jimmy might have encoded. She had found Jimmy hunched in the floorboard of the backseat in the wake of the accident and assumed he had seen very little. I threw a bunch of google eyes in the sandtray, and Jimmy and I played a game of hide-and-seek in which he would tell me the first thing he saw, the next thing he saw, etc., each time he found an eyeball.

Jimmy really enjoyed hide-and-seek, and, once again, the playful experience of finding the hidden eyes induced a surge of competency chemicals that was released in his brain and absorbed as strength in his body. These chemicals mitigated his approach to sharing visual pictures that he had previously kept to himself. Most of these images lined up with his mom's account of the trauma, but at one point Jimmy shared about the moment he peeked over the backseat, and saw a disturbing sight that he had never shared with his mom. The family had been unaware that Jimmy was holding this picture in his head, and they were all relieved when the memory was exposed and could become part of the shared narrative, allowing the parents to hold it with him. As we were ending treatment, Jimmy and his mother came in together to create a "Forward Circle." People often use the phrase "We've come full circle," and sometimes this implies that you are back where you started, but after a process of building coherence, integrating somatic and emotional content, correcting cognitive distortions, and journeying toward healing, families are never right back in the place they were. So these sandtray reflections, often done during the termination phase of treatment, are meant to honor the journey toward healing while moving forward to the next chapter of the family's story.

*Figure 9.18* Future Circle Sandtray

Mom and Jimmy were given a circular sandtray, offered a collection of miniatures, and asked to create a tray depicting their journey from the accident to now. The dyad decided to place fencing pieces in the sand, dividing their story into three sections. Jimmy represented himself and his mom with two swords in the middle of the tray. As events swirled around them, they stuck together. The dyad chose a cage to describe their initial response to the accident. Both felt trapped in feelings of guilt, shame, fear, and sadness. The next quadrant shows the beginning journey in therapy, sifting through all they were carrying (represented by the trash can) and a sense of trying to stay afloat in a sea of chaos (represented by the lifesaver). The last quadrant has several symbols representing growth moments (the flowering vine, the treasure chest). While the image below is an approximate recreation of the original tray, the two black grocery carts had been used by both mom and Jimmy and became important symbols, representing for them the work they had done to understand what feelings and responsibilities belonged to each of them (in terms of the car accident) and the permissions they had learned to give each other for having differences both in how they remember the event and how they move forward in healing (see Figure 9.18).

The family felt this new ability to honor each other's experiences and perceptions while maintaining their own truth would be a valuable skill in managing whatever life brought them in the next season. These creations of Forward

Circle sandtrays encourage shared storytelling and shared story holding while offering left/right hemisphere integration in a way that talk therapy alone would not. They also paint a picture of how the dyad wants to move forward in the wake of trauma recovery.

## Conclusion

Narrative Nuance is really the heart of this text. My hope in this chapter, and in all the preceding chapters, is to applaud the children and families in our care for the myriad ways in which they titrate and allow us to titrate the approach to hard things in therapy. There are as many ways for a child to express their pain, confusion, joy, or love as there are stars in the sky. The trauma narrative examples in this chapter have detailed how the Play Therapist's Palette and the various mitigators offered are harnessed by the client system and supported by the therapist. As we remain present with our families and ask ourselves regularly what the client needs right now, being willing to flex ourselves to meet the therapeutic need, my hope is that we will each continue to have the great privilege of participating in the healing of traumatized children and families.

## Reference

Goodyear-Brown, P. (2010). *The Worry Wars: An Anxiety Workbook for Kids and Their Helpful Adults*. Nashville, TN: P. Goodyear-Brown.

# Index

Note: Page numbers in *italics* indicate a figure.

Adlerian play therapy 4
adopted children 25, 58, 92, 147–148, 156, 196, 212–217; anger at birth parents 217–229, *225*, *226*, *227*
adoptive parents 56, 113, 131–133, 147, 158, 162, 168, 198, 212, 217, 219–220
Adverse Childhood Experiences (ACE) 23–24
amygdala 51, 136; role in sensing alarm, 180–183; as seat of somatosensory memory 28–29
anxiety: in children 38, 56–58, 62, 65–66, 76, 91, 120, 132, 166; in clinicians 48; gaining control over 82, 162–164, 169, 220; in parents 188; responses to 66, 75; symptoms of 165; *see also* anxiety disorder; mindfulness; separation anxiety
anxiety disorder 75, 165
attachment figure 16, 23
attachment relationship 8, 12, *13*, 14, 26, 36, 78, 81, 122
autism 51, 156
autonomic nervous system (ANS) 35–36

Badenoch, Bonnie 22, 27, 31
bipolar disorder 25
body temperature 45–46
brain: development, 23, 45; dysregulation 44; limbic 29, 56
Brown, Stuart 23, 34

Camp Nurture 18, *18*, 52–56, 60, 64, 155, 174–176
child-centered play therapy (CCPT) 4

child-parent relationship therapy (CPRT) 4, 25
Circle of Security project xii, 12, 16
containment 12, *13*, 64, 71–73; for "big feelings" 89–91, *90*; compartmentalization as 80–82; containing perpetrator symbols 94–99, *95*, *96*, *98*; in creative spaces 82–83, *83*; in hiding and joining 75–79, *76*, *77*, *79*; limit-setting 91–94; physical containers 88–89, *89*; in sand spaces 83–87, *85*, *86*; use as physical space 73–75
cortisol 23, 32, *33*, 34, 38, 45, 180

developmental trauma disorder 24–25, 75, 94, 136
distrust *see* trust issues in children
domestic violence 24, 75, *76*, 94, *95*
dopamine 24, 34, 63, *63*, 78, 122, 134, 161, 180
drugs and drug use 23–24, 45; addiction in parents 97, 128, 134, 176, 214–215
dysregulation (brain) 23, 36, 43, 50–51, 91, 136, 156

eating disorder 46, 88
executive functioning 59, 131; in child/adolescent brains 161, 163, 220; cortex 28; development of skills linked to 64, 145; *see also* brain; neocortex
exercise (physical) 47, 160–166
explicit memory 30, 81, 114

filial therapy 4, 25
Flexibly Sequential Play Therapy (FSPT) *see* TraumaPlay™

gardening as therapy *see* healing garden
Gil, Eliana xi–xii, xiii, 1, 15, 23, 87, 107

healing garden 48–49, *50*
hippocampus 32; *see also* brain
humor: as coping mechanism 182; lacking a sense of 184; love connector 180; negative forms of 183; styles of 183; therapeutic 12, *13*, 179–183; *see also* oxytocin
Humor Styles Questionnaire (HSQ) 183
hyperarousal/hypoarousal 2, 6–7, 23, 32, 35, 37, 55
hypothalamic-pituitary-adrenal (HPA) axis 32

immobilization response 10, 35–36, 67, 167
implicit memory 30–31
irrationality: importance of reframing term 58–59

Kestly, Theresa 22, 31
kinesthetic competence 54, *54*
kinesthetic involvement 67, 101, 161
kinesthetic telling 109–114, *111, 112, 113*

Laughing for Acceptance (intervention strategy) 185
laughter, therapeutic uses of 180; *see also* humor
Louv, Richard 144

"making sense of our senses" 154
memory 29–31; *see also* amygdala
mindfulness 2, 27, 66, 70, 160, 164
"more than enough" 56; *see also* "need meeting"

nature: children's diminishing exposure to 145–146; importance to Play Therapist's Palette 176; interaction with 12, 27; metaphors in 155, 166–171; as site of play 144–147, *148*; therapeutic benefits of 145
nature deficit disorder 144
"needing more" 58; *see also* "overflowing"
"need meeting" 45–46, 58–60, *61*
negative self-image 222–226, *223, 226*
neocortex 28, 59, 131, 149

neuroplasticity 22–25
nuance, necessity of 2; in play therapy 6
nurture, therapeutic applications of 17, 18, *18*, 45, 55
Nurture House: goals 6, 12, 19; meeting emotional and bodily needs 52–53, *53*; in nature as play area 146–147; neurochemical impacts of 31–35; as nurturing space 45–51; use of scents at 55; *see also* Camp Nurture
Nurturing Narrations 210–215

"overflowing" (as a perceived state) 58, *59*
oxytocin 32, *38*, 63, *63*, 78, 161, 180, 182; stimulated by humor 34, *34*

Paul, Gordon 6
Perry, Bruce 26
play: core therapeutic powers of 6; intrinsic value 5; as neural exercise 36; neurobiology of 22; outdoors 144–154; pretend 155; *156*; relationship to holistic child development 23; risk-taking in 149–153, *151, 152, 153*; sense of safety in 36; as social engagement system 35–37; trauma-informed 14, 25; *see also* posttraumatic play
play deprivation 23
Play Therapist's Palette 1, 6, 12, 13, *13*, 19–20, 22, 26, 231; humor as mitigator 179; importance of metaphor 118, 177; practical applications of 11; as therapeutic "menu" 14
play therapy 14, 37; effectiveness of 5, 6; forms of 4; goals 6
polyvagal theory 35–37, 67, 167, 186
Porges, Stephen 35–38
posttraumatic play 1, *3*, 9, 20, 97, 207–210

risk-taking *see* play, risking taking in

"safe boss" 12, 15, *16*, 17; healthy attachment with 57; lack of trust in 136–137, 161; need for 62–63, 82, 87, 93; parent as 147, 228; role of 18, *18*, 26–27, 43, 87, 92–93; "sensory savvy" version of 49–54; therapist as 43, 161
safe place 81

safety: building a sense of 1, 3–4, *3*,
17, 29, 69, 97; experiences of 27;
as felt 20, 35–37, 43, *76*, 81, 121,
132–33, 170, 180; as linked to role
of therapist 157; neuroception of
10, 14, 57–58, 62–63, 82, 186;
physical 137, 162, 170; sound
(aural) as 80, 180, 186–187
satiation, sense of 27, 30, 57, 62
Schaefer, Charlie 5, 6
Schore, Alan 37, 39
self: motivation of 165; perception of
87, *89*, 99, 205–207, *208*, 220–223,
*223*; regulation of 163, 166, 190;
in relationship to "other" 56, 131;
sense of 1, 16, 126, *127*, 217; sexual
93; social 173; therapeutic uses of 4,
15; unintegrated *223*
self-actualization 184
self-awareness 167, 173
self-care 17, 48, 201
self-esteem 144, 164
self-harm 121, *122*, 201
self-object 94, 220–223, *221*, *223*;
utility of 44–45
self-soothing 22, 55
self-talk 9
sensory defensiveness 156
sensory differences 49
sensory integration 44
separation anxiety 158, 162
sexual abuse in children and teens 23,
24, 31, 44, 109–110, 220–222, *223*
Siegel, Dan 22, 25, 26, 28, 30
somatic responses 10, 11–12, 27, 38,
52, 145–46; experiencing of 45, 59,
66, 89, 94, 167; expressions of 110,
119–121, *120*, *121*; focusing of 45,
199; mastery of 54, 229; in memory
30–31; reactivation of 200; self-
regulation of 146, 166; as states 52;
trauma stored as 101
SOOTHE strategies 124
soothing *3*; environments related to 48,
80, *84*; examples of 55, 61–62, *63*;
need for 125; *see also* self-soothing

stress hormones *see* cortisol
suicide 11, 103; attempts 24; ideation
120–121

"taming the trauma" 19–20
Theraplay 4, 10, 25, 47, 57, 64, 186,
*187*, 210
titration 13; applied to playroom 8;
definition of 6–7; in relationship to
trauma narratives 9; of trust 44, 57,
137, 145, 147, *148*
touch: as anchoring gesture 65;
benefits of 145; debates over 64; as
"need meeting" 64
trauma: brain-body storage of 43;
chronic 24–25; neurobiology of 22,
31; relationship to bodies 14
trauma history 44
trauma narratives 1, 2, *3*, 9, 12, 107,
114, 131–132, 210, 231; in creative
writing 199–207
TraumaPlay™ 1, *3*; assessment phase
6; core values 8; goals of 2–4, 17,
104; therapeutic role of self in 15;
therapists' role in 16
treatment plans 6–7
Trust-Based Relational Intervention
(TBRI) 47
trust building 47–49, 137, 147–154,
157, 169–170, 184, 220
trust issues in children 9, 19, 57–67,
132–133, 136, 161–162, 215, 217

van der Kolk, Bessel 14, 24, 27, 42, 44
Vygotsky, Lev 12, 26, 149

window of tolerance 6, 13, 36, 64, 91;
awareness of 9–10, 15; edge of 145;
expansion of 8, 23–24, 35, 57, 79,
139, 157; in exposure work 44; going
outside of 14, 87, 91, *92*; pushing
against boundaries of 110–112;
sitting with emotions 133–134

zone of proximal development 12,
126, 149

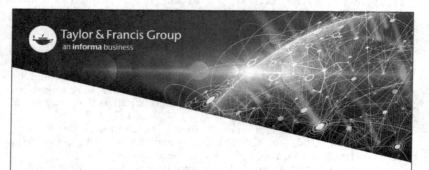